Fundamentals of IMAGE, AUDIO, and VIDEO PROCESSING Using MATLAB®

Fundamentals of IMAGE, AUDIO, and VIDEO PROCESSING Using MATLAB®
With Applications To PATTERN RECOGNITION

Ranjan Parekh

CRC Press
Taylor & Francis Group
Boca Raton London New York

CRC Press is an imprint of the
Taylor & Francis Group, an **informa** business

First edition published 2021
by CRC Press
2 Park Square, Milton Park, Abingdon, Oxon, OX14 4RN

and by CRC Press
6000 Broken Sound Parkway NW, Suite 300, Boca Raton, FL 33487-2742

© 2021 Ranjan Parekh

CRC Press is an imprint of Taylor & Francis Group, LLC

British Library Cataloguing-in-Publication Data
A catalogue record for this book is available from the British Library

Library of Congress Cataloging-in-Publication Data
Names: Parekh, Ranjan, author.
Title: Fundamentals of image, audio, and video processing using MATLAB / Ranjan Parekh.
Description: First edition. | Boca Raton : CRC Press, 2021. | Includes bibliographical references and index.
Identifiers: LCCN 2020050611 | ISBN 9780367895242 (hbk) | ISBN 9780367748340 (pbk) | ISBN 9781003019718 (ebk)
Subjects: LCSH: Digital video. | Image processing—Digital techniques. | Sound—Recording and reproducing—Digital techniques. | MATLAB.
Classification: LCC TK6680.8.M38 P37 2021 | DDC 006.6/96—dc23
LC record available at https://lccn.loc.gov/2020050611

ISBN: 978-0-367-89524-2 (hbk)
ISBN: 978-0-367-74834-0 (pbk)
ISBN: 978-1-003-01971-8 (ebk)

Typeset in Palatino
by codeMantra

Contents

Preface...ix
Author..xv
Abbreviations xvii

1 Image Processing...1
    1.1   Introduction 1
    1.2   Toolboxes and Functions..5
        1.2.1   Basic MATLAB® (BM) Functions 6
        1.2.2   Image Processing Toolbox (IPT) Functions 9
        1.2.3   Signal Processing Toolbox (SPT) Functions 13
        1.2.4   Wavelet Toolbox (WT) Functions 13
    1.3   Import Export and Conversions 14
        1.3.1   Read and Write Image Data 14
        1.3.2   Image-Type Conversion 16
        1.3.3   Image Color 31
        1.3.4   Synthetic Images 44
    1.4   Display and Exploration 50
        1.4.1   Basic Display 50
        1.4.2   Interactive Exploration 54
        1.4.3   Building Interactive Tools 57
    1.5   Geometric Transformation and Image Registration...........................58
        1.5.1   Common Geometric Transformations 58
        1.5.2   Affine and Projective Transformations 64
        1.5.3   Image Registration 67
    1.6   Image Filtering and Enhancement 73
        1.6.1   Image Filtering 73
        1.6.2   Edge Detection 80
        1.6.3   Contrast Adjustment 86
        1.6.4   Morphological Operations 92
        1.6.5   ROI and Block Processing 95
        1.6.6   Image Arithmetic 99
        1.6.7   De-blurring 101
    1.7   Image Segmentation and Analysis..109
        1.7.1   Image Segmentation 109
        1.7.2   Object Analysis 111
        1.7.3   Region and Image Properties 118
        1.7.4   Texture Analysis 125
        1.7.5   Image Quality 129
        1.7.6   Image Transforms 131
    1.8   Working in Frequency Domain 144
    1.9   Image Processing Using Simulink...149
    1.10   Notes on 2-D Plotting Functions...155
    1.11   Notes on 3-D Plotting Functions 182
    Review Questions 190

2 Audio Processing 191
2.1 Introduction 191
2.2 Toolboxes and Functions... 193
    2.2.1 Basic MATLAB® (BM) Functions 193
    2.2.2 Audio System Toolbox (AST) Functions 195
    2.2.3 DSP System Toolbox (DSPST) Functions 196
    2.2.4 Signal Processing Toolbox (SPT) Functions 196
2.3 Sound Waves... 197
2.4 Audio I/O and Waveform Generation 210
2.5 Audio Processing Algorithm Design.. 215
2.6 Measurements and Feature Extraction 225
2.7 Simulation, Tuning and Visualization.. 231
2.8 Musical Instrument Digital Interface (MIDI)............................235
2.9 Temporal Filters.. 237
2.10 Spectral Filters 241
2.11 Audio Processing Using Simulink 254
Review Questions 257

3 Video Processing 259
3.1 Introduction 259
3.2 Toolboxes and Functions... 262
    3.2.1 Basic MATLAB® (BM) Functions................................. 262
    3.2.2 Computer Vision System Toolbox (CVST) Functions ... 263
3.3 Video Input Output and Playback.. 264
3.4 Processing Video Frames 272
3.5 Video Color Spaces... 278
3.6 Object Detection 282
    3.6.1 Blob Detector 282
    3.6.2 Foreground Detector 284
    3.6.3 People Detector 285
    3.6.4 Face Detector 286
    3.6.5 Optical Character Recognition (OCR) 288
3.7 Motion Tracking.. 289
    3.7.1 Histogram Based Tracker 289
    3.7.2 Optical Flow 291
    3.7.3 Point Tracker 293
    3.7.4 Kalman Filter 294
    3.7.5 Block Matcher 296
3.8 Video Processing Using Simulink 297
Review Questions 300

4 Pattern Recognition..303
4.1 Introduction 303
4.2 Toolboxes and Functions...304
    4.2.1 Computer Vision System Toolbox (CVST) 304
    4.2.2 Statistics and Machine Learning Toolbox (SMLT) ... 305
    4.2.3 Neural Network Toolbox (NNT) 306
4.3 Data Acquisition...306
4.4 Pre-processing 311

4.5   Feature Extraction 312
       4.5.1   Minimum Eigenvalue Method 312
       4.5.2   Harris Corner Detector 314
       4.5.3   FAST Algorithm 315
       4.5.4   MSER Algorithm 316
       4.5.5   SURF Algorithm 317
       4.5.6   KAZE Algorithm 320
       4.5.7   BRISK Algorithm 321
       4.5.8   LBP Algorithm 322
       4.5.9   HOG Algorithm 324
4.6   Clustering 324
       4.6.1   Similarity Metrics 324
       4.6.2   *k*-means Clustering 328
       4.6.3   Hierarchical Clustering 332
       4.6.4   GMM-Based Clustering 335
4.7   Classification 337
       4.7.1   *k*-NN Classifiers 338
       4.7.2   Artificial Neural Network (ANN) classifiers 339
       4.7.3   Decision Tree Classifiers 346
       4.7.4   Discriminant Analysis Classifiers 348
       4.7.5   Naive Bayes Classifiers 353
       4.7.6   Support Vector Machine (SVM) Classifiers 354
       4.7.7   Classification Learner App 356
4.8   Performance Evaluation 365
       Review Questions 369

Function Summary 371

References 381

Index 385

Preface

This book introduces the concepts and principles of media processing and its applications in pattern recognition by adopting a hands-on approach using program implementations. The primary objective of this book is to inform the reader about the tools and techniques using which image, audio and video files can be read, modified, and written using the data analysis and visualization tool MATLAB®. This book has been written for graduate and post-graduate students studying courses on image processing, speech and language processing, signal processing, video object detection and tracking, and related multimedia technologies with a focus on practical implementations using programming constructs and skill developments. Benefits of this book will be to provide a concise explanation of software technologies related to media processing and its varied applications along with relevant theoretical backgrounds together with steps for practical implementations, while avoiding lengthy theoretical discourses. This book is also relevant for researchers in the field of pattern recognition, computer vision, and content-based retrieval, and for students of MATLAB courses dealing with media processing, statistical analysis, and data visualization.

MATLAB by MathWorks, Inc. is a data analysis and visualization tool suitable for numerical computations, algorithmic development, and simulation applications. One of the advantages of MATLAB over other contemporary programming tools is the repository of a large number of ready-made functions for a plethora of media processing tasks which can directly be included in custom applications and problem-solving tasks without any additional programming efforts. It is therefore necessary for the prospective programmer and student to be informed about these functions so that they can be applied as and when necessary for rapid application development and reduction in efforts and time. MATLAB functions in general are divided into two broad categories: a basic set and an extended set. The basic set, referred to in this book as basic MATLAB (BM) functions, deals with elementary functionalities like data-type conversions, arithmetic relational and logical operations, algebraic and trigonometrical computations, interpolation and Fourier analysis, a variety of graphical plots and annotations, file I/O operations, and different types of basic programming structures. It is assumed that the reader has rudimentary knowledge of MATLAB and is familiar with basic mathematical operations like matrix algebra and trigonometry. This book therefore is not meant for teaching elementary MATLAB to the beginner, although a short description of the functions has been provided wherever they have been first used within this book.

The second set of more specialized functions are called "toolboxes," and they extend the basic set functionalities customized for specific domains and application areas. This book discusses functions included in a number of toolboxes for illustrating various media processing tasks, divided into three chapters on image processing, audio processing, and video processing. Each function is illustrated using examples, and the program output is shown in the form of graphical plots for easy visualization. The fourth chapter discusses the applications of media processing tasks discussed in the earlier three chapters for solving pattern recognition problems. In the process, this book delves into a selected subset of some of the frequently used functions of this following toolboxes:

- Image Processing Toolbox (IPT): provides a comprehensive set of reference-standard algorithms and workflow apps for image processing, analysis, visualization, and algorithm development

- Audio System Toolbox (AST): provides algorithms and tools for the design, simulation, and desktop prototyping of audio processing systems
- Computer Vision System Toolbox (CVST): provides algorithms, functions, and apps for designing and simulating computer vision and video processing systems
- Digital Signal Processing System Toolbox (DSPST): provides algorithms, apps, and scopes for designing, simulating, and analyzing signal processing systems
- Statistics and Machine Learning Toolbox (SMLT): provides functions and apps to describe, analyze, and model data
- Neural Network Toolbox (NNT): provides algorithms, pretrained models, and apps to create, train, visualize, and simulate neural networks
- Signal Processing Toolbox (SPT): provides functions and apps to analyze, preprocess, and extract features from uniformly and nonuniformly sampled signals
- Wavelet Toolbox (WT): provides functions and apps for analyzing and synthesizing signals and images using wavelet-based modeling

Since the focus of this book is to familiarize the reader about MATLAB functions, the theoretical concepts have been minimized and only those relevant to the functions being discussed have been included. This is in contrast to other media processing books that focus more on explaining theoretical concepts with some of the concepts being illustrated with examples. The current book, on the other hand, focuses primarily on discussing features of the MATLAB package for performing various media processing tasks, and short relevant theoretical sections have been included wherever necessary for understanding the underlying concepts. This enables this book to present the ideas in a compact manner which can quickly be comprehended by the reader without the need for lengthy discourses. Additionally, the features are illustrated through practical examples for solving specific numerical problems. This is expected to help the prospective students and readers to get acquainted with all the relevant features quickly and implement these for solving customized problems. The codes are provided in a simple copy and execute format for the convenience of the novice learner. The outputs of the codes are displayed in the form of graphs and plots to help the learner visualize the results for quick knowledge assimilation. The arrangement and ordering of the topics in each chapter have been done keeping in view the hierarchical structure of the MATLAB functions in the relevant toolboxes. Mentioned below is a list of the primary topics covered in each section of each chapter, most of which, from section 3 onwards, are illustrated with at least one coding example along with a visual plot to display the program output.

Chapter 1 deals with *image processing* and is divided into 11 sections. **Section 1.1** provides an introduction to basic concepts covering pixels, digitization, binary image, grayscale image, color image, image acquisition and output devices, image transformations, image adjustments, color models, compression schemes, and file formats. **Section 1.2** lists the basic and toolbox functions covered in the chapter and provides a list of around 115 basic MATLAB functions divided into five categories mentioned above and a list of 116 functions belonging to the Image Processing Toolbox (IPT) divided into five categories : Import Export and Conversions, Display and Exploration, Transformation and Registration, Filtering and Enhancement, and Segmentation and Analysis. **Section 1.3** deals with import, export, and conversion of images and covers topics like reading and writing image data, image-type conversions, binarization threshold and Otsu's method, image quantization and gray levels, indexed image and dithering, image representations using unsigned integer 8-bits and double precision 64-bits, image color representations, color models and colormaps, device-dependent and device-independent models, tri-stimulus and chromaticity values, sRGB and AdobeRGB color spaces, color conversion between RGB CMY XYZ HSV L*a*b* representations, synthetic images like checkerboard and phantom, image noise representations, and Gaussian function. **Section 1.4** deals with display and exploration of images and covers topics like basic display techniques, image fusion and montage, image sequences and warping surfaces, and interactive exploration tools. **Section 1.5** deals with geometric transformations and image registration and covers topics like common geometric transformations, affine and projective transformations, image registrations, covariance and correlation, and zooming and interpolation. **Section 1.6** deals with image filtering and enhancement and covers topics like kernel, convolution, image blurring, noise filters, order statistic filters, Gabor filter, edge detection operators, image gradients and partial derivatives of images, contrast adjustment and gamma curves, histogram equalization, morphological operations, region-of-interest and block processing, arithmetic and logical operations, point spread function, deconvolution, inverse filter, Wiener deconvolution, Lucy–Richardson deconvolution, and blind deconvolution. **Section 1.7** deals with image segmentation and analysis and covers topics like image segmentation, object analysis, Hough transform, quad-tree decomposition, extracting region properties, pixel connectivity, texture analysis, gray-level co-occurrence matrix, image quality, signal to noise ratio, mean square error, structural similarity index, image transforms, discrete Fourier transform, discrete cosine transform, and discrete wavelet transform. **Section 1.8** deals with working in the frequency domain and covers topics like convolution theorem, ideal and Gaussian low-pass filters, and ideal and Gaussian high-pass filters. **Section 1.9** deals with image processing using Simulink and covers topics like development of Simulink models for tasks like image-type conversion, color conversion, tonal adjustment, edge detection, geometrical transformations, morphological operations, and blob analysis. **Sections 1.10** and **1.11** deal with discussion of syntax, options, and parameters of various 2-D and 3-D plotting functions, respectively, for visualization of data distributions and their customization aspects.

Chapter 2 deals with *audio processing* and is divided into 11 sections. **Section 2.1** provides an introduction to basic concepts covering properties of sound waves like amplitude and frequency and their perceptual representations viz. loudness and pitch, devices for manipulating environmental sound like microphone amplifier and loudspeaker, stereo and mono sounds, digitization of audio, samples and Nyquist sampling theorem, sound card components, characteristics of CD-quality digital audio, audio filtering, synthesizers and MIDI protocol, compression schemes, and file formats. **Section 2.2** lists the basic and toolbox functions covered in the chapter and provides a list of around 52 basic MATLAB functions divided into five categories mentioned above and a list of 29 functions belonging to the Audio System Toolbox (AST) divided into five categories: Audio I/O and Waveform Generation, Audio Processing Algorithm Design, Measurements and Feature Extraction, Simulation, Tuning, and Visualization, and Musical Instrument Digital Interface (MIDI). Apart from AST functions from two other toolboxes have also been listed here viz. the DSP System Toolbox (DSPST) and Signal Processing Toolbox (SPT). **Section 2.3** deals with study and characterization of sound waves and covers topics like waveform, phase, sampling frequency, aliasing, sinusoidal tones, composite notes, Fourier domain representation of audio signals, coefficients, and basis functions. **Section 2.4** deals with audio I/O and waveform generation and covers topics like reading and writing digital audio files, plotting audio waveforms, recording and playback of digital audio, oscilloscope display, and wavetable synthesizer. **Section 2.5** deals with audio processing algorithm design and covers topics like reverberation, noise gate, dynamic range compressor and expander, and cross-over filter. **Section 2.6** deals with measurements and feature extraction and covers topics like pitch, voice activity detection (VAD), momentary short-term and integrated loudness, and mel frequency cepstral coefficients (MFCC). **Section 2.7** deals with simulation tuning and visualization and covers topics like time scope, sine wave generator, spectrum analyzer, and array plots. **Section 2.8** deals with MIDI and covers topics like playback of MIDI notes, using a variety of instrument sounds. **Section 2.9** deals with temporal filters and covers topics like finite impulse filter (FIR), infinite impulse filter (IIR), and window functions. **Section 2.10** deals with spectral filters and covers topics like frequency representations, spectrogram, signals with variable frequencies, low-pass and high-pass FIR filters, low-pass and high-pass IIR filters, band-stop and band-pass FIR filters, and band-stop and band-pass IIR filters. **Section 2.11** deals with audio processing using Simulink and covers topics like development of Simulink models for tasks like reverb, noise gate, cross-over filter, VAD, loudness measurement meter, and frequency spectrum.

   Chapter 3 deals with *video processing* and is divided into eight sections. **Section 3.1** provides an introduction to basic concepts covering video frames, frame rate, illusion of motion, raster scanning, interlacing, component video signals, composite video signals, luminance and chrominance, conversion of RGB to YC signal format, and chroma sub-sampling. **Section 3.2** lists the basic and toolbox functions covered in the chapter and provides a list of around 21 basic MATLAB functions divided into five categories mentioned above and a list of 20 functions belonging to the Computer Vision System Toolbox (CVST) divided into three categories: Input, Output and Graphics, Object Detection and Recognition, and Object Tracking and Motion Estimation. **Section 3.3** deals with video input–output and playback and covers topics like reading and writing video files, display and playback of a subset of video frames, and creating a movie from a collection of images. **Section 3.4** deals with processing video frames and covers topics like creating a 4-D structure for storing video frames, converting images to video frames and vice versa, inserting text at specific locations of video frames, selectively playing frames at specified frame rates, converting video frames from color to grayscale and binary versions, and applying image filters to video frames. **Section 3.5** deals with video color spaces and covers topics like conversion from RGB to YCbCr color space and vice versa, conversion from RGB to NTSC color space and vice versa, and conversion from RGB to PAL color space and vice versa. **Section 3.6** deals with object detection and covers topics like blob detector, foreground detector, people detector, face detector, and optical character recognition. **Section 3.7** deals with motion tracking and covers topics like histogram-based tracker, optical flow, point tracker, Kalman filter, and block matcher. **Section 3.8** deals with video processing using Simulink and covers topics like conversion of video color space, geometrical transformation, color to grayscale and binary conversion, and applying image filters to video frames.

   Chapter 4 deals with *pattern recognition* and is divided into nine sections. **Section 4.1** provides an introduction to basic concepts covering clustering, classification, supervised learning, unsupervised learning, training phase, testing phase, feature vector, feature space, and similarity metric. **Section 4.2** lists the basic and toolbox functions covered in the chapter and provides a list of ten functions belonging to the Computer Vision System Toolbox (CVST) and 26 functions belonging to the Statistics and Machine Learning Toolbox (SMLT). **Section 4.3** deals with data acquisition and covers topics like saving workspace variables to MAT files, loading variables from MAT files to workspace, working with the Fisher-Iris dataset, reading multiple media files from arbitrary folders, and using the image datastore. **Section 4.4** deals with pre-processing and covers topics like media-type conversions, color conversions, geometric transformations, tonal correction, noise filtering, edge detection, morphological operations, object segmentation, and temporal and spectral filtering. **Section 4.5** deals with feature extraction methods and covers topics like minimum eigenvalue method, Harris corner detector, features from accelerated segment test (FAST) algorithm, maximally stable extremal regions (MSER) algorithm, speeded up robust features (SURF) algorithm, KAZE algorithm, binary robust invariant scalable keypoint (BRISK) algorithm, local binary pattern (LBP) algorithm, and histogram of oriented gradients (HOG) algorithm. **Section 4.6** deals with clustering and covers topics like similarity metrics, *k*-means clustering algorithm, *k*-medoids clustering algorithm, hierarchical clustering algorithm, and GMM-based clustering algorithm. **Section 4.7** deals with classification and covers topics like *k*-NN classifiers, artificial neural network (ANN) classifiers, decision tree classifiers, discriminant analysis classifiers, naïve Bayes classifier, support vector machine (SVM) classifiers, and classification learner app. **Section 4.8** deals with performance evaluation and covers topics like silhouette value, Calinski-Harabasz index, and confusion matrix.

There is a huge demand worldwide for courses on image processing, audio processing, and video processing because of their varied applications in pattern recognition, computer vision, object detection, artificial intelligence, speech and speaker recognition, voice activation, video surveillance, face recognition, vehicle tracking, motion estimation, and others, for most of which this book is expected to be useful. A function summary is included at the end of this book, which provides an alphabetic list of around 400 MATLAB functions discussed in this book mentioning their originating toolbox and a one-line description of each, for easy reference of the reader. Around 75 references of books and research papers have been provided for further reading of the various algorithms, standards, and methodologies discussed throughout this book. Each chapter ends with a set of review questions for self-assessment. This book contains around 250 solved examples with their corresponding MATLAB codes. Although the codes have been tested with MATLAB version 2018, most of these will execute properly with version 2015 or later. The bold face is used to highlight the salient theoretical/conceptual terms discussed in each section. There are more than 100 such terms in this book. The red bold texts are used to indicate a MATLAB function name occurring for the first time. There are more than 400 MATLAB functions discussed in this book. Readers are also asked to use MATLAB Help utilities for further information on the functions discussed. More than 300 color figures have been included to help readers in proper visualization of the program outputs.

All readers are encouraged to provide feedback about the content matter of this book as well as any omissions or typing errors. The author can be contacted at *ranjan_parekh@ yahoo.com*

Ranjan Parekh
Jadavpur University
Kolkata 700032, India
October, 2020

MATLAB® is a registered trademark of The MathWorks, Inc. For product information, please contact:

The MathWorks, Inc.
3 Apple Hill Drive
Natick, MA 01760-2098 USA
Tel: 508-647-7000
Fax: 508-647-7001
E-mail: info@mathworks.com
Web: www.mathworks.com

Author

Dr. Ranjan Parekh, PhD (Engineering), is Professor at the School of Education Technology, Jadavpur University, Calcutta, India. He is involved with teaching subjects related to Graphics and Multimedia at the post-graduate level. His research interests include multimedia information processing, pattern recognition, and computer vision. He is also the author of the following books:

- *Principles of Multimedia 2/e* (McGraw Hill, 2012; *http://www.mhhe.com/parekh/ multimedia2*)
- *Fundamentals of Graphics using MATLAB* (Taylor & Francis, 2019; *https://www.crcpress. com/Fundamentals-of-Graphics-Using-MATLAB/Parekh/p/book/9780367184827*)

Abbreviations

1-D	one dimensional
2-D	two dimensional
3-D	three dimensional
ANN	artificial neural network
AST	audio system toolbox
BM	basic MATLAB
BRISK	Binary Robust Invariant Scalable Keypoints
CIE	International Commission on Illumination
CLUT	color look up table
CMYK	cyan magenta yellow black
CVST	computer vision system toolbox
DCT	discrete cosine transform
DFT	discrete Fourier transform
DRC	dynamic range compressor
DSPST	digital signal processing system toolbox
DWT	discrete wavelet transform
FAST	Features from Accelerated Segment Test
FIR	finite impulse response
FLT	fuzzy logic toolbox
GLCM	gray level co-occurrence matrix
GMM	Gaussian mixture models
HOG	histogram of oriented gradients
HPF	high pass filter
HSV	hue saturation value
ICC	International Color Consortium
IIR	infinite impulse response
IPT	image processing toolbox
JPEG	Joint Photographers Expert Group
LBP	local binary pattern
LDA	linear discriminant analysis
LoG	Laplacian of Gaussian
LPF	low pass filter
LSE	least square error
LUFS	loudness unit full scale
MATLAB	matrix laboratory
MFCC	mel frequency cepstral coefficient
MIDI	musical instrument digital interface
MLP	multi layered perceptron
MMSQ	minimum mean square error
MSE	mean square error
MSER	Maximally Stable Extremal Regions
NNT	neural network toolbox
NTSC	national television systems committee
OCR	optical character recognition

PAL	phase alternation lines
PCA	principal component analysis
PSF	point spread function
RGB	red green blue
ROI	region of interest
SE	structural element
SIFT	Scale Invariant Feature Transform
SMLT	statistics and machine learning toolbox
SNR	signal to noise ratio
SPT	signal processing toolbox
SSIM	structural similarity
SURF	Speeded-Up Robust Features
SVM	support vector machine
VAD	voice activity detection
WT	wavelet toolbox

1

Image Processing

1.1 Introduction

An **image** is a snapshot of the real world taken by a camera. The real world being analog (i.e. continuous) in nature the photograph by a conventional camera is also analog. The process of converting it to digital (i.e. discrete) form is called **digitization**. The process of digitization is usually defined with respect to electrical signals, which are also analog in property. The digitization of any electrical signal consists of three steps: sampling, quantization, and code-word generation. **Sampling** consists of examining the value of the signal at equal intervals of time or space and storing these values only while discarding the rest of the values, which essentially discretizes the time or space axis of the signal. **Quantization** involves specifying how many levels or values of the discrete signal amplitude to consider for further processing while discarding the remaining values, which essentially discretizes the amplitude axis of the signal. **Code-word generation** involves assigning binary values called code-words to each of the retained levels occurring at the specific sampling points considered. The number of sampling points per unit time or length is called the **sampling rate**, while the number of amplitude levels retained is called **quantization levels**. The total number of quantization levels determines the number of bits in the binary code-word that would be used to represent them, which is called the **bit-depth** of the digital signal. An n-bit code-word can represent a total of 2^n levels. For example, a 3-bit code-word can represent a total of eight values: 000, 001, 010, 011, 100, 101, 110, 111. Each of the discrete values making up the digital signal is called a **sample**.

A **digital image** is a two-dimensional signal where the samples are spread across the width and height of the image. These samples are called **pixels**, short for picture elements. Pixels are the structural units of a digital image similar to molecules making up real-world objects. Pixels are visualized as rectangular units arranged side by side over the area of the image. Each pixel is identified with two parameters: its location and value. The location of a pixel is measured with respect to a coordinate system. A 2-D **Cartesian coordinate system,** named after the French mathematician Rene Descartes, is used to measure the location of a point with respect to a reference point known as the origin. The location is represented as a pair of offsets measured along two orthogonal directions, the x-axis or the horizontal direction and the y-axis or the vertical direction. The distances from the two axes, known as the primary axes, are written as an ordered pair of numbers within parenthesis as (x,y) coordinates. The numbers denote how many pixels along the horizontal and vertical directions the current pixel is offset from the origin O and hence they are integers as number of pixels cannot be fractions. For image processing applications, the origin O is usually located at the top-left corner, the x-values increase from left to right,

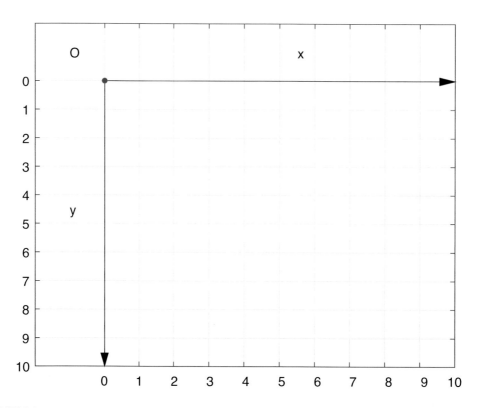

FIGURE 1.1
Image as collection of pixels.

and y-values increase from top to bottom (Figure 1.1). Obviously, the origin itself denoted as O has coordinates (0,0).

The value of a pixel represents the intensity or color of the image at that point. Depending on the value, images can be categorized into three broad types: binary, grayscale, and color. A **binary image** is a digital image, where the pixel values are represented as binary digits, i.e. either 0 or 1. Such images typically contain two types of regions, those containing value 0 appear as black and those containing value 1 appear as white. Since a single bit can be used to denote the pixel values, such images are also called 1-bit images. The second category is called **grayscale image,** where the information is represented using various shades of gray. Although any number of gray shades or levels is possible, the standard number used is 256 because the typical human eye can discern so many shades and this can be represented using 8-bit binary numbers. Starting with 00000000 to denote black at one extreme and 11111111 to denote white at the other extreme, there are 254 levels of gray shades possible in between, taking the total number of values possible to 256. The third category is called **color image,** where the information is represented using various shades of color. Although there is theoretically an infinite number of colors, the standard number used is around 16.7 million based on the capabilities of the human eye to distinguish between so many colors, and this can be represented using 24-bit binary numbers. All colors are represented as a combination of three primary colors, red (R), green (G), and blue (B), which are referred to as color channels in an image (Figure 1.2). Thus, a color image has three channels, while a grayscale image has a single channel. We will talk more about these in the following sections.

binary	grayscale	color
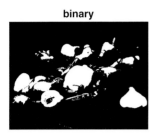		

FIGURE 1.2
Binary, grayscale, and color images.

A digital image is created using an **image acquisition** device, which converts an analog image to electrical signals and then to digital data. Two types of acquisition devices, also called **input devices**, are commonly used: the scanner and the digital camera. A **scanner** is usually used to digitize printed documents and printed photographs. It consists of a glass panel on which the document is placed face down and a scan head below it containing a source of white light. As the scan head moves from left to right and top to bottom, the light is reflected from all parts of the document and falls on an array of electronic sensors called **charge coupled device** (CCD). The CCD converts the light from the document into electrical signals with amplitudes proportional to the intensity of light. The electrical signals are then fed to an **analog-to-digital converter** (ADC), which converts them to digital data using the digitization process described earlier. A software interface can be used to specify parameters associated with digitization viz. the spacing between pixels called the **image resolution**, measured in dots per inch (dps), and the number of bits for representing the pixel values called **bit-depth**. Grayscale images usually have a bit-depth of 8 while color images are represented using a bit-depth of 24, as mentioned previously. Other parameters like compression scheme and file formats are explained later in this section. The **digital camera** also works in a similar principle i.e. the light from surrounding objects is focused through the lens and falls on a CCD array which converts the light into electrical signals. The signals pass through an ADC and gets converted to digital signals, which is then stored as an image file in a storage device within the camera.

Once an image is represented in digital form i.e. as a collection of pixels, editing and programming software tools can be used to change the pixel values and hence modify the image in a variety of ways. A collective term used to denote all these various operations for modifying images is called **image processing**. Image processing operations can be categorized into a number of types. At the most basic level, it can only involve **geometrical transformation** of the image like cropping, translation, rotation, and scaling without actual modification of pixel values. The second category involves changing the pixel values using input–output relationships called **gamma curves**. The third category involves changing pixel values using probability density functions like **histograms**. The fourth category involves using **kernel operations** which modify pixel values based on local neighborhoods. The fifth category involves modification using **arithmetic-, logical-, and morphological-operations**. The sixth category involves **color modification** and conversion. **Color models** enable us to represent and communicate color information. Studies have shown that all colors we can see can be represented as combinations in different proportions of three **primary colors** viz. red (R), green (G), and blue (B). This has led to the development of the **RGB color model** which defines how composite colors can be produced using RGB primaries and is applicable for colored lights. Equal quantities of primary colors when mixed together produce **secondary colors** viz. cyan (C) from G and B, magenta (M) from B and R, and yellow (Y) from R and G. Another color model called the **CMY color model** has been developed to predict behaviors of colored inks on paper and has cyan, magenta, and yellow as its primaries. Both the RGB and CMY color models are visually represented as **color cubes** with each of the three primary colors along one of the orthogonal axes and any point within the cube indicating a particular color specified using three coordinates. While both the RGB and CMYK are **device-dependent** i.e. the actual colors produced depend on the properties of the devices used to display the colors, attempts have been made to develop **device-independent** color models based on the inherent nature of human vision. Studies have shown that presence of rod and cone cells in the human retina is responsible for perceptions of luminance (brightness) and chrominance (color) in people. Accordingly, the **Hue-Saturation-Value (HSV) color model** has been developed which defines a color in terms of three components viz. hue, which is a set of colors the human eye can see, saturation, which is the amount of gray mixed with the pure colors, and value, which is the brightness of each color. Pure colors are henceforth called **saturated colors** with 100% saturation, while colors with varying amount of gray mixed to them are called **de-saturated colors** with lower values of saturation. Hue is visually represented on a circular scale along the circumference of a circle and measured in angles from 0° to 359°, while saturation, measured in percentage values, is represented along the radial direction, decreasing from 100% at the periphery to 0% at the center of the circle. This representation of hue and saturation is commonly referred to as the **color wheel**. Brightness being measured along a direction perpendicular to the plane of the color wheel generates the pictorial representation of the HSV model as an inverted cone. Other contemporary studies on human visual systems have led to the development of the **L*a*b* color model,** where L* denotes the brightness axis similar to the V axis of the HSV model, while the colors are represented using a red–green axis called a* and a blue–yellow axis called b*. The total range of actual colors generated by a color model is called the **color space** of the model. Conversion formulae to convert from one color space to another have also been proposed which are mostly non-linear in nature. One interesting observation is that the RGB color space is larger than the CMY color space, which implies that all colors that can be displayed on screen may not be printable. The L*a*b* color space is a superset of both the RGB and CMY spaces and is represented using the **chromaticity diagram**.

When images are displayed on **output devices** like monitors and printers, the image pixels are mapped to the physical pixels of the devices which enable them to be seen. A monitor can be either a CRT type or an LCD type. A **cathode ray tube** (CRT)-type monitor consists of a vacuum tube with a phosphor-coated screen. An electron beam from the cathode falling on the phosphor dots generates light which is seen as a glowing pixel. The beam moves from left to right and top to bottom of the screen using a **raster scan** pattern activating the phosphors one by one. For a steady picture on the screen, the pattern needs to be completed 60 times per second so that the **persistence of vision** (PoV) of the human eye creates an illusion that all the phosphors are glowing simultaneously. A **liquid crystal display** (LCD) monitor has separate pixel elements which can be controlled independently using transistor circuits. A printer can be either inkjet type or LASER type. An **inkjet** printer produces a drop of liquid ink for each pixel which is activated from a nozzle using either a heating element or a piezo-crystal. A **LASER** printer uses a photo-conductor drum on which a charge image is generated using a LASER beam. Powdered toner particles stick to the charge points on the drum, which are then melted and fused to the paper to create a print version of the image.

When an image is stored as a file on a storage device like a hard disk, it usually takes up a lot of disk space. For example, a 24-bit color image 800 by 600 pixels is the size that requires approximately 1.4 MB disk space. A **compression scheme** is a software application that reduces the file size of an image so that more image files can be stored in the storage device. This is especially useful for portable devices like a digital camera. Scanners can also employ compression software to reduce the file size of images stored in the hard disk of the computer. Compression algorithms are broadly of two types: lossless and lossy. A **lossless algorithm** manipulates the data inside a file so that it takes up less space when stored on a storage device, without making any permanent changes in the file. On the other hand, a **lossy algorithm** actually deletes some information from the file and thereby reduces its file size. The advantage of lossy compression is that it produces a much larger amount of compression than non-lossy compression, although the disadvantage is that it degrades the image quality permanently. Whatever be the compression process, it is only used to store the file on the storage device. When the image is to be viewed again, a reverse **decompression** process is used to restore the original state of the file. This application is therefore commonly referred to as a CODEC (coder/decoder). **File formats** in which the image is saved depend on the compression scheme used. The Windows native image file format is BMP which is typically uncompressed. When image quality is an important factor and space requirement is secondary, non-lossy compression algorithms can be used to save the file in TIFF or PNG formats. When space requirement is more important than image quality, lossy compression algorithms can be used to save the file in JPG format. If a very small number of colors are involved, then an 8-bit format like GIF can also be used to save the file.

1.2 Toolboxes and Functions

MATLAB® functions for image processing can be divided into two broad categories: basic functions and toolbox functions. **Basic MATLAB** (BM) functions provide a set of supporting functionalities to the core media processing tasks like I/O operations, matrix manipulations, and plotting routines. The toolbox functions provide a set of more advanced features for specialized processing tasks like geometrical transformations, filtering, enhancement, and segmentation. For image processing tasks, the main toolbox is the **Image Processing Toolbox** (IPT) which provides a comprehensive set of reference-standard algorithms and workflow apps for image processing, analysis, visualization, and algorithm development. Most of the toolbox functions discussed in this chapter are part of the IPT, however, a small number of functions have been taken from two other toolboxes: the **Signal Processing Toolbox** (SPT), which provides functions and apps to analyze, preprocess, and extract features from uniformly and nonuniformly sampled signals, and the **Wavelet Toolbox** (WT), which provides functions and apps for analyzing and synthesizing signals and images using wavelet based modeling. It is to be noted that some functions, like basic I/O functions, can be common to both the BM set and the toolbox sets. The MATLAB features for media processing tasks are illustrated as solutions to specific examples throughout this book. For a particular task at hand, functions from both the BM set and multiple toolboxes might be necessary, and the sources of these functions are mentioned as and when they are used for the solutions. The image file formats supported by MATLAB include BMP (Windows Bitmap), GIF (Graphics Interchange Format),

ICO (Icon File), JPEG (Joint Photographic Experts Group), JPEG 2000, PBM (Portable Bitmap), PCX (Windows Paintbrush), PGM (Portable Graymap), PNG (Portable Network Graphics), PPM (Portable Pixmap), RAS (Sun Raster), and TIFF (Tagged Image File Format). Most of the images used in the examples are included in the MATLAB package and do not need to be supplied externally. The image samples are available in the following folder: *(matlab-root)/toolbox/images/imdata/*.

1.2.1 Basic MATLAB® (BM) Functions

The BM functions fall into five different categories: Language Fundamentals, Mathematics, Graphics, Data Import and Analysis, and Programming Scripts and Functions. A list of the BM functions used in this chapter is provided below along with their hierarchical structure and a one-line description of each. The BM set actually consists of thousands of functions, out of which a subset has been used in this chapter keeping in view the scope and coverage of this book.

1. *Language Fundamentals*
1. ceil: round toward positive infinity
2. clc: clear command window
3. double: convert to double precision data type
4. eps: epsilon, a very small value equal to 2^{-52} or 2.22×10^{-16}
5. find: find indices and values of nonzero elements
6. format: set display format in command window
7. hex2dec: convert hexadecimal to decimal
8. length: length of largest array dimension
9. linspace: generate vector of evenly spaced values
10. meshgrid: generates 2-D grid coordinates
11. NaN: not a number, output from operations which have undefined numerical results
12. num2str: convert numbers to strings
13. numel: number of array elements
14. ones: create array of all ones
15. prod: product of array elements
16. strcat: concatenate strings horizontally
17. uint8: unsigned integer 8-bit (0 to 255)
18. unique: unique values in array
19. whos: list variables in workspace, with sizes and types
20. zeros: create array of all zeros

2. *Mathematics*
1. abs: absolute value
2. area: filled area 2-D plot
3. atan: inverse tangent in radians

4. boundary: boundary of a set of points
5. cos: cosine of argument in radians
6. exp: exponential
7. fft: fast Fourier transform
8. fft2: 2-D fast Fourier transform
9. fftshift: shift zero-frequency component to center of spectrum
10. filter2: 2-D digital filter
11. imag: imaginary part of complex number
12. log: natural logarithm
13. magic: magic square with equal row and column sums
14. polyshape: creates a polygon defined by 2-D vertices
15. real: real part of complex number
16. rng: control random number generation
17. sin: sine of argument in radians
18. sqrt: square root
19. rand: uniformly distributed random numbers
20. randi: uniformly distributed random integers
21. sum: sum of elements

3. Graphics

1. axes: Specify axes appearance and behavior
2. axis: Set axis limits and aspect ratios
3. bar, bar3: Bar graph, 3-D Bar graph
4. colorbar: Displays color scale in colormap
5. colormap: Color look-up table
6. comet: 2-D animated plot
7. compass: Plot arrows emanating from origin
8. contour, contourf: Contour plot of a matrix
9. cylinder: 3-D cylinder
10. datetick: date formatted tick labels
11. ellipsoid: generates 3-D ellipsoid
12. errorbar: Line plot with error bars
13. ezplot: Plots symbolic expressions
14. feather: Plot velocity vectors
15. figure: Create a figure window
16. fill: Fill 2-D polygons
17. fmesh: 3-D mesh
18. fsurf: 3-D surface
19. gca: Get current axis for modifying axes properties
20. gcf: Current figure handle for modifying figure properties

21. grid: Display grid lines
22. histogram: Histogram plot
23. hold: Hold on the current plot
24. hsv2rgb: Convert colors from HSV space to RGB space
25. im2double: Convert image to double precision
26. image, imagesc: Display image/scaled image from array
27. imapprox: Approximate indexed image by reducing number of colors
28. imread: Read image from file
29. imshow: Display image
30. imwrite: Write image to file
31. imfinfo: Display information about file
32. ind2rgb: Convert indexed image to RGB image
33. legend: Add legend to axes
34. line: Create line
35. mesh, meshc: Mesh/mesh with contour plot
36. peaks: Sample function of two variables
37. pie, pie3: Pie chart, 3-D pie chart
38. plot, plot3: 2-D line plot, 3-D line plot
39. plotyy: Plot using two y-axis labelings
40. polar: Polar plot
41. polarhistogram: Polar histogram
42. quiver: Arrow plot
43. rgb2gray: Convert RGB image to grayscale
44. rgb2hsv: Convert colors from RGB space to HSV space
45. rgb2ind: Convert RGB image to indexed image
46. rgbplot: Plot colormap
47. set: Set graphics object properties
48. sphere: Generate sphere
49. stairs: Stairs plot
50. stem: Stem plot
51. subplot: Multiple plots in a single figure
52. surf: Surface plot
53. text: Text descriptions in graphical plots
54. title: Plot title
55. view: Viewpoint specification
56. xlabel, ylabel: Label x-axis, y-axis
57. scatter, scatter3: 2-D scatter plot, 3-D scatter plot

4. Data Import and Analysis

1. clear: Remove items from workspace memory
2. disp: Display value of variable
3. imageDatastore: Create datastore for image data
4. imread: Read image from graphics file
5. imwrite: Write image to graphics file
6. imfinfo: Information about graphics file
7. load: Load variables from file into workspace
8. max: Maximum value
9. mean: Average value
10. min: Minimum value

5. Programming Scripts and Functions

1. dir: list folder contents
2. fullfile: Build full file name from parts
3. if, elseif, else: Execute statements if condition is true
4. continue: Pass control to next iteration of loop
5. end: Terminate block of code
6. function: Create user-defined function
7. pause: Stop MATLAB execution temporarily

1.2.2 Image Processing Toolbox (IPT) Functions

The IPT functions fall into five different categories: Import Export and Conversions, Display and Exploration, Transformation and Registration, Filtering and Enhancement, and Segmentation and Analysis. Each of these is again divided into several sub-categories, and each sub-category can contain multiple functions. A list of the IPT functions used in this chapter is provided below along with their hierarchical structure and a one-line description for each. The IPT set actually consists of more than 600 functions, out of which a subset has been selected and included in this chapter keeping in view the scope and coverage of this book.

1. Import, Export, and Conversion
   Image-Type Conversions:
   1. gray2ind: Convert grayscale or binary image to indexed image
   2. ind2gray: Convert indexed image to grayscale image
   3. mat2gray: Convert matrix to grayscale image
   4. imbinarize: Convert grayscale image to binary image
   5. imquantize: Quantize image using specified quantization levels
   6. graythresh: Global image threshold using Otsu's method

7. otsuthresh: Global histogram threshold using Otsu's method
8. multithresh: Multilevel image thresholds using Otsu's method

Color:

9. rgb2xyz: Convert RGB to CIE 1931 XYZ
10. xyz2rgb: Convert CIE 1931 XYZ to RGB
11. rgb2lab: Convert RGB to CIE 1976 L*a*b*
12. lab2rgb: Convert CIE 1976 L*a*b* to RGB
13. makecform: Create color transformation structure
14. applycform: Apply device-independent color space transformation

Synthetic Images:

15. checkerboard: Create checkerboard image
16. imnoise: Add noise to image
17. phantom: Create head phantom image

2. Display and Exploration

Basic Display:

1. montage: Display multiple image frames as rectangular montage
2. implay: Play movies, videos, or image sequences
3. warp: Display image as texture-mapped surface
4. imshowpair: Compare differences between images
5. imfuse: Composite of two images

Interactive Exploration:

6. imtool: Open Image Viewer app
7. imageinfo: Image Information tool
8. impixelinfo: Pixel Information tool
9. impixelregion: Pixel Region tool
10. imdistline: Distance tool
11. imcontrast: Adjust Contrast tool
12. imcolormaptool: Choose Colormap tool
13. immagbox: Magnification box to the figure window

Build Interactive Tools:

14. imcontrast: Adjust Contrast tool
15. imcolormaptool: Choose Colormap tool
16. imrect

3. Geometric Transformation and Image Registration

Common Geometric Transformations:

1. imcrop: Crop image
2. imresize: Resize image

3. imrotate: Rotate image
4. imtranslate: Translate image

Generic Geometric Transformations:

5. imwarp: Apply geometric transformation to image
6. affine2d: 2-D affine geometric transformation
7. projective2d: 2-D projective geometric transformation

Image Registration:

8. imregister: Intensity-based image registration
9. normxcorr2: Normalized 2-D cross-correlation

4. Image Filtering and Enhancement

Image Filtering:

1. imfilter: Multidimensional filtering of images
2. fspecial: Create predefined 2-D filter
3. roifilt2: Filter region of interest (ROI) in image
4. wiener2: 2-D adaptive noise-removal filtering
5. medfilt2: 2-D median filtering
6. ordfilt2: 2-D order-statistic filtering
7. gabor: Create Gabor filter
8. imgaborfilt: Apply Gabor filter to 2-D image
9. bwareafilt: Extract objects from binary image by size
10. entropyfilt: Filter using local entropy of grayscale image

Contrast Adjustment:

11. imadjust: Adjust image intensity values or colormap
12. imsharpen: Sharpen image using unsharp masking
13. histeq: Enhance contrast using histogram equalization

Morphological Operations:

14. bwmorph: Morphological operations on binary images
15. imclose: Morphologically close image
16. imdilate: Morphologically dilate image
17. imerode: Morphologically erode image
18. strel: Morphological structuring element
19. imtophat: Top-hat filtering
20. imbothat: Bottom-hat filtering

Deblurring:

21. deconvblind: Deblur image using blind deconvolution
22. deconvlucy: Deblur image using the Lucy–Richardson method
23. deconvwnr: Deblur image using the Wiener filter

ROI-Based Processing:

24. roipoly: Specify polygonal ROI

25. imrect: Create draggable rectangle

Neighborhood and Block Processing:

26. blockproc: Distinct block processing for image

27. col2im: Rearrange matrix columns into blocks

28. im2col: Rearrange image blocks into columns

Image Arithmetic:

29. imadd: Add two images or add constant to image

30. imabsdiff: Absolute difference of two images

31. imcomplement: Complement image

32. imdivide: Divide one image into another

33. imlincomb: Linear combination of images

34. immultiply: Multiply two images

35. imsubtract: Subtract one image from another

5. Image Segmentation and Analysis

Image Segmentation:

1. imoverlay: Burn binary mask into 2-D image

2. superpixels: 2-D superpixel oversegmentation of images

3. boundarymask: Find region boundaries of segmentation

4. labeloverlay: Overlay label matrix regions on 2-D image

5. grayconnected: Select contiguous image region with similar gray values

Object Analysis:

6. bwboundaries: Trace region boundaries in binary image,

7. bwtraceboundary: Trace object in binary image

8. edge: Find edges in intensity image

9. imgradient: Gradient magnitude and direction of an image

10. hough: Hough transform

11. houghlines: Extract line segments based on Hough transform

12. houghpeaks: Identify peaks in Hough transform

13. qtdecomp: Quadtree decomposition

14. imfindcircles: Find circles using circular Hough transform

15. viscircles: Create circle

Region and Image Properties:

16. regionprops: Measure properties of image regions

17. bwarea: Area of objects in binary image

18. bwconncomp: Find connected components in binary image
19. bwdist: Distance transform of binary image
20. bwconvhull: Generate convex hull image from binary image
21. bweuler: Euler number of binary image
22. bwperim: Find perimeter of objects in binary image
23. imhist: Histogram of image data
24. corr2: 2-D correlation coefficient
25. mean2: 2-D average or mean of matrix elements
26. std2: 2-D standard deviation of matrix elements
27. bwlabel: Label connected components in 2-D binary image

Texture Analysis:

28. graycomatrix: Create gray-level co-occurrence matrix (GLCM) from image
29. graycoprops: Compute properties of GLCM
30. entropy: Entropy of grayscale image

Image Quality:

31. psnr: Peak Signal-to-Noise Ratio (PSNR)
32. immse: Mean-squared error
33. ssim: Structural Similarity Index (SSIM) for measuring image quality

Image Transforms:

34. dct2: 2-D discrete cosine transform
35. dctmtx: Discrete cosine transform matrix
36. idct2: 2-D inverse discrete cosine transform

1.2.3 Signal Processing Toolbox (SPT) Functions

1. dct : Discrete cosine transform
2. idct : Inverse discrete cosine transform

1.2.4 Wavelet Toolbox (WT) Functions

1. appcoef2 : 2-D approximation coefficients
2. detcoef2 : 2-D detail coefficients
3. dwt2 : Single level discrete 2-D wavelet transform
4. idwt2 : Single level inverse discrete 2-D wavelet transform
5. wavedec2 : 2-D wavelet decomposition
6. waverec2 : 2-D wavelet reconstruction

1.3 Import Export and Conversions

1.3.1 Read and Write Image Data

The first step of image processing is to read the pixel values of an image and store them in memory. This is called image acquisition. The image can now be viewed in a figure window and information about the image like its dimensions, file size, file format, and bit-depth can be displayed. It is also possible to write an image back to the disk. The file formats supported by MATLAB are mentioned in the previous section. In the example below, the BM function **imread** is used to read data from an image and the data are stored using a variable name. The image is displayed in a figure window by referring to the variable name using the BM function **imshow**. The BM function **whos** displays the height and width of the image in pixels along with the number of channels, the byte size of the file, and the data type. The string *uint8* indicates **unsigned integer 8-bits**, which means that the pixels are represented using 1 byte each and hence can range in value from 0 to 255. The BM function **imfinfo** displays information about the image file like the file-path, file modification date, file-format, bit-depth, and so on. The BM function **clear** clears the memory of previous variable assignments, and the BM function **clc** clears the workspace of existing text and command lines. The BM function **imwrite** allows us to save an image in the clipboard onto a file. The file format used is JPEG which being a lossy format, a quality factor can be specified during saving. The smaller the quality value, the more degraded the image becomes and smaller is its file size. The BM function **figure** creates a new figure window to display the images, while the BM function **subplot** partitions a single figure window into partitions or cells by specifying the number of rows and columns for displaying multiple images. The BM function **title** displays a title on top of the individual cells (Figure 1.3).

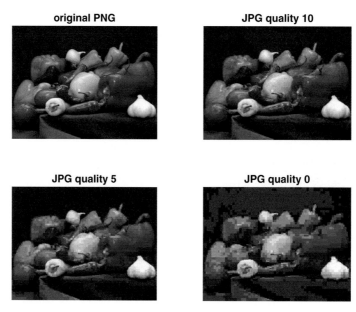

FIGURE 1.3
Image in PNG and JPEG formats with different quality factors.

Example 1.1: Write a program to read an image, display information about it and write it back in a different file format.

```
clear; clc;
x = uint8(1000)        % returns 255
a = imread('peppers.png');
whos
imfinfo('peppers.png')
imwrite(a, 'p10.jpg', 'quality', 10);
imwrite(a, 'p5.jpg', 'quality', 5);
imwrite(a, 'p0.jpg', 'quality', 0);
b = imread('p10.jpg');
c = imread('p5.jpg');
d = imread('p0.jpg');

figure,
subplot(221), imshow(a); title('original PNG');
subplot(222), imshow(b); title('JPG quality 10');
subplot(223), imshow(c); title('JPG quality 5');
subplot(224), imshow(d); title('JPG quality 0');
```

A general 2-D matrix can also be displayed as an image. Elements of the matrix are used as indices into the current colormap, which is a table containing a list of colors to be used for displaying the image. The BM function **colormap** specifies the name of a color-map called **jet** containing 64 colors. The BM function **colorbar** displays all the colors in the current colormap. Colormaps are discussed in detail in Section 1.3. In the following example, the BM function **image** is used to display a matrix as an image by using the values of the matrix to specify the colors they are to be displayed in. Thus, the colors 0–6.4 in the colormap are used to display the matrix. The BM function **imagesc** scales the values so as to cover the entire range of values in the colormap. In this case, the values are scaled by about 10 so that value 1 is displayed using color 10, 2 using color 20, and so on until 6.4 is displayed using color 64. The BM function **axis** makes the current axis box square in shape (Figure 1.4).

Example 1.2: Write a program to display a 2-D matrix as an image by specifying a colormap.

```
clear; clc;
a = [1, 3, 4 ; 2, 6.4, 0 ; 0.5, 5, 5.5];
colormap jet
subplot(121), image(a); colorbar; axis square;
subplot(122), imagesc(a); colorbar; axis square;
```

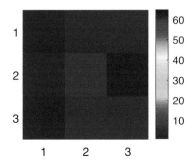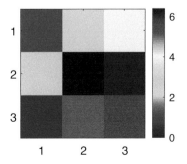

FIGURE 1.4
Output of Example 1.2.

1.3.2 Image-Type Conversion

A color image has three **color channels** which contain the red (R), green (G), and blue (B) information. These three colors are called **primary colors**, as all other colors can be generated by combining various percentages of these colors. It is possible to separate the three color channels of an image and display them separately. Since for a color image each pixel takes up 24 bits, the three color channels are each allotted 8-bits of space. This implies that each color channel behaves like a grayscale image with intensities ranging from 0 (color absent) to 255 (color present in full intensity). The condition which means full intensity of 255 is referred to by the term **saturation**. Thus, saturated red is represented by the triad (255, 0, 0), saturated green by (0, 255, 0), and saturated blue by (0, 0, 255). The superposition of three color channels of varying intensities creates the perception of color. The portions of the image, where a specific primary color is present, are reflected by a brighter area in the corresponding color channel, while areas which are devoid of a primary color are reflected by darker areas in the corresponding channel. For example, red portions of an image are displayed as white regions in the red color channel (which appears like a grayscale image). Equal amounts of two primary colors generate colors called **secondary colors**. Red and green produce yellow Y (255, 255, 0), green and blue produce cyan C (0, 255, 255), and red and blue produce magenta M (255, 0, 255). In MATLAB, a color image is represented by three dimensions: height (number of rows in the image), width (number of columns in the image), and channels (primary colors R, G, B). In the following example, a color image along with its three color channels are displayed. Note that each of the color channels appears as a grayscale image because they contain 8-bits of information. The **:** **(colon)** operator signifies the range of values from start to end, for each channel. The colon only by itself indicates all current values. After the three dimensional color image is read and stored in a variable, a value of 1 for the third dimension indicates the R channel, a value of 2 indicates the G channel, and a value of 3 indicates the B channel. The **%** **(percent)** symbol indicates comments (Figure 1.5).

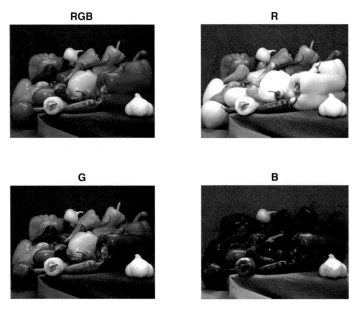

FIGURE 1.5
Output of Example 1.3.

Example 1.3: Write a program to read a color image and display its color channels separately

```
clear; clc;
a = imread('peppers.png');
ar = a(:,:,1);   % red channel
ag = a(:,:,2);   % green channel
ab = a(:,:,3);   % blue channel
subplot(221), imshow(a); title('RGB');
subplot(222), imshow(ar); title('R');
subplot(223), imshow(ag); title('G');
subplot(224), imshow(ab); title('B');
```

As mentioned before, images can generally be classified into three types: color, grayscale, and binary. There are well-established methods to convert one image type to another. A color image has three matrices of primary colors R, G, B, each taking up 8-bits per pixel. A grayscale image, on the other hand, consists of only a single matrix of 8-bits per pixel. To convert a color image to a grayscale image therefore consists of merging the three primary matrices into a single matrix. The simplest way of doing this would be to take an average of the R, G, B values for each pixel. However, there are two problems in this simplistic approach: (1) First, there is a chance that multiple colors may map to the same grayscale intensity. In fact all the three primary colors would be converted to the same intensity of $(1/3)*255$, and all secondary colors would be converted to the same intensity of $(2/3)*255$. This would lead to a degraded version of the image, where details might be lost. (2) Second, the human eye does not place equal emphasis on all the three primary colors. Studies have shown that the human vision is most sensitive to the yellow–green range of the visible spectrum and least sensitive to the blue range. Keeping the above factors into consideration, the accepted scheme of color to grayscale conversion, as proposed by the **International Commission on Illumination** (CIE), abbreviated based on its French name *Commission Internationale de l'Éclairage*, in 1931 (CIE, 1931), models the intensity Y as a weighted sum of the RGB components in the ratio 3: 6: 1

$$Y = 0.2989 * R + 0.5870 * G + 0.1140 * B$$

For converting a grayscale image to binary, a specific gray-level intensity needs to be specified as the **binarization threshold** (T) below which all pixel values would be converted to black and above which all pixel values would be converted to white. Lower the threshold a larger portion of the image is converted to white, higher the threshold more pixels will be black. The threshold value is usually specified as a fraction between 0 and 1 or as a percentage value from 0 to 100. A threshold value of k indicates that all values from 0 to $k* 255$ are converted to black and remaining values from $k * 255 + 1$ to 255 are converted to white. In the following example, the IPT function **rgb2gray** converts an RGB color image to grayscale intensity image using the above equation. The IPT function **imbinarize** is used to convert the grayscale image to binary image by specifying a threshold value. Two different threshold values $T = 0.6$ and $T = 0.5$ are used i.e. in the first case, all pixel values from 0 to $0.3 \times 255 = 76$ are converted to black and all values from 77 to 255 are converted to white. Likewise for the second case, all values from 0 to 127 are converted to black and values from 128 to 255 are displayed as white (Figure 1.6).

Example 1.4: Write a program to convert a color image to grayscale and binary

```
clear; clc;
a = imread('peppers.png');
b = rgb2gray(a);
```

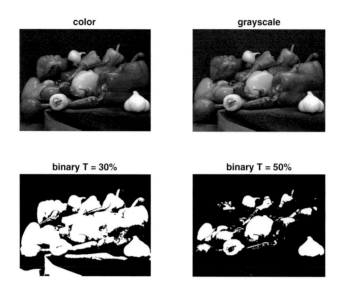

FIGURE 1.6
Output of Example 1.4.

```
c = imbinarize(b, 0.3);
d = imbinarize(b, 0.5);

figure,
subplot(221), imshow(a); title('color');
subplot(222), imshow(b); title('grayscale');
subplot(223), imshow(c); title('binary T = 30%');
subplot(224), imshow(d); title('binary T = 50%');
```

Since different thresholds generate different images, the question of a best or optimum threshold naturally arises. Nobuyuki Otsu proposed a method (Otsu, 1979) henceforth referred to as **Otsu's method**, to automatically calculate the threshold from grayscale images which minimizes the intra-class variance of black and white pixels from image histograms. Histograms are discussed in Section 1.6. If t be the threshold intensity level and σ_1 and σ_2 be the standard deviation of the histogram values on either side of the threshold, then Otsu's method fixes the threshold at a point, where the weighted sum of the variances $(q_1 \cdot \sigma_1 2 + q_2 \cdot \sigma_2 2)$ is minimum. Here, q_1 and q_2 are the weights, and if $p(i)$ be the pixel count at the i-th intensity level, and L be the total number of intensity levels, then:

$$q_1 = \sum_{i=1}^{t} \left[p(i) \right]$$

$$q_2 = \sum_{i=t+1}^{L} \left[p(i) \right]$$

The class means are defined as follows:

$$\mu_1 = \sum_{i=1}^{t} \left[i.p(i)/q_1 \right]$$

$$\mu_2 = \sum_{i=t+1}^{L} \left[i.p(i)/q_2 \right]$$

The class variances are defined as follows:

$$\sigma_1^2 = \sum_{i=1}^{t} \left[(i - \mu_1)^2 \cdot p(i)/q_1 \right]$$

$$\sigma_2^2 = \sum_{i=t+1}^{L} \left[(i - \mu_2)^2 \cdot p(i)/q_2 \right]$$

Apart from single thresholds which leads to a binary image, multiple thresholds can also be computed which generates an image with two or three gray levels. In the following example, the IPT function **graythresh** is used to calculate an optimum threshold from a grayscale image for conversion to binary using Otsu's method. Alternatively, the IPT function **otsuthresh** can also be used to calculate the threshold by Otsu's method using the histogram counts. The IPT function **multithresh** can be used to create a specified number of multiple thresholds using Otsu's method. These thresholds can be fed to the IPT function **imquantize** to generate images with multiple gray levels. In the following example, Otsu's method is used to calculate a single threshold in the first two cases, which uses black for levels 0–100 and white for levels 101–255. In the third case, two thresholds are computed which represent the image using three gray shades: black for levels 0–85, 50% gray or 128 for levels 86–171, and white for levels 172–255. In the fourth case, three thresholds are computed which are used to split the image into four gray shades: black for levels 0–49, 33.33% gray ($0.3333 \times 255 = 85$) for levels 50–104, 66.66% gray ($0.6666 \times 255 = 170$) for levels 105–182, and white for levels 183–255 (Figure 1.7). We would discuss how to read the actual pixel values from a displayed image, in the next section.

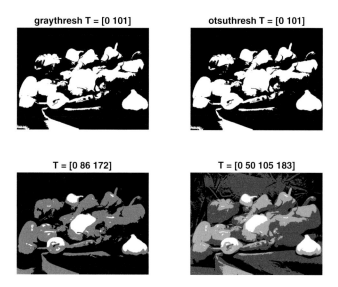

FIGURE 1.7
Output of Example 1.5.

Example 1.5: Write a program to convert a greyscale image to binary using automatically calculated single and multiple thresholds

```
clear; clc;
a = imread('peppers.png');
b = rgb2gray(a);
t1 = graythresh(b);
h = imhist(b);
t2 = otsuthresh(h);
c = imbinarize(b, t1);
d = imbinarize(b, t2);
t1, t2

t3 = multithresh(a, 2);
t4 = multithresh(a, 3);
e = imquantize(b, t3);
f = imquantize(b, t4);

figure,
subplot(221), imshow(c); title('graythresh T = [0 101]');
subplot(222), imshow(d); title('otsuthresh T = [0 101]');
subplot(223), imshow(e, []); title('T = [0 86 172]');
subplot(224), imshow(f, []); title('T = [0 50 105 183]');
```

Quantization is the process of substituting a continuously varying function by a set of discrete values also called levels. For images, the quantization levels are also called gray levels and are usually represented by specifying the number of bits required to generate the requisite number of gray levels. Typical values include 8-bit (256 levels), 6-bit (64 levels), 4-bit (16 levels), and so on up to 1-bit, where the image is represented using two levels, black and white, and hence called binary image. In the following example, the BM function **round** rounds the values of the pixels to the nearest integer. When the image is divided by 255, all pixel values from 0 to 127 are rounded to 0, and values from 128 to 255 are rounded to 1. On multiplying these rounded values by 255, the entire set of pixel values is divided into two discrete values: 0 and 255, resulting in an image with two gray levels (binary). When the image is divided by 127, all pixel values from 0 to 63 are rounded to 0, values from 64 to 190 are rounded to 1, and values from 191 to 255 are rounded to 2. On multiplying these rounded values by 127, the entire set of pixel values is divided into three discrete values: 0, 127, and 255, resulting in an image with three gray levels. When the image is divided by 85, all pixel values from 0 to 42 are rounded to 0, values from 43 to 127 are rounded to 1, values from 128 to 212 are rounded to 2, and values from 213 to 255 are rounded to 3. On multiplying these rounded values by 85, the entire set of pixel values is divided into four discrete values: 0, 85, 170, and 255, resulting in an image with four gray levels. In a similar fashion, the image when divided by 51, is divided into five discrete values: 0, 63, 126, 189, and 252. Also when divided by 63, it is divided into six discrete values: 0, 51, 102, 153, 204, and 255. The BM function **unique** returns these unique discrete values from the modified images (Figure 1.8).

Example 1.6: Write a program to reduce the quantization levels of a grayscale image

```
clear; clc;
a = imread('peppers.png');
b = rgb2gray(a);
c = round(b/51)*51;      %6 levels
d = round(b/63)*63;      %5 levels
e = round(b/85)*85;      %4 levels
f = round(b/127)*127;    %3 levels
g = round(b/255)*255;    %2 levels

subplot(231), imshow(b); title('original');
```

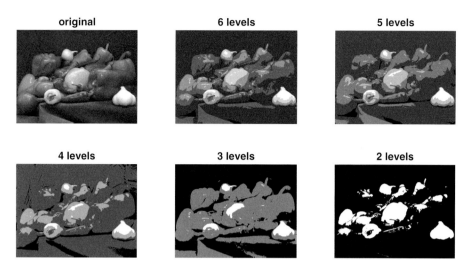

FIGURE 1.8
Output of Example 1.6.

```
subplot(232), imshow(c);  title('6 levels');
subplot(233), imshow(d);  title('5 levels');
subplot(234), imshow(e);  title('4 levels');
subplot(235), imshow(f);  title('3 levels');
subplot(236), imshow(g);  title('2 levels');

unique(c), unique(d), unique(e), unique(f), unique(g)
```

An **indexed image** is a color image, where the pixel values do not contain the actual color information but only index numbers. The index numbers are converted to color values by looking up a colormap or a table in memory with a set of rows and four columns, called the **color look up table** (CLUT). The first column contains the index number of the row in a sequential manner, while the other three columns contain the RGB values for each index number. When the image is to be displayed, the index numbers from the image are converted to color values using the table, and corresponding pixels are generated. To display an indexed image, the corresponding colormap needs to be specified. The colormap usually contains 64 or 128 or 256 colors, but a smaller subset of colors can also be used to display an approximate version of the image. When an image is displayed using a CLUT containing a smaller number of colors than the original, then a process called dithering is used to approximate the colors absent in the CLUT. **Dithering** is a process which mixes available colors to produce new colors. It varies the distribution of the existing colors to simulate colors not present in the color table, in a way similar to how distribution of black dots can be varied to simulate shades of gray, usually seen in the print medium (Figure 1.9).

An indexed image can be displayed using a colormap different from the original containing a different number of colors. If the number of colors is more in the new map, then the original set of colors are displayed in the image. If the number of colors is less in the new map, the image is displayed using a subset of the original colors. Dithering can be used to improve the quality of image by simulating colors not present in the new map. The BM function **imapprox** is used to display an approximate version of the image with reduced number of colors. Two options are available: the *dither* option produces dithering to change the distribution of the available colors for simulating colors absent in the colormap, while the *nodither* option displays the image only using the colors available in

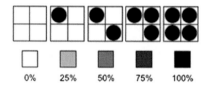

FIGURE 1.9
Dithering.

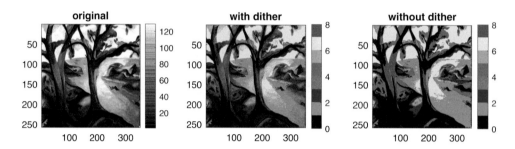

FIGURE 1.10
Output of Example 1.7.

the colormap. In the following example, an indexed image is first displayed using the colormap with a full set of 128 colors. Next, the number of colors in the colormap is reduced to eight and the same image is displayed using the new map, first with dithering enabled and next with dithering disabled. A colorbar at the side shows which colors are available in the colormaps. While in the first case the colorbar shows all 128 colors, in the second and third cases, the same set of eight colors are displayed (Figure 1.10).

> **Example 1.7: Write a program to display an indexed image using a CLUT with a reduced number of colors.**
>
> ```
> clear; clc;
> load trees;
> [Y, newmap] = imapprox(X, map, 8, 'dither');
> [Z, newmap] = imapprox(X, map, 8, 'nodither');
> a = subplot(131), image(X);
> colormap(a, map); colorbar; axis square; title('original');
> b = subplot(132), image(Y);
> colormap(b, newmap); colorbar; axis square; title('with dither');
> c = subplot(133), image(Z);
> colormap(c, newmap); colorbar; axis square; title('without dither');
> ```

A color RGB image can be converted to an indexed image. The pixel values are replaced by index values of the CLUT, and the color table is created in memory to reference those values. The CLUT contains as many entries as there are colors in the image and each entry contains the R, G, B components of the corresponding color. The IPT function **rgb2ind** is used to convert an RGB image to indexed format and create the corresponding colormap. In the following example, the first image *image_1* contains only two most representative colors and hence is a matrix containing two values 0 and 1 (Figure 1.11). The corresponding colormap *map_1* contains two rows containing RGB values of the two colors expressed within the range 0–1:

original

2 colors

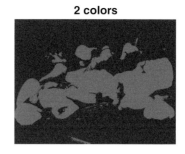

5 colors

10 colors

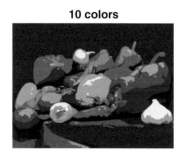

FIGURE 1.11
Output of Example 1.8.

```
0.3137    0.1882    0.2314
0.8118    0.4039    0.2078
```

Converted to the range 0 to 255, these colors are c_1 (80, 48, 59) and c_2 (207, 103, 53). The second image *image_2* contains numbers 0–4 to represent five most representative colors specified in *map_2* which are:

```
0.2784    0.1373    0.2353
0.7608    0.1686    0.1373
0.8902    0.7255    0.6353
0.4275    0.3765    0.2235
0.8471    0.5569    0.1020
```

The *nodither* option ensures that the original colors of the image are inserted into the map and not modified through mixing. Similarly the third *image_3* contains numbers 0–9 representing ten colors specified in *map_3*:

```
0.3333    0.0902    0.0745
0.6745    0.1176    0.1216
0.8118    0.5608    0.5333
0.4078    0.3961    0.0941
0.6510    0.5451    0.1255
0.4824    0.3216    0.5412
0.2706    0.1451    0.2627
0.9686    0.5647    0.0824
0.9647    0.8941    0.7451
0.8784    0.2431    0.1608
```

Example 1.8: Write a program to convert an RGB image to indexed format and display using varying number of colors

```
clear; clc;
a = imread('peppers.png');
[image_1, map_1] = rgb2ind(a, 2, 'nodither');
[image_2, map_2] = rgb2ind(a, 5, 'nodither');
[image_3, map_3] = rgb2ind(a, 10, 'nodither');

subplot(221), imshow(a); title('original');
subplot(222), imshow(image_1, map_1); title('2 colors');
subplot(223), imshow(image_2, map_2); title('5 colors');
subplot(224), imshow(image_3, map_3); title('10 colors');
```

The index values and the actual colors they represent can be directly viewed by using the *Data Cursor* tool of the figure window and clicking on a point in the image. The index values, the RGB values, and the location are displayed within text boxes over the image. Figure 1.12 below shows *image_2*, and the five colors specified in *map_2*.

Since the image only contains index values, it is possible to change the colors displayed by changing the colormap. Thus, if *image_1* is displayed using a map other than *map_1*, then the overall color scheme can be changed. Similar is the case when the *image_2* is displayed using a map other than *map_2* (Figure 1.13).

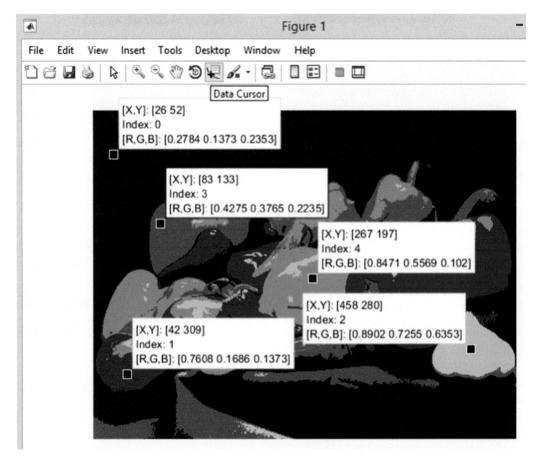

FIGURE 1.12
Image along with color details.

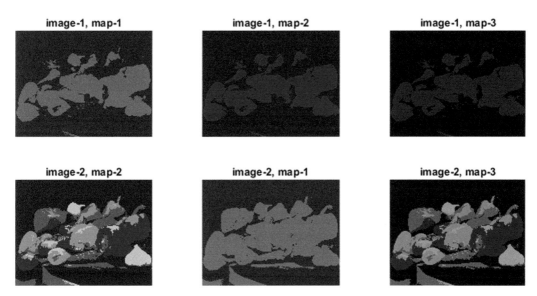

FIGURE 1.13
Displaying images using multiple colormaps.

When *image_1* is displayed using *map_1*, then the colors reflect the two entries in the CLUT (top-left). When *image_1* is displayed using *map_2*, then the first two entries in the CLUT are used to display the two colors in the image (top-right) while the other colors are ignored. When *image_2* is displayed using *map_1*, then the two entries in the CLUT are used to display the image (bottom-left). When *image_2* is displayed using *map_3*, then the first five entries in the CLUT are used to display the image (bottom-right) while the remaining colors are ignored (Figure 1.14).

The process of conversion from RGB to indexed format can also be reversed by specifying an indexed color image and a map to transform it back to RGB format. In this case, the index values in the image are replaced by actual RGB pixel values read from the colormap. The colormap itself can also be visualized as a graphical plot. As already mentioned, the colormap contains RGB values represented in the range 0–1. For example, the *map_1* of the previous example containing two colors is represented as below:

```
              R          G          B
color 1    0.3137     0.1882     0.2314
color 2    0.8118     0.4039     0.2078
```

The three columns indicate R, G, B components of each color, and the two rows indicate two colors in the map. This can be graphically represented as a red colored line joining the first R value (0.3137) with the second R value (0.8118), a green colored line joining the first G value (0.1882) with the second G value (0.4039), and a blue line joining the first B value (0.2314) with the second B value (0.2078). If the colormap contains more color values, then each of these pairs are joined by segments of colored lines. In the following example, the IPT function **ind2rgb** converts an indexed color image to RGB format by reading the color values from the specified colormaps and substituting them back into the image. The BM function **rgbplot** is used to represent the colormaps graphically, as explained in the previous paragraph (Figure 1.15).

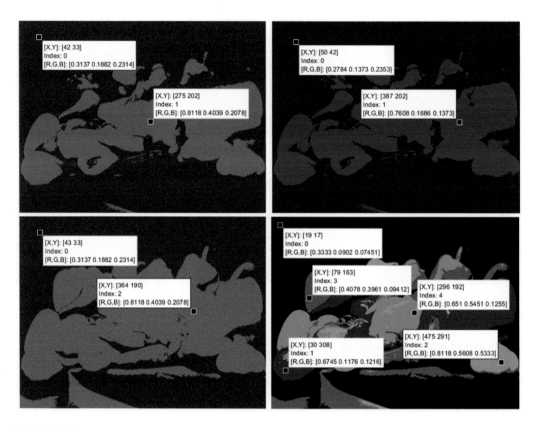

FIGURE 1.14
Displaying images with customized colormaps.

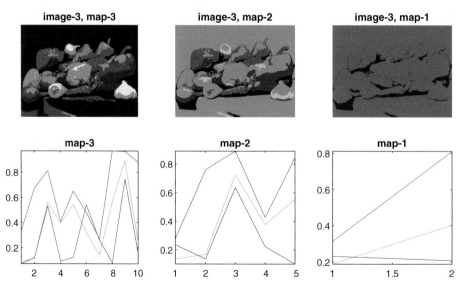

FIGURE 1.15
Output of Example 1.9.

**Example 1.9: Write a program to convert an indexed
color image to RGB format by using different colormaps
and represent each colormap graphically.**

```
clear; clc;
a = imread('peppers.png');
[b1, map1] = rgb2ind(a, 2, 'nodither');
[b2, map2] = rgb2ind(a, 5, 'nodither');
[b3, map3] = rgb2ind(a, 10, 'nodither');

p = ind2rgb(b3, map3);
q = ind2rgb(b3, map2);
r = ind2rgb(b3, map1);

subplot(231), imshow(p); title('image-3, map-3');
subplot(232), imshow(q); title('image-3, map-2');
subplot(233), imshow(r); title('image-3, map-1');
subplot(234), rgbplot(map3); title('map-3'); axis tight;
subplot(235), rgbplot(map2); title('map-2'); axis tight;
subplot(236), rgbplot(map1); title('map-1'); axis tight;
```

MATLAB has a set of 18 inbuilt predefined colormaps: parula (default), jet, hsv, hot, cool, spring, summer, autumn, winter, gray, bone, copper, pink, lines, colorcube, prism, flag, and white. Each of these maps consist of 64 colors. An indexed color image can be displayed using any of the predefined maps. The number of colors for a specific colormap, say *winter*, can be obtained using the command: `size(winter)`. The following example shows the same image being displayed using different colormaps, which are specified as arguments to the imshow function. A colormap can also be invoked by using its name as a function name (Figure 1.16).

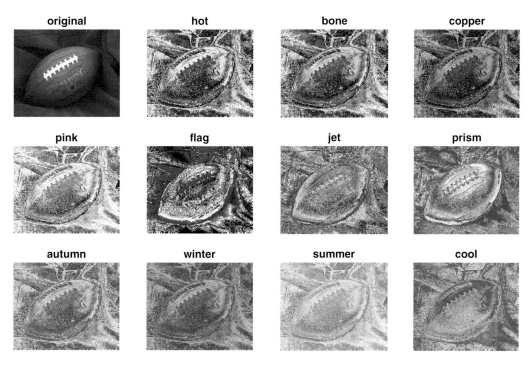

FIGURE 1.16
Output of Example 1.10.

Example 1.10: Write a program to display an indexed color image using various inbuilt colormaps

```
clear; clc;
rgb = imread('football.jpg');
[g, map] = rgb2ind(rgb, 64);
subplot(341), imshow(g, map); title('original');
subplot(342), imshow(g, hot); title('hot');
subplot(343), imshow(g, bone); title('bone');
subplot(344), imshow(g, copper); title('copper');
subplot(345), imshow(g, pink); title('pink');
subplot(346), imshow(g, flag); title('flag');
subplot(347), imshow(g, jet); title('jet');
subplot(348), imshow(g, prism); title('prism');
subplot(349), imshow(g, autumn); title('autumn');
subplot(3,4,10), imshow(g, winter); title('winter');
subplot(3,4,11), imshow(g, summer); title('summer');
subplot(3,4,12), imshow(g, cool); title('cool');
```

In some cases, specially when displaying graphical plots, we might require limited number of colors to be displayed from the colormaps. We can specify how many colors to be displayed by creating a customized map. In the example below, *map1* contains eight colors from the *hot* colormap which can be displayed by typing map1. The BM function **colormap** sets the specified figure's colormap to the specified customized map. In the following example, the BM function **contourf** generates a filled contour plot and the BM function **peaks** is a sample function of two variables, obtained by translating and scaling Gaussian distributions. The BM function **rand** generates a set of random numbers within the interval (0, 1). The BM functions **hot**, **spring**, **summer**, and **jet** are used to generate colormap arrays with the specified number of colors. The BM function **rng** is a random number generator used with a seed number to produce a predictable sequence of numbers. Instead of using colors from a predefined map, it is also possible to create a set of custom colors in a user-specified map viz. *mymap1* and *mymap2* (Figure 1.17).

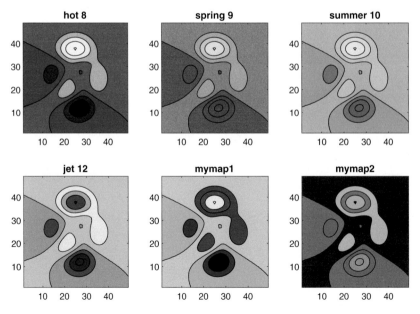

FIGURE 1.17
Output of Example 1.11.

Example 1.11: Write a program to display a graphics plot using limited colors from predefined and customized colormaps

```
clear; clc;

ax1 = subplot(231); map1 = hot(8);
contourf(peaks); colormap(ax1, map1); title('hot 8');
ax2 = subplot(232); map2 = spring(9);
contourf(peaks); colormap(ax2, map2); title('spring 9');
ax3 = subplot(233); map3 = summer(10);
contourf(peaks); colormap(ax3, map3); title('summer 10');
ax4 = subplot(234); map4 = jet(12);
contourf(peaks); colormap(ax4, map4); title('jet 12');
mymap1 = [  0 0 0
            0 0 1
            0 1 0
            0 1 1
            1 0 0
            1 0 1
            1 1 0
            1 1 1];
ax5 = subplot(235); map5 = mymap1;
contourf(peaks); colormap(ax5, map5);  title('mymap1');
rng(2); mymap2 = rand(10,3);
ax6 = subplot(236); map6 = mymap2;
contourf(peaks); colormap(ax6, map6); title('mymap2');
```

It is possible to extend the number of colors in a colormap beyond the default value of 64 or to use a specific subset of colors from the default colorset. In the following example, the BM function **meshgrid** generates a 2-D rectangular grid of equally spaced points from –10 to +10 and the BM function **surf** generates a surface by plotting a specified *Z* value for each of the (*X*, *Y*) coordinates of the mesh grid. The surface is then colored using the colormap named *colorcube*. Initially, a default version of the colormap is used consisting of 64 colors, then an extended version of the colormap consisting of 75 colors is used and finally a specific set of colors from the colormap i.e. colors 45–50, both inclusive, is used (Figure 1.18).

Example 1.12: Write a program to generate a surface and use the default, a subset and a superset of a colormap to color the surface

```
clear; clc;
[X,Y] = meshgrid(-10:1:10);
Z = X + Y;
a = subplot(131),surf(X,Y,Z); colormap(a, colorcube);
title('colorcube(64)');
```

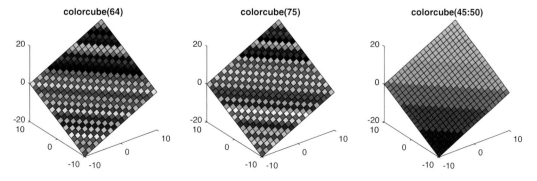

FIGURE 1.18
Output of Example 1.12.

```
b = subplot(132),surf(X,Y,Z); colormap(b, colorcube(75));
title('colorcube(75)');
cc = colorcube; cc = cc(45:50,:);
c = subplot(133),surf(X,Y,Z); colormap(c, cc);
title('colorcube(45:50)');
```

When an image is read into the system using the *imread* function, the image is represented as a 2-D matrix having datatype *uint8* which stands for **unsigned integer 8-bits**. This means that the values of the matrix range from 0 to 255. Sometimes for the sake of visualization, it becomes necessary to display an arbitrary 2-D matrix as an image. However, when a matrix is created in MATLAB, such as by using A = [−1, 2, 3; 4, −5, 6], it is by default allotted a datatype of *double* which stands for **double precision 64-bits**. Now, if this matrix needs to be displayed as an image, then its values should be converted to fractional numbers within the range [0, 1]. This is because when *imshow* is used to display the image, it expects one of the two situations to occur (a) either the matrix is of datatype *uint8* and its values range over [0, 255] (b) or the matrix is of datatype *double* and its values range over [0, 1]. Additionally, an image of type *uint8* can also be required to be converted to type *double* when its pixel values need to be manipulated in such a way that it exceeds the maximum value of 255. In the following example, the grayscale image *I* has an intensity ranging from 0 to 255; however, when it is passed through an edge detection filter, the resultant matrix *J* has values ranging from approximately −700 to +700. Filters are discussed in Section 1.6. To represent these values *J* is saved as datatype *double*. Now to display this as a grayscale image, the range of values is mapped to the range [0,1] because of the reason mentioned above. The IPT function **mat2gray** is used to convert a 2-D matrix of type double into a grayscale image by representing its values within the range [0,1]. As another example, *L* is a matrix of random integer values ranging from −300 to +300. To display this as a grayscale image, the values are mapped to the range [0,1] using the function *mat2gray*. As a third example, *X* is an RGB image saved as type *uint8*. One of its channels is isolated and its values are increased by 100. In the *uint8* mode, most of the values would freeze at 255 and the modification would not be visually perceptible. So its datatype is first changed to *double*, the channel values are increased and finally its values are mapped to the range [0, 1] using the function *mat2gray* (Figure 1.19).

Example 1.13: Write a program to illustrate how a 2-D matrix of arbitrary values can be displayed as an image

```
clear; clc;
I = imread('coins.png');
J = filter2(fspecial('sobel'), I);
K = mat2gray(J);
rng('default');
L = randi([-300 300], 4, 5);
M = mat2gray(L);
X = imread('peppers.png');
n = 2;
Xd = double(X);
Y = Xd;
Yn = Y(:,:,n);
Yn = Yn + 100;
Y(:,:,n) = Yn;
Y = mat2gray(Y);
figure,
subplot(231), imshow(I); title('I');
subplot(232), imshow(J); title('J');
subplot(233), imshow(K); title('K');
subplot(234), imshow(M); title('M');
subplot(235), imshow(X); title('X');
subplot(236), imshow(Y); title('Y');
```

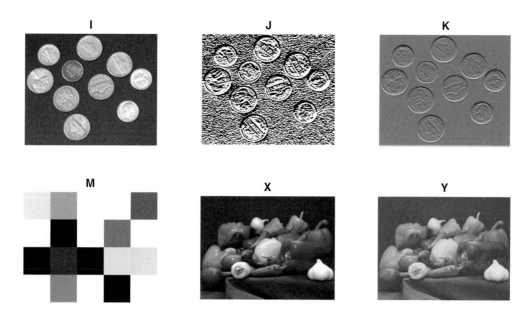

FIGURE 1.19
Output of Example 1.13.

1.3.3 Image Color

Although we have been discussing concept of color channels of RGB images and color-maps of indexed images, in this section we formally lay the foundation of color representation and color models. The phenomenon of **color perception** depends on three factors: nature of light, physiology of human vision, and interaction of light with matter. Out of the total range of electromagnetic spectrum, a small part called **visible spectrum**, spanning around 400–700 nm in the wavelength, causes the sensation of vision in our eyes. The photoreceptors in the retina of the human eyes respond to three different type of colors – red, green, and blue, depending on the frequencies of light. When light strikes an opaque object, part of it gets absorbed and part of it gets reflected back to our eyes. The frequencies present in the reflected light imparts the color of the object from which it is reflected. Studies have revealed that the human eye is capable of distinguishing more than 16 million different colors (Young, 1802). **Color models** enable us to represent and communicate information regarding color in a meaningful and computer recognizable way. All colors that we see around us can be derived by mixing a few elementary colors in different proportions, called **primary colors**. Primary colors mixed together gives rise to **composite colors**, while two primary colors mixed in equal proportions generate **secondary colors**. The total range of primary, secondary, and composite colors is defined by the **color space** of the model (Wright, 1929; Guild, 1931). A number of color models have been developed over the years, and formulae for conversion between the color spaces defined by these models have been standardized by the global standard body **International Color Consortium** (ICC) and documented in their specifications, the latest of which is version 4.3 (*http://www.color.org/specification/ICC1v43_2010-12.pdf*). These specifications are technically identical to ISO 15076-1:2010 standard. A few of these models and inter-relations between them are discussed below.

The **RGB color model** is used to represent all colors produced by red (R), green (G), and blue (B) primaries. Each primary color is quantized to 256 levels so that the total number of colors in the RGB model is $256 \times 256 \times 256$ or 16.7 million. The RGB model takes the form of a color cube, where red, green, and blue are three orthogonal axes along which the RGB components of a color are plotted in the range 0–255. Any point inside the cube therefore represents a specific color, for example, 'orange' is composed of RGB components (245, 102, 36). The RGB color model is applicable for emission of colored lights and is additive in nature i.e. the intensities increase when primary colors are mixed. When all three primaries are added in full intensities, the resultant color is white (255, 255, 255), whereas black (0, 0, 0) is the absence of all colors (Figure 1.20).

The **CMY color model** is used to represent all colors produced by cyan (C), magenta (M), and yellow (Y) primaries. Each primary color is expressed in percentage values and quantized to 100 levels so that the total number of colors in the CMY model is $100 \times 100 \times 100$ or 1 million. The CMY model also takes the form of a color cube, where cyan, magenta, and yellow are three orthogonal axes along which the CMY components of a color are plotted in the range 0–100. Any point inside the cube therefore represents a specific color, for example 'orange' is composed of CMY components (4%, 60%, 86%). The CMY color model is applicable for colored inks printed on paper and is based on absorption of light instead of emission. The model is therefore subtractive in nature i.e. the intensities decrease when primary colors are mixed due to absorption. When all three primaries are mixed in equal proportions, the resultant color is theoretically black (0%, 0%, 0%). However, since chemical inks cannot be absolutely pure in their composition, the color actually obtained is usually dark brown, thus a fourth component of pure black (K) ink is added along with the three primaries to generate the CMYK model. The following example illustrates how to draw polygonal shapes and fill them with specified colors. The colors are expressed as RGB triplets with values in the range [0, 1]. If values are in hexadecimal units, the BM function **hex2dec** is used to convert them to decimal. The BM function **fill** is used to fill 2-D polygons given their vertex coordinates and specified color. The BM function **rectangle** is used to draw 2-D rectangles with sharp or curved boundaries. Boundaries of the rectangle are sharp when the *Curvature* parameter is 0 and curved when the parameter value is increased; with a maximum value of 1, the rectangle is converted to a circle. The BM function **polyshape** is used to draw 2-D polygons with specified vertex coordinates. The BM function **boundary** returns the boundary of a set of data points. The BM function **sphere** is used to draw a spherical surface. A new colormap using three specified colors is created and used to colorize the sphere (Figure 1.21).

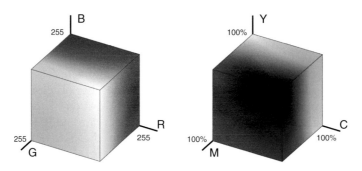

FIGURE 1.20
RGB and CMY color models.

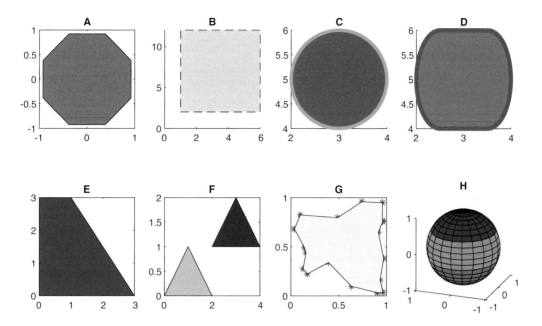

FIGURE 1.21
Output of Example 1.14.

Example 1.14: Write a program to draw some polygonal shapes and surfaces and fill them with specified colors.

```
clear; clc;

subplot(241),
t = (1/16:1/8:1)'*2*pi;
x = cos(t);
y = sin(t);
orange = [245, 102, 36]/255;
fill(x,y, orange);
axis square; title('A');

subplot(242),
aquamarine = [0.4980, 1, 0.8314];
rectangle('Position',[1,2,5,10],'FaceColor',aquamarine, 'LineStyle', '--');
axis square; title('B');

subplot(243),
olivegreen = [hex2dec('55'), hex2dec('6b'), hex2dec('2f')]/255;
goldenrod = [hex2dec('da'), hex2dec('a5'), hex2dec('20')]/255;
pos = [2 4 2 2];
rectangle('Position',pos,'Curvature',[1 1], 'FaceColor', ...
    olivegreen, 'EdgeColor', goldenrod, 'LineWidth',3);
axis square; title('C');

subplot(244),
tomato = [hex2dec('ff'), hex2dec('63'), hex2dec('47')]/255;
pos = [2 4 2 2];
rectangle('Position',pos,'Curvature',[0.5 1], 'FaceColor', tomato, ...
    'LineWidth',4, 'EdgeColor', orange-0.1);
axis square; title('D');

subplot(245),
pgon = polyshape([0 0 1 3], [0 3 3 0]);
```

```
plot(pgon);
fill([0 0 1 3], [0 3 3 0], 'b');  axis square; title('E');

subplot(246),
x1 = [0 1 2];
y1 = [0 1 0];
x2 = [2 3 4];
y2 = [1 2 1];
darkgreen = [00, 64/255, 00];
polyin = polyshape({x1,x2},{y1,y2});
plot(polyin);
fill(x1,y1, 'g', x2, y2, darkgreen); axis square; title('F');

subplot(247),
rng('default')
kx = rand(1,30);
ky = rand(1,30);
plot(kx, ky, 'r*')
k = boundary(kx',ky');
hold on;
plot(kx(k),ky(k));
fill(kx(k), ky(k), 'y'); axis square; title('G');

subplot(248),
[x,y,z] = sphere;
surf(x,y,z); view(-70,20)
crimson = [hex2dec('DC'), hex2dec('14'), hex2dec('3C')]/255;
chocolate = [hex2dec('d2'), hex2dec('69'), hex2dec('1e')]/255;
coral = [hex2dec('ff'), hex2dec('7f'), hex2dec('50')]/255;
redmap = [chocolate ; coral ; crimson ];
colormap(redmap); axis square; title('H');
```

Both the RGB and CMY color models have one similarity – they are **device-dependent** models. This means that the actual colors produced are dependent on the physical characteristics of the devices using which the colors are displayed. RGB-based colors depend on the LED/CRT properties of the monitors and CMY colors depend on the chemical properties of the printing inks. The colors therefore are not absolute or constant and would tend to vary from device to device. To define **device-independent** color models, the global standards body CIE defined the **CIE XYZ** color model in 1931. Through a series of experiments, it was found that the human eye had photoreceptors called cone cells, which had peak sensitivities at wavelengths around 700 nm for red, 550 nm for green, and 440 nm for blue (Wright, 1929; Guild, 1931). These parameters are called **tristimulus** values, and they describe a color sensation. To model the color sensation, the CIE proposed a color model based on the tristimulus values denoted as X, Y, Z and the color space was henceforth known as the CIE XYZ color space. The colors of the CIE XYZ model are based on the human sensation of color from a physiological aspect and hence not dependent on any device characteristics. Due to the nature of distribution of the cone cells within the human eye, the XYZ values depend on the observer's angular field of view. To standardize this, CIE in 1931 defined the standard angular view as 2°. Since colors look different under various lighting conditions, two standard **white point illuminant** specifications were used: (1) D50 which has a color temperature of 5003 K and (2) D65 which has a color temperature of 6504 K. The letter 'D' indicates *daylight* meaning that the standard illuminants were proposed to simulate daylight conditions (CIE, 2013). The **color temperature** denotes the color of light radiated by an ideal black body radiator when heated to the specified temperature in degrees Kelvin (K). Since each color was defined using three parameters X, Y, Z, it could not be effectively represented on a planar surface. To make this possible, **chromaticity values** *x,y,z* were defined as proportions of the tristimulus values as follows:

$$x = X/(X + Y + Z)$$

$$y = Y/(X + Y + Z)$$

$$z = Z/(X + Y + Z)$$

Since $z = 1 - x - y$, only two parameters x and y values are sufficient to define a specific color which can now be plotted on a planar surface. By plotting x and y for all visible colors, we obtain the **CIE chromaticity diagram**. The values around the periphery indicate wavelengths in nanometers (Figure 1.22).

Since the CIE XYZ color space is too large for a single device or application to encompass, a subset color space called standard RGB or **sRGB** was created by Hewlett-Packard and Microsoft for use on devices like monitors and printers. sRGB uses color primaries defined in the ITU-T Rec. 709 (*https://www.itu.int/rec/R-REC-BT.709/en*). The primary colors in sRGB are defined in terms of following chromaticity values and use the D65 white point.

Chromaticity	Red	Green	Blue	White (D65)
x	0.64	0.30	0.15	0.3127
y	0.33	0.6	0.06	0.3290

The **AdobeRGB** color space was developed by Adobe Systems Inc. for use in software development and extended upon the sRGB space in the cyan–green region. The primary colors in AdobeRGB are defined in terms of following chromaticity values and use the D65 white point.

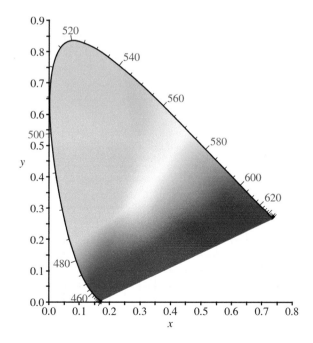

FIGURE 1.22
CIE chromaticity diagram.

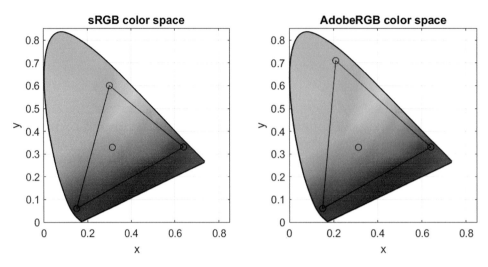

FIGURE 1.23
The sRGB and AdobeRGB color spaces.

```
Chromaticity  Red    Green  Blue   White (D65)
x             0.64   0.21   0.15   0.3127
y             0.33   0.71   0.06   0.3290
```

In the following example, the IPT function **plotChromaticity** is used to plot the CIE chromaticity diagram and the BM function **line** is used to draw the boundaries of the sRGB and the AdobeRGB color spaces (Figure 1.23).

> **Example 1.15: Write a program to show the primaries and boundaries of the sRGB and the AdobeRGB color spaces overlaid on top of the CIE chromaticity diagram and using the D65 white point**
>
> ```
> clear; clc;
> subplot(121)
> plotChromaticity;
> hold on;
> r = [0.64, 0.33];
> g = [0.3, 0.6];
> b = [0.15, 0.06];
> w = [0.3127, 0.3290];
> px = [r(1), g(1), b(1), w(1)];
> py = [r(2), g(2), b(2), w(2)];
> qx = [r(1), g(1), b(1), r(1)];
> qy = [r(2), g(2), b(2), r(2)];
> plot(px, py, 'ko');
> L = line(qx,qy);
> L.Color = [0 0 0];
> hold off;
> title('sRGB color space');
>
> subplot(122)
> plotChromaticity;
> hold on;
> r = [0.64, 0.33];
> g = [0.21, 0.71];
> b = [0.15, 0.06];
> w = [0.3127, 0.3290];
> px = [r(1), g(1), b(1), w(1)];
> py = [r(2), g(2), b(2), w(2)];
> ```

```
qx = [r(1), g(1), b(1), r(1)];
qy = [r(2), g(2), b(2), r(2)];
plot(px, py, 'ko');
L = line(qx,qy);
L.Color = [0 0 0];
hold off;
title('AdobeRGB color space');
```

The conversions between the XYZ and the sRGB and AdobeRGB color spaces are governed according to the following relations as specified by the ICC, where R, G, B denote unsigned integer 8-bit values in the range [0, 255] and D65 is used as the reference white point:

- sRGB to XYZ

Step 1	$R' = \dfrac{R}{255}; G' = \dfrac{G}{255}; B' = \dfrac{B}{255}$
Step 2a	*If* $R', G', B' \leq 0.04045$
	$R'' = \dfrac{R'}{12.92}; G'' = \dfrac{G'}{12.92}; B'' = \dfrac{B'}{12.92}$
Step 2b	*If* $R', G', B' > 0.04045$
	$R'' = \left\{ \dfrac{R'+0.055}{1.055} \right\}^{2.4}; G'' = \left\{ \dfrac{G'+0.055}{1.055} \right\}^{2.4}; B'' = \left\{ \dfrac{B'+0.055}{1.055} \right\}^{2.4}$
Step 3	$X = 0.4212R'' + 0.3576G'' + 0.1805B''$
	$Y = 0.2126R'' + 0.7152G'' + 0.0722B''$
	$Z = 0.0193R'' + 0.1192G'' + 0.9505B''$

- XYZ to sRGB

Step 1	$R' = 3.2406X - 1.5372Y - 0.4986Z$
	$G' = -0.9689X + 1.8758Y + 0.0415Z$
	$B' = 0.0557X - 0.2040Y + 1.0570Z$
Step 2a	*If* $R', G', B' \leq 0.0031308$
	$R'' = 12.92 * R'; G'' = 12.92 * G'; B'' = 12.92 * B'$
Step 2b	*If* $R', G', B' \leq 0.0031308$
	$R'' = 1.055 * (R')^{\frac{1}{2.4}} - 0.055; G'' = 1.055 * (G')^{\frac{1}{2.4}} - 0.055; B'' = 1.055 * (B')^{\frac{1}{2.4}} - 0.055$
Step 3	$R = round(255 * R''); G = round(255 * G''); B = round(255 * B'')$

- AdobeRGB to XYZ

Step 1	$R' = \dfrac{R}{255}; G' = \dfrac{G}{255}; B' = \dfrac{B}{255}$
Step 2	$R'' = (R')^{2.199}; G'' = (G')^{2.199}; B'' = (B')^{2.199}$
Step 3	$X = 0.5766R'' + 0.1855G'' + 0.1882B''$
	$Y = 0.2973R'' + 0.6273G'' + 0.0752B''$
	$Z = 0.0270R'' + 0.0707G'' + 0.9913B''$

- XYZ to AdobeRGB

Step 1	$R' = 2.0416X - 0.5650Y - 0.3447Z$
	$G' = -0.9692X + 1.8759Y + 0.0415Z$
	$B' = 0.0134X - 0.1183Y + 1.015Z$
Step 2	$R" = (R')^{0.4547}; G" = (G')^{0.4547}; B" = (B')^{0.4547}$
Step 3	$R = round(255*R"); G = round(255*G"); B = round(255*B")$

In the following example, the IPT function **rgb2xyz** converts RGB values to XYZ values and the IPT function **xyz2rgb** converts XYZ values back to RGB values. In both cases, there are options to use the D50 and D65 white point references and for translating to the sRGB and AdobeRGB color spaces.

Example 1.16: Write a program to convert RGB values of white and green to XYZ values and chromaticity values xyz, using the sRGB and AdobeRGB color spaces and D65 white point reference. Revert the XYZ values back to the corresponding RGB values.

```
clear; clc; format compact;

fprintf('sRGB White D65: \n');
XYZ = rgb2xyz([1 1 1],'WhitePoint','d65', 'ColorSpace','srgb')
X = XYZ(1); Y = XYZ(2); Z = XYZ(3);
xyz = [X/(X+Y+Z), Y/(X+Y+Z), Z/(X+Y+Z)]
RGB = xyz2rgb([X Y Z],'WhitePoint','d65','ColorSpace','srgb')
fprintf('\n')

fprintf('sRGB Green D65: \n');
XYZ = rgb2xyz([0 1 0],'WhitePoint','d65', 'ColorSpace','srgb')
X = XYZ(1); Y = XYZ(2); Z = XYZ(3);
xyz = [X/(X+Y+Z), Y/(X+Y+Z), Z/(X+Y+Z)]
RGB = xyz2rgb([X Y Z],'WhitePoint','d65','ColorSpace','srgb')
fprintf('\n')

fprintf('AdobeRGB White D65: \n');
XYZ = rgb2xyz([1 1 1],'WhitePoint','d65', 'ColorSpace','adobe-rgb-1998')
X = XYZ(1); Y = XYZ(2); Z = XYZ(3);
xyz = [X/(X+Y+Z), Y/(X+Y+Z), Z/(X+Y+Z)]
RGB = xyz2rgb([X Y Z],'WhitePoint','d65','ColorSpace','adobe-rgb-1998')
fprintf('\n')

fprintf('AdobeRGB Green D65: \n');
XYZ = rgb2xyz([0 1 0],'WhitePoint','d65', 'ColorSpace','adobe-rgb-1998')
X = XYZ(1); Y = XYZ(2); Z = XYZ(3);
xyz = [X/(X+Y+Z), Y/(X+Y+Z), Z/(X+Y+Z)]
RGB = xyz2rgb([X Y Z],'WhitePoint','d65','ColorSpace','adobe-rgb-1998')
fprintf('\n')
```

The program output displays the following:
```
sRGB White D65:
XYZ = 0.9505    1.0000    1.0888
xyz = 0.3127    0.3290    0.3583
RGB = 1.0000    1.0000    1.0000
sRGB Green D65:
XYZ = 0.3576    0.7152    0.1192
xyz = 0.3000    0.6000    0.1000
RGB = 0.0000    1.0000         0
AdobeRGB White D65:
XYZ = 0.9505    1.0000    1.0888
xyz = 0.3127    0.3290    0.3583
```

```
RGB = 1.0000    1.0000    1.0000
AdobeRGB Green D65:
XYZ = 0.1856    0.6273    0.0707
xyz = 0.2100    0.7100    0.0800
RGB = 0.0000    1.0000         0
```

The CMY and CMYK values are related to the RGB values according to the following relations (Poynton, 1995), where all values R, G, B, C, M, Y, and K are expressed in the range [0, 1]:

- RGB to CMY

$$C = 1 - R$$
$$M = 1 - G$$
$$Y = 1 - B$$

- CMY to RGB

$$R = 1 - C$$
$$G = 1 - M$$
$$B = 1 - Y$$

- RGB to CMYK

$$K = 1 - max(R, G, B)$$
$$C = (1 - R - K)/(1 - K)$$
$$M = (1 - G - K)/(1 - K)$$
$$Y = (1 - B - K)/(1 - K)$$

- CMYK to RGB

$$R = (1 - C)(1 - K)$$
$$G = (1 - M)(1 - K)$$
$$B = (1 - Y)(1 - K)$$

- CMY to C'M'Y'K'

$$K' = min(C, M, Y)$$
$$C' = (C - K)/(1 - K)$$
$$M' = (M - K)/(1 - K)$$
$$Y' = (Y - K)/(1 - K)$$

- C'M'Y'K' to CMY

$$C = C'(1 - K') + K'$$
$$M = M'(1 - K') + K'$$
$$Y = Y'(1 - K') + K'$$

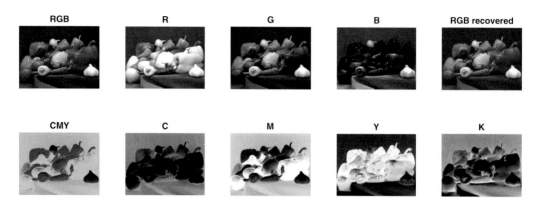

FIGURE 1.24
Output of Example 1.17.

In the following example, the IPT function **makecform** creates a color transformation structure for color space conversion defined by the specified argument, in this case, *srgb-2cmyk* option is used for RGB to CMYK conversion. A reverse transformation of CMYK to RGB can be specified by the argument *cmyk2srgb*. The IPT function **applycform** converts the color values in the specified image to the specified color space defined by the color transformation structure. The BM function **pause** pauses program execution for 5 seconds to allow the TIFF file to be written down to disk. The BM function **imfinfo** can be used to verify that the TIFF file has a color type of CMYK. Since MATLAB can only display a three channel color image, the first three channels are used for displaying the CMYK image (Figure 1.24).

Example 1.17: Write a program to convert an RGB image to CMYK format and display the 4 color channels individually.

```
clear; clc;
I = imread('peppers.png');
S = makecform('srgb2cmyk');
J = applycform(I, S);
r = I(:,:,1);
g = I(:,:,2);
b = I(:,:,3);
imwrite(J, 'test.tiff');
pause (5);
imfinfo('test.tiff')

M = imread('test.tiff');
c = M(:,:,1);
m = M(:,:,2);
y = M(:,:,3);
k = M(:,:,4);
cmy = M(:,:,1:3);

T = makecform('cmyk2srgb');
N = applycform(M, T);

subplot(251), imshow(I); title('RGB');
subplot(252), imshow(r); title('R');
subplot(253), imshow(g); title('G');
subplot(254), imshow(b); title('B');
subplot(256), imshow(cmy); title('CMY');
```

```
subplot(257), imshow(c); title('C');
subplot(258), imshow(m); title('M');
subplot(259), imshow(y); title('Y');
subplot(2,5,10), imshow(k); title('K');
subplot(255), imshow(N); title('RGB recovered');
```

The **CIE L*a*b*** model has been derived in 1976 from the CIE XYZ model to design a more perceptually uniform color space, which means that equal changes in color values produce equal changes in color as perceived by the human visual system. It expresses color as a set of three values: L^* for lightness or brightness ranging from black (0) to white (100), a^* for color ranging from green (–128) to red (+127), and b^* for color ranging from blue (–128) to yellow (+127); the color values being represented as signed 8-bit integers. Thus, when the color channels need to be displayed on the screen, the values need to be mapped to the range [0, 255] by adding 128 to them. With respect to a given white point, like D50 or D65, the L*a*b* model is device-independent i.e. it defines colors as they are perceived by the human vision irrespective of how they are created by devices (Figure 1.25). The asterisk (*) after the letters were inserted to distinguish them from another similar color space called Hunter Lab defined by Richard Hunter in 1948 (Hunter, 1948).

An alternative color model named HSV is also based on human perception of color. The HSV color space takes the form of an inverted cone, the base of the cone being referred to as the **color wheel**. *Hue* defines the color itself and is measured along the circumference of the color wheel in degrees. By convention, red is assigned an angle of 0° which increases to 120° for green and 240° for blue. Intermediate colors like yellow, cyan, and magenta are located in between the primary colors. *Saturation* is a measure of the purity of the color and indicates the amount of gray added to the hue, specified as percentage value between 0 and 100. If no gray is added, then the color is called pure or saturated color and has a saturation of 100%. As more and more gray is added to the color, its saturation decreases until at 0% the actual color is fully converted to gray. On the color wheel, saturation decreases radially from periphery of the circle toward the center. The third parameter *value* indicates the brightness or level of illumination of the color. Colors on the color wheel have a value of 100% which however decreases in a perpendicular direction toward the apex of the cone, where all colors become black with a value of 0% (Figure 1.26).

The color conversion relations between sRGB, L*a*b* and HSV color space as defined by ICC are detailed below, where R, G, B denote unsigned integer 8-bit values in the range [0, 255], L* ranges over [0, 100], a* and b* each ranges over [–128, +127], H ranges over [0°, 359°], S and V each ranges over [0, 100], and $X\omega$, $Y\omega$, $Z\omega$, denote tri-stimulus values of reference white e.g. D65:

FIGURE 1.25
L*a*b* color model.

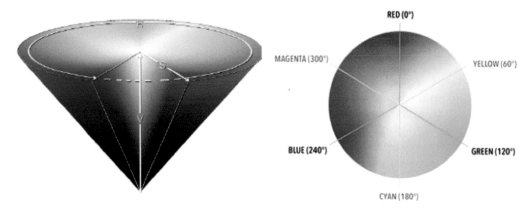

FIGURE 1.26
HSV color model and color wheel.

- sRGB to L*a*b*

Step 1	$sRGB$ to XYZ (as detailed previously)
Step 2	$X' = \dfrac{X}{X\omega}; Y' = \dfrac{Y}{Y\omega}; Z' = \dfrac{Z}{Z\omega}$
Step 3a	$If\ X', Y', Z' > \left(\dfrac{6}{29}\right)^3$
	$X'' = (X')^{1/3}; Y'' = (Y')^{1/3}; Z'' = (Z')^{1/3}$
Step 3b	$If\ X', Y', Z' \le \left(\dfrac{6}{29}\right)^3$
	$X'' = \dfrac{1}{3}\left(\dfrac{29}{6}\right)^2 X' + \dfrac{4}{29}; Y'' = \dfrac{1}{3}\left(\dfrac{29}{6}\right)^2 Y' + \dfrac{4}{29}; Z'' = \dfrac{1}{3}\left(\dfrac{29}{6}\right)^2 Z' + \dfrac{4}{29}$
Step 4	$L = 116 * Y'' - 16;\ a = 500 * (X'' - Y'');\ b = 200 * (Y'' - Z'')$

- L*a*b* to sRGB

Step 1	$Y' = \dfrac{L+16}{116}; X' = \dfrac{a}{500} + Y'; Z' = \dfrac{-b}{200} + Y'$
Step 2a	$If\ X', Y', Z' > \dfrac{6}{29}$
	$X'' = (X')^3; Y'' = (Y')^3; Z'' = (Z')^3$
Step 2b	$If\ X', Y', Z' \le \dfrac{6}{29}$
	$X'' = 3\left(\dfrac{6}{29}\right)^2\left(X' - \dfrac{16}{116}\right); Y'' = 3\left(\dfrac{6}{29}\right)^2\left(Y' - \dfrac{16}{116}\right); Z'' = 3\left(\dfrac{6}{29}\right)^2\left(Z' - \dfrac{16}{116}\right)$
Step 3	$X = X'' \cdot X\omega; Y = Y'' \cdot Y\omega; Z = Z'' \cdot Z\omega$
Step 4	XYZ to $sRGB$ (as detailed previously)

- sRGB to HSV

Step 1	$R' = \dfrac{R}{255}; G' = \dfrac{G}{255}; B' = \dfrac{B}{255}$
Step 2	$N = min(R', G', B'); X = max(R', G', B'); D = X - N$
	$V' = X;$
Step 3a	*If $D = 0$ then $S' = 0$ and $H' = 0$;*
Step 3b	*If $D \neq 0$ then $S' = \dfrac{D}{X}$*
	$R'' = 0.5 + \dfrac{X - R'}{6D}; G'' = 0.5 + \dfrac{X - G'}{6D}; B'' = 0.5 + \dfrac{X - B'}{6D}$
	If $R' = X$ then $H' = B'' - G''$
	elseif $G' = X$ then $H' = 1/3 + R'' - B''$
	elseif $B' = X$ then $H' = \dfrac{2}{3} + G'' - R''$
	If $H' < 0$ then $H' = H' + 1$
	If $H' > 1$ then $H' = H' - 1$
Step 4	$H = round(360 * H'); S = round(100 * S'); V = round(100 * V');$

- HSV to sRGB

Step 1	$C = V * S; D = V - C$
	$H' = \dfrac{H}{60}; X = C\left(1 - \left|H' mod 2 - 1\right|\right)$
Step 2	*If $0 \leq H' < 1$ then $R' = C; G' = X; B' = 0$;*
	If $1 \leq H' < 2$ then $R' = X; G' = C; B' = 0$;
	If $2 \leq H' < 3$ then $R' = 0; G' = C; B' = X$;
	If $3 \leq H' < 4$ then $R' = 0; G' = X; B' = C$;
	If $4 \leq H' < 5$ then $R' = X; G' = 0; B' = C$;
	If $5 \leq H' < 6$ then $R' = C; G' = 0; B' = X$;
Step 3	$R'' = R' + D; G'' = G' + D; B'' = B' + D$
Step 4	$R = round(255 * R''); G = round(255 * G''); B = round(255 * B'')$

The IPT functions **rgb2lab** and **lab2rgb** convert colors from RGB to L*a*b* color spaces and vice versa. The BM functions **rgb2hsv** and **hsv2rgb** convert colors from RGB color space to HSV color space and vice versa. The following example shows an RGB image being converted to HSV and L*a*b* color spaces and displays the individual channels (Figure 1.27).

Example 1.18: Write a program to convert an RGB image to HSV and Lab formats and display the channels individually.

```
clear; clc;

rgb1 = imread('peppers.png');
r = rgb1(:,:,1);
```

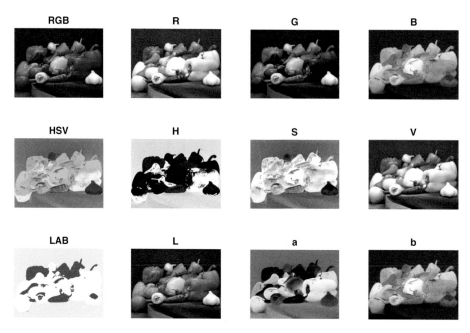

FIGURE 1.27
Output of Example 1.18.

```
g = rgb1(:,:,2);
b = rgb1(:,:,3);

hsv = rgb2hsv(rgb1);
h = hsv(:,:,1);
s = hsv(:,:,2);
v = hsv(:,:,3);

lab = rgb2lab(rgb1);
l = lab(:,:,1);
a = lab(:,:,2);
b = lab(:,:,3);

subplot(341), imshow(rgb1); title('RGB');
subplot(342), imshow(r, []); title('R');
subplot(343), imshow(g, []); title('G');
subplot(344), imshow(b, []); title('B');
subplot(345), imshow(hsv); title('HSV');
subplot(346), imshow(h, []); title('H');
subplot(347), imshow(s, []); title('S');
subplot(348), imshow(v, []); title('V');
subplot(349), imshow(lab); title('LAB');
subplot(3,4,10), imshow(l, []); title('L');
subplot(3,4,11), imshow(a, []); title('a');
subplot(3,4,12), imshow(b, []); title('b');

rgb2 = hsv2rgb(hsv);
rgb3 = lab2rgb(lab);
```

1.3.4 Synthetic Images

Synthetic images are those which are not created by reading pixel values but generated directly by manipulation of values of 2-D matrices. The IPT function **checkerboard** creates a checkerboard pattern which is made up of tiles. Each tile contains four squares,

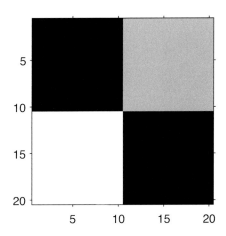

FIGURE 1.28
Checkerboard pattern.

each having a default size of ten pixels per side. The light squares on the left half of the checkerboard are white. The light squares on the right half of the checkerboard are gray (Figure 1.28).

Calling the function without arguments generates a 4×4 tile i.e. 8×8 squares, with a default size of ten pixels per square. The function can be called with three parameters: the first specifying size of each square in pixels, the second specifying the number of rows measured in tiles (not squares), and the third parameter specifying number of columns in tiles. To create pure black and white tiles, instead of gray, the function should be called by specifying intensity greater than 50%. The following example illustrates variants of the pattern (Figure 1.29).

Example 1.19: Write a program to generate checkerboard patterns using the default configuration as well as by modifying side length, number of rows and number of columns. Specify how to build pure black and white tiles without including grays.

```
clear; clc;
I = checkerboard;
subplot(221), imshow(I); axis on; title('4 x 4 @ 10')
s = 20;      % side length of each square in pixels
r = 1;       % number of rows
c = 3;       % number of columns
J = checkerboard(s, r, c);
subplot(222), imshow(J); axis on;  title('1 x 3 @ 20');
K = (checkerboard (40, r+2, c-1) > 0.5);
subplot(223), imshow(K); axis on;  title('3 x 2 @ 40');
L = checkerboard (100, 1, 1);
subplot(224), imshow(L); axis on;  title('1 x 1 @ 100');
```

The IPT function **phantom** generates the image consisting of one large ellipse and a number of smaller ellipses, referred to as the **phantom head**. The size of the image can be specified as an argument (default 256). For any given pixel in the image, pixel's value is equal to the sum of the additive intensity values of all ellipses that the pixel is a part of (Jain, 1989). If a pixel is not part of any ellipse, its value is 0. Two image types can be specified: Shepp–Logan and modified Shepp–Logan in which the contrast is improved for better viewing. These figures are attributes to Larry Shepp and Benjamin Logan (Shepp, 1974).

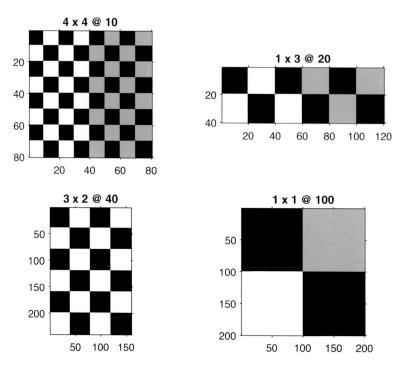

FIGURE 1.29
Output of Example 1.19.

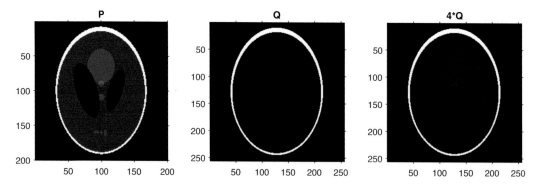

FIGURE 1.30
Output of Example 1.20.

The function describing the plot is defined as the sum of ten ellipses inside a 2×2 square, with varying centers, major axes, minor axes, inclination angles, and gray levels. The following example illustrates the variants of the function and their relationship (Figure 1.30).

> **Example 1.20: Write a program to generate a phantom head image using the Shepp-Logan and the modified Shepp-Logan algorithms.**
>
> ```
> clear; clc;
>
> P = phantom('Modified Shepp-Logan',200);
> Q = phantom('Shepp-Logan');
> ```

```
figure,
subplot(131), imshow(P); axis on; title('P');
subplot(132), imshow(Q); axis on; title('Q');
subplot(133), imshow(4*Q); axis on; title('4*Q');
```

Noise in an image is referred to as the presence of unwanted dots arising out of various reasons like defects in camera sensors, electrical interference, and inadequate illumination. The IPT function **imnoise** synthetically generates noise in the image of specified type and intensity. In the following example, three types of noise have been shown. *Salt and pepper noise* consists of black and white dots, with a default density of 0.05 which affects approximately 5% of the pixels. The density can be specified as an argument after the name of the noise. *Speckle noise* adds multiplicative noise using the equation $L = I + n*1$, where n is uniformly distributed noise with default values of mean 0 and variance 0.04. The variance value can be added as an additional argument after the name of the noise. *Gaussian noise* adds white noise with mean 0 and variance 0.01. The mean and variance can be specified as additional arguments (Figure 1.31). Further details about the Gaussian function are added after the example.

Example 1.21: Write a program to add three different types of noise to a grayscale image and display the resultant noisy images

```
clear; clc;
I = imread('eight.tif');
J = imnoise(I,'salt & pepper'); % salt and pepper noise
K = imnoise(I,'gaussian');      % Gaussian noise
L = imnoise(I,'speckle');       % speckle noise
figure,
subplot(221), imshow(I); title('I');
subplot(222), imshow(J); title('J');
subplot(223), imshow(K); title('K');
subplot(224), imshow(L); title('L');
```

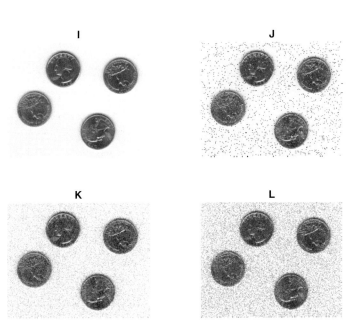

FIGURE 1.31
Output of Example 1.21.

In mathematics, the **Gaussian function**, named after German mathematician Carl Friedrich Gauss

$$f(x) = a \cdot e^{\left\{-(x-\mu)^2\right\}/(2\sigma^2)}$$

The graph of a Gaussian function is a characteristic bell-shaped curve, also known as the **normal distribution** curve. The parameter a (amplitude) controls the height of the curve, μ (mean) controls the position of the peak, and σ (standard deviation) controls the width of the curve. The following example plots a 1-D Gaussian curve with mean 2.5 and standard deviation 4.37. For a Gaussian curve, area under curve from $x = -\sigma$ to $x = +\sigma$ is 68.26%, area under curve from $x = -2\sigma$ to $x = +2\sigma$ is 95.44%, and area under curve from $x = -3\sigma$ to $x = +3\sigma$ is 99.74% of the total area. Alternatively, the **Fuzzy Logic Toolbox** (FLT) function **gaussmf** can also be used to draw a Gaussian curve for a specified mean and variance (Figure 1.32).

Example 1.22: Write a program to generate Gaussian curves with varying mean and variance

```
clear; clc;
x = -5:0.1:10;
mu = mean(x);
sigma = std(x);
den = 2*sigma^2;
for i = 1:numel(x)
    num = -(x(i) - mu)^2;
    frac = num/den;
    y(i) = exp(frac);
end
subplot(121), plot(x, y); grid;

x = -5:0.1:5;
a = 0.9; m = 0; s = sqrt(0.2);
g = gaussmf(x, [s, m]);
subplot(122), plot(x, a*g, 'b-', 'LineWidth', 2);
hold on; grid;
```

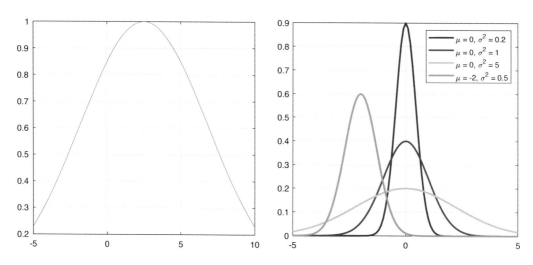

FIGURE 1.32
Output of Example 1.22.

```
a = 0.4; m = 0; s = sqrt(1);
g = gaussmf(x,[s, m]);
plot(x, a*g, 'r-', 'LineWidth', 2);

a = 0.2; m = 0; s = sqrt(5);
g = gaussmf(x,[s, m]);
plot(x, a*g, 'Color', [0.8 0.8 0], 'LineWidth', 2);

a = 0.6; m = -2; s = sqrt(0.5);
g = gaussmf(x,[s, m]);
plot(x, a*g, 'Color', [0 0.7 0], 'LineWidth', 2);
hold off;

legend('\mu = 0, \sigma^2 = 0.2', '\mu = 0, \sigma^2 = 1', ...
    '\mu = 0, \sigma^2 = 5', '\mu = -2, \sigma^2 = 0.5');
```

A **2-D Gaussian** curve is a function of two variables x and y and is defined as follows, where μx and μy are the means along the x- and y- directions and σx and σy are the corresponding standard deviations:

$$f(x, y) = a \cdot e^{\dfrac{-(x - \mu x) \cdot (y - \mu y)}{2 \cdot \sigma x \cdot \sigma y}}$$

A plot of a 2-D Gaussian curve can be generated by varying the values of x and y and using the values of the mean and variance along orthogonal directions. Alternatively, the IPT function **fspecial** can be used to generate a 2-D Gaussian matrix by specifying the size along the x- and y- directions of the matrix. The following example illustrates both the methods with symmetrical Gaussians (Figure 1.33).

Example 1.23: Write a program to display 2-D Gaussian function be specifying varying means and variances

```
clear; clc;
x = -5:5; y = -5:5;
mux = mean(x); muy = mean(y);
sigmax = std(x); sigmay = std(y);
den = 2*sigmax*sigmay;
for i = 1:11
    for j = 1:11
        a = (x(i) - mux)^2;
        b = (y(j)  -muy)^2;
        num = -(a*b);
        frac = num/den;
```

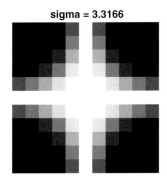 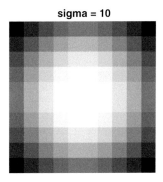

FIGURE 1.33
Output of Example 1.23.

```
            z(i,j) = exp(frac);
        end
    end
end
subplot(121),imshow(z); title(strcat('sigma = ', num2str(sigmax)));
hsize = 10; sigma = 10;
G = fspecial('gaussian', hsize, sigma);
subplot(122), imshow(G, []); title(strcat('sigma = ', num2str(sigma)));
```

1.4 Display and Exploration

1.4.1 Basic Display

After an image is read into the system, referred to as **image acquisition**, the image needs to be displayed to the user. We have already seen that the figure window can be used to display an image. In this section, we will discuss other ways of displaying images. The BM function **imshow** is also included in the IPT for displaying images which can be color, grayscale, binary, or indexed. For grayscale images, the function can stretch the range covering existing minimum and maximum values to the entire range [0, 255] by including a pair of empty square brackets []. Alternatively, it can take two user-specified intensity ranges *low* and *high* and stretch the intensities between them to the entire set of values [0, 255]. This way of stretching grayscale values is called **contrast stretching** as the contrast of an image is the difference between the minimum and maximum gray values, and the contrast is stretched to the maximum range of 0 to 255. In the following example, the empty square values [] map the existing minimum (74) and maximum (224) gray values of the image to 0 and 255, respectively, while the argument [50, 150] maps 50 to 0 and 150 to 255. Any values less than 50 are also mapped to 0, while any values greater than 150 are all mapped to 255. For displaying an indexed color image, the map needs to be included as the function argument (Figure 1.34).

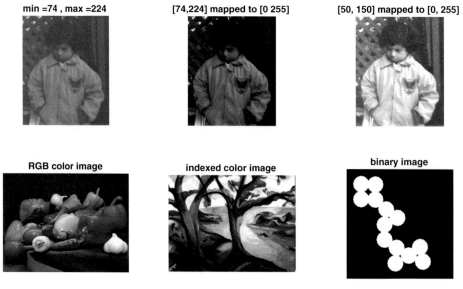

FIGURE 1.34
Output of Example 1.24.

Example 1.24: Write a program to display binary, grayscale, RGB color and indexed color images. For grayscale image demonstrate how contrast stretching can be implemented

```
clear; clc;
I = imread('pout.tif'); title('grayscale image');
Imin = min(min(I)); Imax = max(max(I));
subplot(231), imshow(I);
title(strcat('min = ', num2str(Imin),' , max = ', num2str(Imax)));
subplot(232), imshow(I, []);
title(strcat('[',num2str(Imin),',',num2str(Imax),'] ',' mapped to [0 255]'));
% all existing values mapped to [0, 255]
subplot(233), imshow(I, [50 150]);
title('[50, 150] mapped to [0, 255]');
subplot(234), imshow('peppers.png'); title('RGB color image');
load trees; subplot(235), imshow(X, map); title('indexed color image');
subplot(236),  imshow('circles.png'); title('binary image');
```

To display multiple images, the BM function **subplot** can be used as we have seen previously. Alternatively, the IPT function **imshowpair** can be used to display two images in a number of ways. Using the option *montage* places them side by side in the same image, while using the option *blend* overlays the two images on each other using alpha blending, and using the option *checkerboard* creates an image with alternate rectangular regions from both images. In the example below, an edge detector is used to highlight the edges in the image (Figure 1.35). Edge detectors are discussed in Section 1.6.

montage

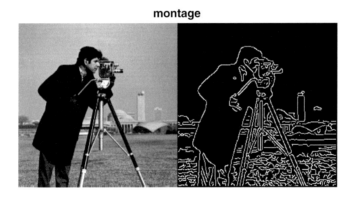

blend

checkerboard

FIGURE 1.35
Output of Example 1.25.

Example 1.25: Write a program to display a pair of images using different variations

```
clear; clc;
a = imread('cameraman.tif');
b = edge(a, 'canny');
figure, imshowpair(a, b, 'montage');
figure, imshowpair(a, b, 'blend');
figure, imshowpair(a, b, 'checkerboard');
```

Image fusion is a process of gathering information from multiple cameras or sensors and combining them into a single image. The composite image is created by using spatial referencing information for aligning areas with similar intensities. The final results is a blended overlay image usually with false coloring i.e. separate colors for areas of the individual images and areas common to them, The IPT function **imfuse** creates a composite from two images by fusing their intensities. The following example creates a blended image using red for image 1, green for image 2, and yellow for areas of similar intensity between the two images. The IPT function **imresize** resizes the dimensions of an image by a specified percentage (Figure 1.36).

Example 1.26: Write a program to fuse a pair of images together

```
clear; clc;
a = imread('coins.png');
b = imresize(a, 0.8);
c = imfuse(a, b, 'scaling', 'joint', 'ColorChannels',[1 2 0]);
figure, imshowpair(a, c, 'montage');
```

The IPT function **montage** can be used to display multiple images together in a rectangular grid. The size of the grid can be specified by including the number of rows and number of columns. A cell array is used to specify the filenames of the images. If the images have been read and stored in variable names, then these can be included in a character vector. In the following example, the first six images are displayed in a 2×3 grid and the next four images are displayed in a 1×4 grid (Figure 1.37).

FIGURE 1.36
Output of Example 1.26.

2 × 3 grid

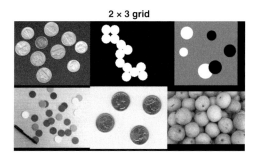

1 × 4 grid

FIGURE 1.37
Output of Example 1.27.

Example 1.27: Write a program to display multiple images in a rectangular grid

```
clear; clc;

% part 1
f = {'coins.png',
     'circles.png',
     'circlesBrightDark.png',
     'coloredChips.png',
     'eight.tif',
     'pears.png'};
montage(f, 'Size', [2 3]);  % cell array, 2 rows 3 columns

% part 2
a = imread('football.jpg');
b = imread('flamingos.jpg');
c = imread('fabric.png');
d = imread('foggysf1.jpg');
montage({a,b,c,d}, 'Size', [1 4]);    % character vector, 1 row 4 column
```

In the above examples, multiple images are displayed by using spatial distribution. Multiple images can also be displayed one after another along a **temporal timeline**. The IPT function **implay** is used to display a sequence of images one after another at a specified frame rate, by invoking the **Movie Player**. The default frame rate is 20 fps. The BM function **load** copies variables from a specified file into workspace. The following example loads two image sequences and plays them back using the Movie Player (Figure 1.38).

**Example 1.28: Write a program to display multiple images
in a temporal sequence one after another**

```
clear; clc;
load cellsequence;
fps = 10;
implay(cellsequence, fps);
load mristack;
implay(mristack);
```

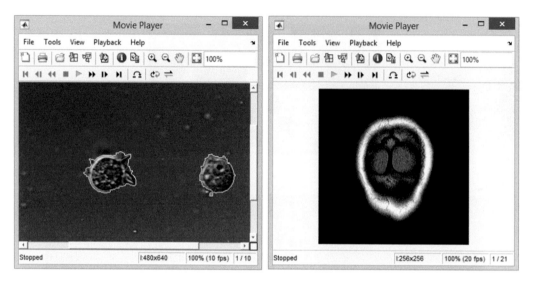

FIGURE 1.38
Output of Example 1.28.

Apart from displaying images simply as a 2-D rectangular plane, images can also be displayed on a non-planar 3-D surface, which is referred to as **image warping**. The IPT function **warp** is used to map an image as a texture on to a graphical surface. In the following example, an indexed image X is loaded from a file and displayed. Then, a rectangular grid of equally spaced points are created using the BM function **meshgrid** and for each point on the meshgrid, non-planar functions are used to create a vertical elevation along the z-axis. The image is then mapped to the surface created. Two surfaces are created, one using the function $f(x, y) = -(x^3 + y^3)$ and the other using the function $f(x, y) = -(x^2 + y^4)$. The BM function **view** creates a viewpoint specification by specifying the azimuth and elevation (Figure 1.39).

Example 1.29: Write a program to display images on a non-planar surface

```
clear; clc;
load trees;
[x, y] = meshgrid(-100:100, -100:100);
z1 = -(x.^3 + y.^3);
z2 = -(x.^2 + y.^4);
figure,
subplot(131), imshow(X, map);
subplot(132), warp(x, y, z1, X, map); axis square;
view(-20, 30); grid; xlabel('x'); ylabel('y'); zlabel('z');
subplot(133), warp(x, y, z2, X, map); axis square;
view(-50, 30); grid; xlabel('x'); ylabel('y'); zlabel('z');
```

1.4.2 Interactive Exploration

Interactive exploration involves not only displaying the image but also performing common image processing tasks by the user using menu-based options provided by the graphical user interface. The IPT function **imtool** provides an integrated environment **Image Viewer** application for displaying images and performing some common image processing tasks.

 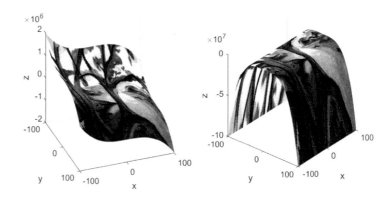

FIGURE 1.39
Output of Example 1.29.

Apart from providing all image display capabilities of the *imshow* function it provides additional tools for displaying pixel values, specifying magnification factors, adjusting contrast and color, measuring distances between points, cropping the image, and so on. The output of these tools can also be invoked using functions from the IPT. The **Pixel Region tool** displays a selected region of pixels along with pixel values, which can also be invoked using the IPT function **impixelregion**. The **Image Information tool** displays information about the image and file metadata, which can also be invoked using the IPT function **imageinfo**. The **Measure Distance tool** measures distances from one point to another point within the image, which can also be invoked using the IPT function **imdistline**. The **Pixel Information tool** is used for getting information about the pixel under the pointer, which can also be invoked using the IPT function **impixelinfo**. The **Crop Image tool** is used for defining a crop region on the image and cropping the image, which can also be invoked using the IPT function **imcrop**. Cropping is discussed in more details in Section 1.5. The **Overview tool** is used for determining what part of the image is currently visible in the Image Viewer and for changing this view using a sliding rectangle, which can also be invoked using the IPT function **imoverview**. The **Pan tool** is used for moving the image to view other parts of the image, and the **Zoom tool** is used for getting a closer view of any part of the image (Figure 1.40).

> **Example 1.30: Write a program to display an image along with pixel values for a selected region**
>
> ```
> clear; clc;
> h = imtool ('peppers.png');
> impixelregion(h)
> ```

Magnification and scrollbars to the figure window can be controlled using the IPT function **immagbox** in an interactive way by adding a magnification box to the image in a specified location having a specified size. The following example adds a magnification box at location (10, 50) with dimensions (100, 50) using which magnification of the image can be specified, and accordingly scrollbars can automatically appear along the sides of the figure window (Figure 1.41). The last line magnifies the figure window itself so that it occupies the full screen by specifying a normalized size of (0,0) to (1,1).

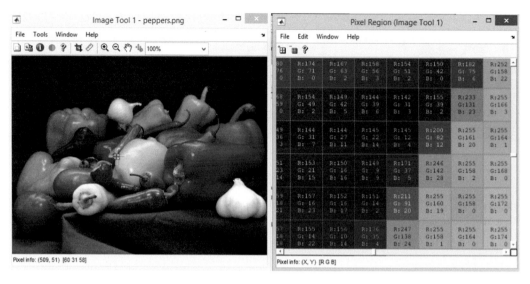

FIGURE 1.40
Output of Example 1.30.

FIGURE 1.41
Output of Example 1.31.

Example 1.31: Write a program to display an image which can be interactively magnified. The figure window should also be enlarged to occupy the full screen.

```
clear; clc;
f = figure;
i = imshow('pears.png');
s = imscrollpanel(f,i);
m = immagbox(f,i);
set(m,'Position',[10 50 100 50])
set(gcf, 'Units', 'Normalized', 'OuterPosition', [0, 0, 1, 1]);
```

1.4.3 Building Interactive Tools

The **Adjust Contrast tool** is used for adjusting the contrast of a grayscale image displayed in the Image Viewer and modifying the contrast using a histogram, which can also be invoked using the IPT function **imcontrast**. Histograms plot the number of pixels occurring for each gray level and have been discussed in Section 1.6. As the sliders are moved from an initial position to a final position, the intensity at the final position is mapped to the intensity at the initial position, thereby stretching the histogram and producing tonal correction of the image. For example, dragging the left slider from 0 to 50 maps all values in the range [0, 50] to 0 effectively converting dark gray to pure black. Similarly dragging the right slider from 250 to 200 maps all values in the range [200, 250] to 250 effectively converting light gray to pure white. Such mappings help to improve the brightness and contrast of the image by utilizing all possible gray levels. In the following example, the histogram is displayed to adjust the tonal contrast of the image (Figure 1.42).

> **Example 1.32: Write a program to display an image along with a histogram to modify the contrast**
>
> ```
> clear; clc;
> h = imtool ('coins.png');
> imcontrast(h)
> ```

The **Choose Colormap tool** allows the user to choose an in-built colormap for viewing a grayscale image, which can also be invoked using the IPT function **imcolormaptool**. In-built colormaps have been discussed later in this section. The gray levels of the image are mapped to the colors in the specified colormap, and the image can be viewed in color (Figure 1.43).

> **Example 1.33: Write a program to display a grayscale image along with an option of choosing colormaps to display the image in color**
>
> ```
> clear; clc;
> h = imtool ('coins.png');
> imcolormaptool(h)
> ```

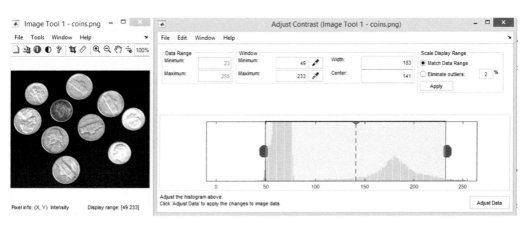

FIGURE 1.42
Output of Example 1.32.

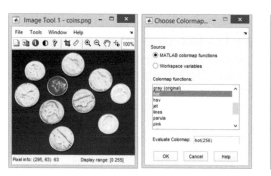
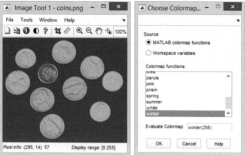

FIGURE 1.43
Output of Example 1.33.

1.5 Geometric Transformation and Image Registration

1.5.1 Common Geometric Transformations

Geometrical transformation is a family of image processing techniques, where the pixel values of an image are usually not modified, rather the location, orientation, and number of pixels of the original image are changed. **Cropping** is one of the simplest image processing operations and involves displaying a small portion of the whole image i.e. a subset of the entire pixel set. To crop an image a rectangle needs to be specified smaller than the whole image so that only the portion of the image within the rectangle will be displayed. Two parameters of the rectangle need to be specified viz. start point (upper-left corner) and size (number of rows and columns within it). The IPT function **imcrop** is used for the purpose. The crop rectangle can be specified either interactively or parametrically. The following example illustrates **interactive cropping**. The user is asked to draw the cropping rectangle over the image in the figure window, right-click on it, and choose option *Crop Image* from the pop-up menu. The portion of the image within the rectangle is retained and displayed beside the original image (Figure 1.44).

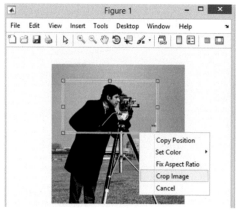

FIGURE 1.44
Output of Example 1.34.

Example 1.34: Write a program to crop an image by specifying the crop rectangle interactively

```
clear; clc;
I = imread('cameraman.tif');
[J, rect] = imcrop(I);
imshowpair(I, J, 'montage');
```

The following example illustrates **parametric cropping**. Here, the cropping rectangle is specified in terms of four parameters: coordinates of the top-left corner of the rectangle (xmin, ymin), its width and its height (width, height) in that order. In the following example, an image is cropped into four quadrants which are then displayed in a different order i.e. each quadrant is replaced by its diagonally opposite quadrant. To avoid fractional number of pixels in each quadrant, the BM function **floor** is used to round off to the nearest integer less than or equal to the specified value (Figure 1.45).

Example 1.35: Write a program to crop an image into four quadrants and display each separately. Then re-assemble the image by swapping each quadrant with its diagonally opposite quadrant.

```
clear; clc;
I = imread('cameraman.tif');
H = floor(size(I,1)); W = floor(size(I,2));
h = floor(H/2); w = floor(W/2);       % dimensions of crop rectangle
% syntax: imcrop(I, [xmin, ymin, width, height])
Q1 = imcrop(I, [0, 0,    w, h]);
Q2 = imcrop(I, [w+1, 0,   w, h]);
Q3 = imcrop(I, [0, h+1,   w, h]);
Q4 = imcrop(I, [w+1, h+1, w, h]);
figure,
subplot(221), imshow(Q1); title('Q1');
subplot(222), imshow(Q2); title('Q2');
subplot(223), imshow(Q3); title('Q3');
subplot(224), imshow(Q4); title('Q4');
J = [Q4 Q3 ; Q2 Q1];
figure, imshowpair(I, J, 'montage');
```

FIGURE 1.45
Output of Example 1.35.

Translation can be done by moving segments or the entire image on a 2-D plane by specifying translation amounts along the *x*- and *y*- directions. If the translation amounts are positive, then the image segments move from left to right and top to bottom i.e. along the positive directions of the *x*- and *y*- axes, if they are negative, then the movements are in the opposite direction. A pixel with coordinates $P(x_1, y_1)$ when translated by amounts (tx, ty) has new coordinates $Q(x_2, y_2)$ given by the following:

$$x_2 = x_1 + tx$$

$$y_2 = y_1 + ty$$

The IPT function **imtranslate** is used to translate an image by specifying a translation vector (tx, ty) i.e. *x* and *y* values by which the image should be moved. In the following example, the *FillValues* option is used to fill the region outside the image by a graylevel intensity between 0 and 255. The *OutputView* option is used to specify whether the translated image will be viewed as cropped or full. In case of a full view, the image canvas area is increased to accommodate the image and the extra portion by which the image has moved. In case of a cropped view, the image canvas remains at its original size so that the cropped portion of the image is viewed along with the extra portion by which the image has moved. In the following example, a translated image is viewed using the full and cropped option (Figure 1.46).

> **Example 1.36: Write a program to translate an image viewing the result as a full view and a cropped view**
>
> ```
> clear; clc;
> I = imread('coins.png');
> J = imtranslate(I, [25, -20],   'FillValues', 200, 'OutputView', 'full');
> K = imtranslate(I, [-50.5, 60.6],   'FillValues',   0, 'OutputView', 'same');
> subplot(131), imshow(I); title('I'); axis on;
> subplot(132), imshow(J); title('J'); axis on; % full view
> subplot(133), imshow(K); title('K'); axis on; % cropped view
> ```

Rotation implies moving the pixels of an image along the circumference of a circle with the origin of the coordinate system as the pivot point. Counterclockwise rotations are considered positive while clockwise rotations are negative. A pixel $P(x_1, y_1)$ when rotated by angle θ about the origin, has new coordinates $Q(x_2, y_2)$ given by the following:

$$x_2 = x_1 \cdot \cos(\theta) - y_1 \cdot \sin(\theta)$$

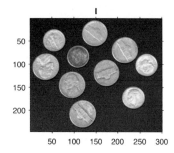
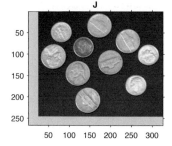
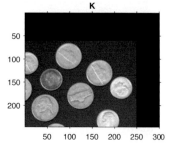

FIGURE 1.46
Output of Example 1.36.

$$y_2 = x_1 \cdot \sin(\theta) - y_1 \cdot \cos(\theta)$$

The IPT function **imrotate** is used to rotate an image by specifying the angle in degrees. The *crop* option makes the output image size same as the input image cropping the additional portions of the image, while the *loose* option makes the image canvas large enough to view the entire image. The following example displays a rotated image using the crop and loose options (Figure 1.47).

> **Example 1.37: Write a program to rotate an image by +30 and −40 degrees specifying the 'crop' and 'loose' options.**
>
> ```
> clear; clc;
> I = imread('saturn.png');
> J = imrotate(I, 30, 'crop');
> K = imrotate(I, -40, 'loose');
> figure,
> subplot(131), imshow(I); title('I');
> subplot(132), imshow(J); title('J'); % crop option
> subplot(133), imshow(K); title('K'); % loose option
> ```

Scaling occurs when the dimensions of an image are multiplied by constants, which are known as scaling factors. There are two scaling factors one for the height and the other for the width. If both the factors are equal then the scaling is called **uniform**, as the ratio of the height to width (aspect ratio) remains the same. If the factors are unequal then the scaling is **non-uniform** which usually results in a distorted image. If the factors are greater than 1 then it leads to an increase in the image size, if less than 1 then it reduces the image size. Obviously if the scaling factors are 1 then the image remains as it is as there is no change in the image dimensions. A pixel $P(x_1, y_1)$ when scaled by amounts (sx, sy) has new coordinates $Q(x_2, y_2)$ given by the following:

$$x_2 = x_1 \cdot sx$$

$$y_2 = y_1 \cdot sy$$

The IPT function **imresize** is used to resize the image dimensions with one or two following arguments indicating the scaling factors. The following example illustrates the

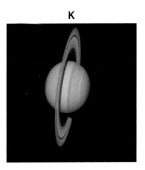

FIGURE 1.47
Output of Example 1.37.

various ways in which scaling parameters can be specified. If there is a single argument, it indicates the scaling factor by which both the height and width are to be scaled. If there are two arguments, it indicates the number of rows and number of columns, respectively. If one of the arguments is a number and the other argument is set as *NaN* (not a number), then both the dimensions are set to the single value (Figure 1.48).

Example 1.38: Write a program to scale an image uniformly and non-uniformly

```
clear; clc;
I = imread('testpat1.png');
h = size(I, 1); w = size(I, 2);
J = imresize(I, 0.5);
K = imresize(I, [h/2, 1.5*w]);
L = imresize(I, [500, 300]);
M = imresize(I, [100, NaN]);
subplot(151), imshow(I); title('I'); axis on;
subplot(152), imshow(J); title('J'); axis on;
subplot(153), imshow(K); title('K'); axis on;
subplot(154), imshow(L); title('L'); axis on;
subplot(155), imshow(M); title('M'); axis on;
```

Sampling is the process of examining the values of a function at discrete intervals along the appropriate dimensions. Sampling rate measures how often the sampling is done. For a temporal signal, like audio, sampling rate is measured in samples per second (Hertz) and for a spatial signal, like image, sampling rate is measured in pixels per inch (ppi). Increasing the sampling rate increases the number of pixels and hence encapsulates more information in the image. This has the result of increasing the image size and improving the image quality. Decreasing the sampling rate reduces the pixel information and hence reduces image size and quality. Sampling is therefore related to the scale and quality of an image. Sampling is frequently associated with two other operations viz. zooming and interpolation. **Zooming** is an operation of increasing or decreasing the display size of an image without changing its pixel structure. When we reduce the size of an image we can zoom-in so that the image still appears at its original size. During size reduction, the total number of pixels in the image decreases so that zooming makes the pixels appear larger and beyond a certain point the smooth image takes on a blocky appearance. This is called **pixelation** and usually degrades the image quality. To restore the quality we can either zoom-out or incorporate **interpolation** which introduces new pixels in the reduced image so that its pixel count becomes same as it was originally before reduction. However, these pixels are synthetically created using an interpolation algorithm which

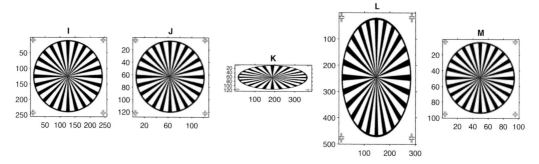

FIGURE 1.48
Output of Example 1.38.

uses average values of the neighboring pixels to create a new value, and which in general do not correspond to the actual pixel value of the original image at that location. Thus, although the total pixel count is restored, the image quality improves only slightly. The interpolation can be done using one of the three methods: *nearest neighbor*, where the new pixel is assigned the value of the next pixel; *bilinear*, where the new pixel is assigned a value of the weighted average of a 2×2 neighborhood; and *bicubic*, where the new pixel is assigned a value of the weighted average of a 4×4 neighborhood. The following example shows an image scaled down to half its size along each dimension appears nearly the same when zoomed (top-middle) and interpolated (top-right) to its original size, but when the scaling factor is one-tenth along each dimension, there is a perceptible degradation in quality both when zoomed (bottom-left) and interpolated using nearest neighbor method (bottom-middle). The quality is slightly better when interpolated using bicubic method (bottom-right) but nowhere near the original quality (top-left). So even though the top-left and bottom-right images have the same number of pixels, one can clearly distinguish the difference in quality (Figure 1.49).

Example 1.39: Write a program to sample down an image and then zoom and interpolate to its original size.

```
clear; clc;
a = imread('peppers.png');
b = rgb2gray(a);
h = size(b,1);
w = size(b,2);
c1 = imresize(b, 0.5);
c2 = imresize(c1, 2);
d1 = imresize(b, 0.1);
d2 = imresize(d1, 10, 'nearest');
d3 = imresize(d1, 10, 'bicubic');

subplot(231), imshow(b); title('original'); axis on;
```

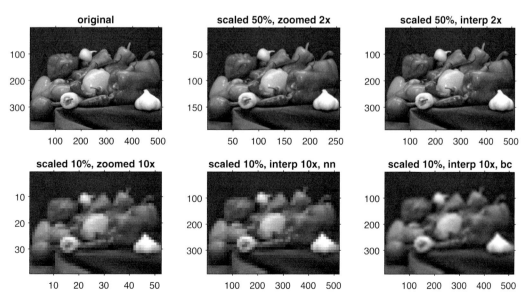

FIGURE 1.49
Output of Example 1.39.

```
subplot(232), imshow(c1);  axis on; title('scaled 50%, zoomed 2x');
subplot(233), imshow(c2);  axis on; title('scaled 50%, interp 2x');
subplot(234), imshow(d1);  axis on; title('scaled 10%, zoomed 10x');
subplot(235), imshow(d2);  axis on; title('scaled 10%, interp 10x, nn');
subplot(236), imshow(d3);  axis on; title('scaled 10%, interp 10x, bc');
```

A **reflection** operation about an axis reverses the sign of the coordinate of the pixels which are perpendicular to the axis. A reflection about the y-axis reverses the x-coordinates of the image pixels while a reflection about the x-axis reverses the y-axis coordinates. The concerned axis behaves like a mirror, and the reflected image behaves as a flipped mirror image of the original. A reflection can also take place about the origin in which case both x- and y- coordinates are reversed. A pixel $P(x_1, y_1)$ when subjected to y-axis reflection has new coordinates $Q(x_2, y_2)$ given by the following:

$$x_2 = -x_1$$
$$y_2 = y_1$$

When subjected to x-axis reflection the new coordinates are given by the following:

$$x_2 = x_1$$
$$y_2 = -y_1$$

When subjected to reflection about the origin has new coordinates are given by the following:

$$x_2 = -x_1$$
$$y_2 = -y_1$$

The BM function **fliplr** is used to flip an image left to right which is equivalent to a y-axis reflection, while the BM function **flipud** flips an image up to down which is equivalent to an x-axis reflection. The following example illustrates the three types of reflection (Figure 1.50).

Example 1.40: Write a program to illustrates the three types of reflection of an image

```
clear; clc;
I = imread('peppers.png');
J = fliplr(I);  % reflection about Y-axis
K = flipud(I);  % reflection about X-axis
L = flipud(J);  % reflection about origin
figure,
subplot(221), imshow(I); title('original');
subplot(222), imshow(J); title('reflected about Y-axis');
subplot(223), imshow(K); title('reflected about X-axis');
subplot(224), imshow(L); title('reflected about origin');
```

1.5.2 Affine and Projective Transformations

Generic transformations can be applied using a specified transformation matrix which can be a single operation or a composite of several operations. A composite transformation involving a combination of translation, rotation, scaling, reflection, and shear is referred to as **affine** transformation. The following example uses shear, scale, rotation, and reflection

original

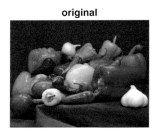

reflected about Y-axis

reflected about X-axis

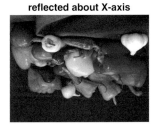

reflected about origin

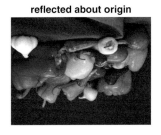

FIGURE 1.50
Output of Example 1.40.

in various combinations and uses the corresponding transformation matrices to display the output result. The IPT function **affine2d** creates a affine transformation object using a mapping between the initial and final position of a pixel given a specified transformation matrix, while the IPT function imwarp applies the transformation object to a specified image (Figure 1.51).

Example 1.41: Write a program to transform images using transformation matrices.

```
clear; clc;
C = imread('peppers.png');
sx = 3; sy = 1.5;
cost = 0.7; sint = 0.7;
hx = 1.5; hy = 3;
S = [sx 0 0 ; 0 sy 0 ; 0 0 1];           % scaling
R = [cost -sint  0 ; sint cost 0 ; 0 0 1];  % rotation
H = [1 hx 0 ; 0 1 0 ; 0 0 1];            % shear
F = [-1 0 0 ; 0 -1 0 ; 0 0 1];           % reflection
M1 = F*R;                                % rotation & reflection
M2 = R*S;                                % scaling & rotation
M3 = H;                                  % shear
A1 = affine2d(M1');
A2 = affine2d(M2');
A3 = affine2d(M3');

I1 = imwarp(C, A1);
I2 = imwarp(C, A2);
I3 = imwarp(C, A3);

subplot(221),imshow(C); axis on; title('original');
subplot(222),imshow(I1); axis on; title('after rotation & reflection');
subplot(223),imshow(I2); axis on; title('after rotation & scaling');
subplot(224),imshow(I3); axis on; title('after shear');
```

The constraint to be satisfied by an affine transformation is that the opposite sides of a rectangle should change by equal amounts i.e. a rectangle is converted to a parallelogram.

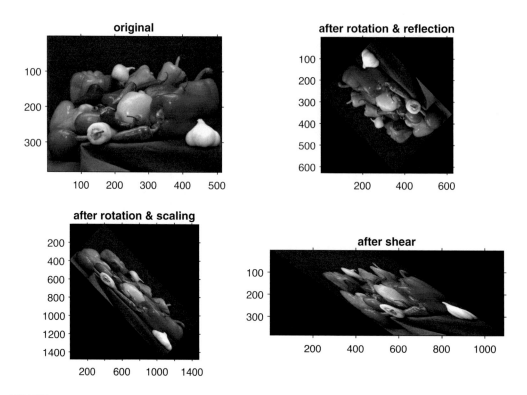

FIGURE 1.51
Output of Example 1.41.

A pixel $P(x_1,y_1)$ when subjected to affine transformation has new coordinates $Q(x_2,y_2)$ given by the following, where a,b,c,d,e,f are constants:

$$x_2 = ax_1 + by_1 + c$$

$$y_2 = dx_1 + ey_1 + f$$

However if the above constraint is not applied i.e. the opposite sides are not changed by equal amounts, then a rectangle is converted to an arbitrary quadrilateral, and the transformation is referred to as **projective** or **perspective** transform. A pixel $P(x_1,y_1)$ when subjected to projective transformation has new coordinates $Q(x_2,y_2)$ given by the following, where a,b,c,d,e,f,g,h are constants:

$$x_2 = \frac{ax_1 + by_1 + c}{gx_1 + hy_1 + 1}$$

$$y_2 = \frac{dx_1 + ey_1 + f}{gx_1 + hy_1 + 1}$$

The following example illustrates the difference between an affine transformation and a projective transformation. The IPT function **projective2d** creates a projective transformation object using a mapping between the initial and final position of a pixel given a

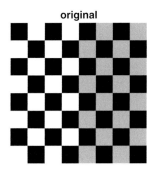

FIGURE 1.52
Output of Example 1.42.

specified transformation matrix M, while the IPT function imwarp applies the transformation object to a specified image (Figure 1.52).

> **Example 1.42: Write a program to apply affine transformation and projective transformation to an image**
>
> ```
> clear; clc;
> C = checkerboard(10);
> M = [5 2 5 ; 2 5 5 ; 0 0 1];
> A = affine2d(M');
> I = imwarp(C, A);
>
> M = [5 2 5 ; 2 5 5 ; 0.01 0.01 1];
> B = projective2d(M');
> J = imwarp(C, B);
>
> figure,
> subplot(131), imshow(C); title('original');
> subplot(132), imshow(I); title('affine transform');
> subplot(133), imshow(J); title('perspective transform');
> ```

1.5.3 Image Registration

Image registration is a process of aligning two or more same or similar images. One of the images is called the **fixed** or reference image, and the other image called the **moving** image is aligned to the former by applying geometric transformations. Mis-alignment usually originates from a change in camera orientation or different sensor characteristics or distortions. The process of image registration requires a metric, an optimizer, and a transformation type. The **metric** defines a quantitative measure of the difference/similarity between the two images, the **optimizer** defines a methodology to minimize or maximize the difference/similarity value, and the **transformation type** defines the 2-D transformation required to align the images. The image registration process involves three steps: (1) the specified transformation type generates an internal transformation matrix which is applied to the moving image with bilinear interpolation; (2) the metric compares the transformed moving image with the reference image and computes a similarity value; and (3) the optimizer adjusts the transformation matrix to increase the similarity value. These three steps are repeated in iterations until a stop condition is reached, which is either the maximum similarity/minimum difference value or the maximum number of iterations. The IPT function imregister is used to register two mis-aligned images based on relative intensity patterns. The optimizer has a number of parameters like *growth factor* which controls the rate at which the search radius grows in parameter space and decides whether the process is slow or fast, *epsilon* which specifies the size of the search radius, *InitialRadius* which specifies the initial search radius value, and *MaximumIterations* which specifies that maximum number of iterations that may be performed by the optimizer. IPT function **imregconfig** returns an optimizer and metric with default settings to provide a basic registration configuration. The *multimodal* option specifies that the two images have different brightness and contrast. In the following example, the second image is rotated by 5° and then registered with respect to the unrotated image. The images are then viewed together using the IPT function **imshowpair**, where the *Scaling* option scales the intensity values of the two images together as a single data set instead of processing them independently (Figure 1.53).

Example 1.43: Write a program to demonstrate image registration by aligning a rotated version of an image with the fixed version by using a metric and an optimizer

```
clear; clc;
I = imread('coins.png');
J = imrotate(imread('coins.png'), 5);
[optimizer, metric] = imregconfig('multimodal');
K = imregister(J, I, 'affine', optimizer, metric);

figure,
subplot(121), imshowpair(I, J, 'Scaling', 'joint'); title('before registration');
subplot(122), imshowpair(I, K, 'Scaling', 'joint'); title('after registration');
```

For measuring the similarity between two images or portions of images, two measures that are frequently used are the covariance and the correlation. The **covariance** of two vectors A and B with N elements is given by the following, where μA is the mean of A and μB is the mean of B and i is an index of an element (Kendall, 1979).

$$\mathrm{cov}(A,A) = \frac{1}{N-1}\sum_{i=1}^{N}\{A(i)-\mu A\}\{A(i)-\mu A\}$$

$$\mathrm{cov}(A,B) = \frac{1}{N-1}\sum_{i=1}^{N}\{A(i)-\mu A\}\{B(i)-\mu B\}$$

before registration

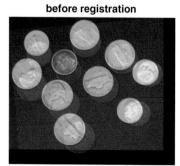

after registration

FIGURE 1.53
Output of Example 1.43.

$$\mathrm{cov}(B,A) = \frac{1}{N-1} \sum_{i=1}^{N} \{B(i) - \mu B\}\{A(i) - \mu A\}$$

$$\mathrm{cov}(B,B) = \frac{1}{N-1} \sum_{i=1}^{N} \{B(i) - \mu B\}\{B(i) - \mu B\}$$

Here, cov (*A, A*) is called the variance of *A* i.e. var(*A*), and cov (*B,B*) is called the variance of *B* i.e. var (*B*). The mean and variance are as defined below:

$$\mu A = \frac{1}{N} \sum_{i=1}^{N} A(i); \mu B = \frac{1}{N} \sum_{i=1}^{N} B(i)$$

$$\mathrm{var}(A) = \frac{1}{N-1} \sum_{i=1}^{N} \{A(i) - \mu A\}^2 \; ; \mathrm{var}(B) = \frac{1}{N-1} \sum_{i=1}^{N} \{B(i) - \mu B\}^2$$

The **covariance matrix** is the matrix of pairwise covariance between each variable.

$$C = \begin{bmatrix} \mathrm{cov}(A, A) & \mathrm{cov}(A, B) \\ \mathrm{cov}(B, A) & \mathrm{cov}(B, B) \end{bmatrix}$$

Using vector notations, if the normalized vectors be $m = A - \mu A$ and $n = B - \mu B$, then the covariance matrix is given by the following, where the superscript T indicates the transpose matrix:

$$C = \frac{1}{N-1} \begin{bmatrix} m \cdot m^T & m \cdot n^T \\ n \cdot m^T & n \cdot n^T \end{bmatrix}$$

The following example calculates covariance matrix of two vectors using both of the above two expressions to compare the results. The BM function **cov** returns the covariance of a vector or matrix, the BM function **mean** returns the average value of a set of numbers, and the BM function **numel** returns the total number of array elements. The ' (apostrophe) operator calculates the transpose of a vector or matrix.

Example 1.44: Write a program to calculate covariance of two vectors

```
clear; clc;
a = [2 4 6];
b = [6 9 -3];
cov(a,b)

% Alternative computation
n = numel(a);
ma = mean(a);
mb = mean(b);
d = 1/(n-1);
c1 = a-ma; c2 = b-mb;
C = d*[c1*c1', c1*c2' ; c2*c1', c2*c2']
```

To calculate the **covariance of a matrix**, each column is treated as a vector and the above procedure is followed for each pair of columns. If the number of columns of the data matrix is N, then the dimensions of the covariance matrix is $N \times N$. The following example illustrates how to calculate covariance of a matrix with two columns and three columns. Calculations are performed both using the in-built function and the above expressions to cross validate the results.

Example 1.45: Write a program to calculate covariance of a matrix

```
clear; clc; format compact;
% Case 1
A = [2 6 ; 4 9 ; 6 -3]
C1 = cov(A)
% Alternative computation
c1 = A(:,1); c2 = A(:,2);
C2 = cov(c1, c2)

% Case 2
B = A'
C1 = cov(B)
% Alternative computation
a = B(:,1); b = B(:,2); c = B(:,3);
ma = mean(a);
mb = mean(b);
mc = mean(c);
c1 = a-ma; c2 = b-mb; c3 = c-mc;
n = numel(a);
d = 1/(n-1);
a11 = c1'*c1;
a12 = c1'*c2;
a13 = c1'*c3;
a21 = c2'*c1;
a22 = c2'*c2;
a23 = c2'*c3;
a31 = c3'*c1;
a32 = c3'*c2;
a33 = c3'*c3;
C2 = d*[a11 a12 a13 ; a21 a22 a23 ; a31 a32 a33]
```

To find the **covariance of two matrices** of equal size, each matrix is converted to a vector by appending all its columns together and then covariance of the two vectors are calculated as shown above. The following example illustrates calculation of covariance of two matrices of equal size using two methods to compare the results. The : **(colon)** operator here converts a matrix into a vector by appending all its columns together.

Example 1.46: Write a program to calculate covariance of two matrices of equal size

```
clear; clc;

A = [2 4 6 ; 6 9 -3];
B = fliplr(A); % flip the matrix left to right
C1 = cov(A,B)

% Alternative computation

a = A(:);     % convert matrix to vector
b = B(:);     % convert matrix to vector
C2 = cov(a,b)
```

The **correlation coefficient** of two vectors is their covariance divided by their standard deviations. If vectors A and B consists of N elements each and σA, σB represent their standard deviations which is the square root of their variance values, then correlation coefficient r is given by:

$$r(A, B) = \frac{\text{cov}(A, B)}{\sigma A * \sigma B}$$

The **correlation coefficient matrix** is the matrix of correlation coefficients for each pairwise combination. Since $r(A,A) = \text{var}(A)/(\sigma A * \sigma A) = 1$ and $r(B,B) = \text{var}(B)/(\sigma B * \sigma B) = 1$, the diagonal elements are 1.

$$R(A,B) = \begin{bmatrix} r(A, A) & r(A, B) \\ r(B, A) & r(B, B) \end{bmatrix} = \begin{bmatrix} 1 & r(A, B) \\ r(B, A) & 1 \end{bmatrix}$$

The IPT function **corr2** computes 2-D correlation between two images A and B and returns the correlation coefficient $r(A,B)$ as given above. The following example calculates 2-D correlation between pairs of images and prints the correlation coefficient values above the corresponding images (Figure 1.54).

> **Example 1.47: Write a program to calculate correlation between an image and its variations produced by adding noise, filtering with a median filter and tonal correction using histogram equalization**
>
> ```
> clear; clc;
> I = imread('pout.tif');
> J = imnoise(I, 'salt & pepper', 0.1);
> K = medfilt2(J);
> L = histeq(I);
>
> R1 = corr2(I,J); R2 = corr2(I,K); R3 = corr2(I,L);
> subplot(141), imshow(I); title('I');
> subplot(142), imshow(J); title(strcat('J: R =  ', num2str(R1)));
> subplot(143), imshow(K); title(strcat('K: R =  ', num2str(R2)));
> subplot(144), imshow(L); title(strcat('L: R =  ', num2str(R3)));
> ```

Normalized 2-D cross correlation is a technique to find out if one image is contained within another image. The smaller image called the **template** is treated as a kernel and slides over the larger image calculating the correlation value at each point. If the template is a part of the second image, then the computed correlation will be maximum when the two images coincide. The location of the match is identified by computing the location of the maximum correlation (Lewis, 1995). The IPT function **normxcorr2** is used to compute whether the template is part of an image by looking at the location of the maximum correlation value. To demonstrate this the following example shows a 3×3 matrix A and

I   J : R =0.45757   K : R =0.99116   L : R =0.97825

FIGURE 1.54
Output of Example 1.47.

a smaller 2×2 matrix T, such that the pattern of values in T is contained within A. The normalized cross correlation operation indicates a match by a value of 1 at the matching location. Matrix A is padded with zero values before the sliding of the template T begins.

Example 1.48: Write a program to demonstrate normalized cross correlation by using a 3×3 data matrix and 2×2 templates

```
clear; clc;

A = [1 -2 3 ; -4 5 -6 ; 7, -8, 9]
T1 = [-4 5 ; 7, -8]
T2 = [-2 3 ; 5, -6]

C1 = normxcorr2(T1, A)
C2 = normxcorr2(T2, A)
```

The program output generates the following cross correlation scores in which the maximum value of 1 indicates where the templates have matched with the data matrix i.e. at locations (3,2) and (2,3).

```
C1 =
    -0.7444     0.1551    -0.4026     0.6513
     0.1416    -0.9743     0.9930    -0.2658
    -0.6563     1.0000    -0.9994     0.6282
     0.4652    -0.3722     0.4652    -0.3722
C2 =
    -0.8054     0.1342    -0.4546     0.6712
     0.0471    -0.9941     1.0000    -0.1533
    -0.7385     0.9930    -0.9881     0.6613
     0.4027    -0.2685     0.4027    -0.2685
```

The following example shows how templates can be specified interactively. The user is first asked to create the template by drawing a rectangle around an object in an image, right-click on it and choose *Crop Image* option, after which it is cropped and used to identify its matching position by computing the maximum correlation value. The IPT function **imrect** is used to draw a rectangle over the image signifying the location of the maximum correlation match (Figure 1.55).

Example 1.49: Write a program to demonstrate how a matching region can be identified in an image

```
clear; clc;
I = imread('coins.png');
[J, rect] = imcrop(I);
```

original image and template

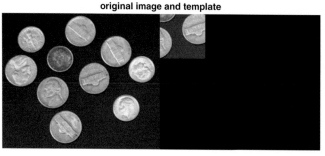

template location identified

FIGURE 1.55
Output of Example 1.49.

```
subplot(1,3,[1,2]),imshowpair(I, J, 'montage');
title('original image and template')
c = normxcorr2(J, I);
[ypeak, xpeak] = find(c == max(c(:)));
yloc = ypeak - size(J,1);
xloc = xpeak - size(J,2);
subplot(133), imshow(I);
imrect(imgca, [xloc + 1, yloc + 1, size(J,2), size(J,1)]);
title('template location identified')
```

1.6 Image Filtering and Enhancement

1.6.1 Image Filtering

Kernel-based operations modify the value of a pixel, not in an isolated manner but in a manner dependent on the values of the neighboring pixels. A kernel or mask is a small array, typically of size 3×3 or 5×5, and contains some coefficient values (Figure 1.56). Starting from the top-left corner of the image the kernel is slid over the image left to right and top to bottom until it reaches the bottom-right corner of the image. At each position of the kernel, the coefficients of the kernel are multiplied with the pixel value just below it and all such products are summed up. Obviously the total number of products would be equal to the total number of coefficients in the kernel. The sum is now made to replace the pixel value just under the center of the kernel. For a 3×3 kernel, let the coefficients be represented as $w(-1, -1)$ through $w(0,0)$ to $w(1,1)$. At any position of the kernel let the pixel values just below the kernel be represented as $f(x - 1, y - 1)$ through $f(x, y)$ to $f(x+1, y+1)$. Then, the sum of the products, shown below, is used to replace the central pixel $f(x,y)$:

$$S = w(-1,-1) \cdot f(x-1,y-1) + \cdots + w(0,0) \cdot f(x, y) + \cdots + w(1,1) \cdot f(x+1,y+1)$$

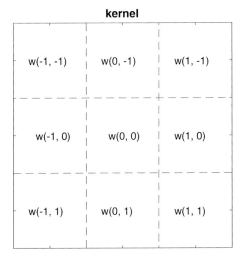
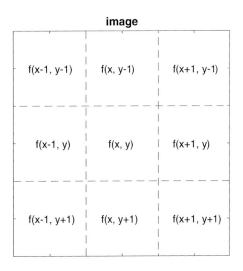

FIGURE 1.56
Kernel and image segment.

In general, for a kernel of size (a, b) we have the sum of products as:

$$S = \sum_{a=-1}^{1}\sum_{b=-1}^{1} w(a, b) \cdot f(x + a, y + b)$$

The pixel value under the center of the kernel is originally $f(x, y)$ and after the operation, the new value of the central pixel would be $g(x, y) = S$. This operation is repeated for all positions of the kernel, and the entire operation is called **correlation**. The same operation is referred to as **convolution** if the kernel is first rotated by 180° before carrying out the process. If the values of the coefficients of the kernel are equal, the kernel is called an **average kernel**. To prevent the pixel values from exceeding 255, the average kernel is divided by the sum of the coefficients, hence the name. Alternatively, the kernel is called a **Gaussian kernel**, where the coefficient values represent a 2-D Gaussian distribution. The effect of correlation or convolution is **image blurring**, the amount of blur depending on the size of the kernel, the larger the kernel greater is the blurring. The blurring is a result of the fact that each pixel of the input image is replaced by a weighted average of the neighboring pixels, due to which the filter behaves like a **low-pass** filter. The foreground object thus gets diffused with some of the background pixels thereby making the line of demarcation between the foreground and background fuzzy. If the kernel is symmetrical then both convolution and correlation produce the same results. The IPT function **imfilter** is used to filter a data matrix using a specified kernel. In the following example, the data matrix A is filtered using the kernel H once using correlation and once using convolution. The filtered matrices along with the original matrix are shown as grayscale images. H_1 is symmetrical average kernal all of whose values are equal to (1/9), and hence both the correlation and convolution results are identical. H_2 is a non-symmetrical kernel which produces different results for the two operations (Figure 1.57).

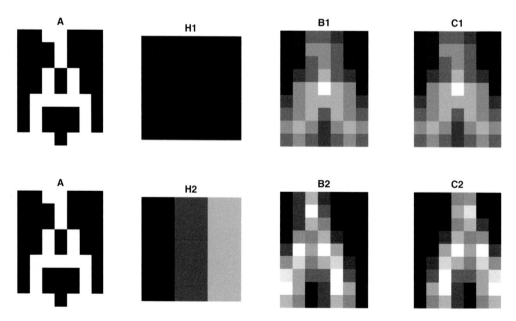

FIGURE 1.57
Output of Example 1.50.

Example 1.50: Write a program to demonstrate the results of correlation and convolution on a binary image using a symmetrical and non-symmetrical kernel

```
clear; clc;

A = [ 0, 0, 1, 1, 0, 0, 0 ; 0, 0, 0, 1, 0, 0, 0 ; 0, 0, 0, 1, 0, 0, 0 ; ...
      0, 0, 1, 0, 1, 0, 0 ; 0, 0, 1, 0, 1, 0, 0 ; 0, 1, 1, 1, 1, 1, 0 ; ...
      0, 1, 0, 0, 0, 1, 0 ; 0, 1, 0, 0, 0, 1, 0 ; 1, 1, 1, 0, 1, 1, 1 ];

H1 = (1/9)*[1 1 1; 1 1 1; 1  1 1];
B1 = imfilter(A, H1, 'corr');
C1 = imfilter(A, H1, 'conv');

subplot(241), imshow(A, []); title('A');
subplot(242), imshow(H1, []); title('H1');
subplot(243), imshow(B1, []); title('B1');
subplot(244), imshow(C1, []); title('C1');

H2 = [0 0.3 0.7; 0 0.3 0.7; 0 0.3 0.7];
B2 = imfilter(A, H2, 'corr');
C2 = imfilter(A, H2, 'conv');

subplot(245), imshow(A, []); title('A');
subplot(246), imshow(H2); title('H2');
subplot(247), imshow(B2, []); title('B2');
subplot(248), imshow(C2, []); title('C2');
```

The IPT function **fspecial** is used to create a customized 2-D kernel filter of the specified type and size. The option *average* creates a averaging filter, the option *disk* creates a circular averaging filter, the option *gaussian* creates a Gaussian filter of the specified standard deviation, the option *motion* creates a motion blur filter. The following example shows four types of kernel filters and their blurring effect on images (Figure 1.58).

Example 1.51: Write a program to demonstrate the results of correlation and convolution on a grayscale image using customized kernels

```
clear; clc;
a = imread('cameraman.tif');
k1 = fspecial('average', 10);
```

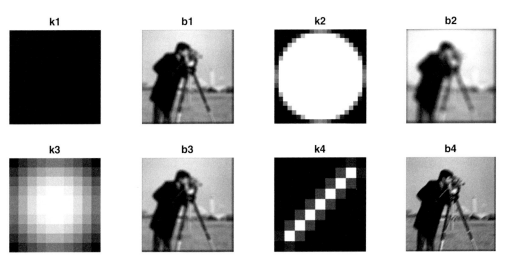

FIGURE 1.58
Output of Example 1.51.

```
k2 = fspecial('disk', 10);
k3 = fspecial('gaussian', 10,10);
k4 = fspecial('motion', 10, 45);
b1 = imfilter(a, k1, 'conv');
b2 = imfilter(a, k2, 'conv');
b3 = imfilter(a, k3, 'conv');
b4 = imfilter(a, k4, 'conv');
figure
subplot(241), imshow(k1, []); title('k1');
subplot(242), imshow(b1, []); title('b1');
subplot(243), imshow(k2, []); title('k2');
subplot(244), imshow(b2, []); title('b2');
subplot(245), imshow(k3, []); title('k3');
subplot(246), imshow(b3, []); title('b3');
subplot(247), imshow(k4, []); title('k4');
subplot(248), imshow(b4, []); title('b4');
```

A major application of convolution is to reduce the effect of noise in an image, as noise is frequently represented as small dots in the background and blurring can effectively reduce this effect to some extent. The blurred image can subsequently be binarized with an appropriate threshold to eliminate the smaller dots. Applied this way, the kernel is referred to as a **noise filter**. Usually an average filter or a Gaussian filter is used as a noise filter to improve or enhance the quality of a general noisy image corrupted by small white dots. Gaussian functions are discussed in detail in Section 1.3. In the following example, an image is filtered using a 5 × 5 and a 10 × 10 Gaussian kernel and then converted to binary using a specific threshold. The effect of the kernel is to blur the image, the larger the size of the kernel the more pronounced is the blurring effect. On blurring the smaller spots is reduced in the intensities due to the convolution with the black background. On thresholding the smaller dots are effectively removed leaving on the larger spots in the image. The IPT function **im2bw** converts an image into a binary format using a specified threshold (Figure 1.59). A 2-D Gaussian function is given by the following form, where σ determines the width or spread of the function.

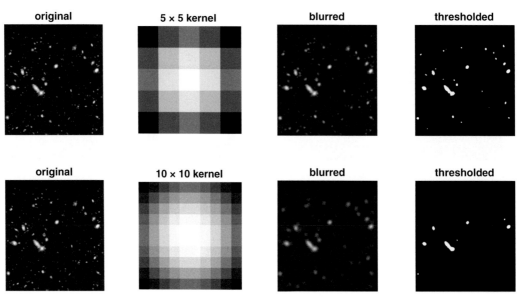

FIGURE 1.59
Output of Example 1.52.

$$G(x, y) = \exp\left(-\frac{x^2 + y^2}{2\sigma^2}\right)$$

Example 1.52: Write a program to perform noise reduction in an image using a Gaussian kernel.

```
clear; clc;
a = imread('sky.png');
k1 = fspecial('gaussian', 5, 5);
k2 = fspecial('gaussian', 10, 10);
b1 = imfilter(a, k1, 'conv');
b2 = imfilter(a, k2, 'conv');
c1 = im2bw(b1, 0.5);
c2 = im2bw(b2, 0.5);

subplot(241), imshow(a); title('original');
subplot(242), imshow(k1, []); title('5 × 5 kernel');
subplot(243), imshow(b1); title('blurred');
subplot(244), imshow(c1); title('thresholded');

subplot(245), imshow(a); title('original');
subplot(246), imshow(k2, []); title('10 × 10 kernel');
subplot(247), imshow(b2); title('blurred');
subplot(248), imshow(c2); title('thresholded');
```

Another type of noise consisting of both white and black dots is called **salt-and-pepper** noise. Such noise can more effectively be removed by a **median filter**. A median filter replaces the central pixel value by the median of the pixel values under the kernel. Other similar filters include the **min-filter** and **max-filter**, where the minimum and maximum of the pixel values under the kernel is used to replace the central pixel value. The min-filter can remove bright spots from an image while the max-filter can remove dark spots. All these types of filters are generally referred to as **order statistic** filters since the pixel values need to be ordered prior to the filtering operation. In the following example, the IPT function **imnoise** adds a specified type of noise to an image, in this case, *salt-and-pepper* noise. The IPT function **ordfilt2** implements a 2-D order statistic filtering of the image, the order being determined by the second argument. The third argument specifies the size of the kernel. In this case, a 3×3 kernel is used which consists of a total of nine elements. When order 9 is specified, it implies that the 9-th or maximum value is retained in the sorted list after the convolution operation. When order 5 is specified, it implies that the 5-th or middle value is retained in the sorted list after the convolution operation. When order 1 is specified, it implies that the 1-st or minimum value is retained in the sorted list after the convolution operation (Figure 1.60). Alternatively, a 2-D median filter can also be implemented using the IPT function **medfilt2**

Example 1.53: Write a program to implement a minimum, maximum and median filtering for removing noise in an image.

```
clear; clc;
a = imread('peppers.png'); a = rgb2gray(a);
b = imnoise(a, 'salt & pepper', 0.1);   % add salt-&-pepper noise
c = ordfilt2(b, 9, ones(3,3));          % max-filter
d = ordfilt2(b, 5, ones(3,3));          % median filter
e = ordfilt2(b, 1, ones(3,3));          % min-filter
f = medfilt2(b);

subplot(231), imshow(a); title('a');
subplot(232), imshow(b); title('b');
subplot(233), imshow(c); title('c');
```

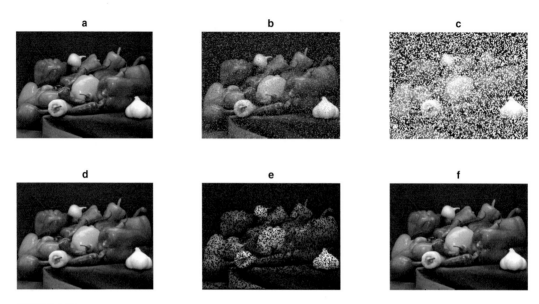

FIGURE 1.60
Output of Example 1.53.

```
subplot(234), imshow(d); title('d');
subplot(235), imshow(e); title('e');
subplot(236), imshow(f); title('f');
```

Another filter frequently used in image processing application is the **Gabor filter**. Named after Dennis Gabor, such a filter analyzes (Gabor, 1946) presence of specific frequency components in specific directions around localized points in an image. In the spatial domain, a 2-D Gabor filter is generated by multiplication of a Gaussian function with a sinusoidal function.

$$g(x, y, \lambda, \theta, \varphi, \sigma) = \exp\left(-\frac{X^2 + Y^2}{2\sigma^2}\right) \cdot \exp\left\{i\left(\frac{2\pi X}{\lambda} + \varphi\right)\right\}$$

Here, λ is the wavelength of the sinusoidal function, θ is the orientation of the Gabor function, φ is the phase offset, σ is the standard deviation of the Gaussian function, and $X = x.\cos(\theta) + y.\sin(\theta)$ and $Y = -x.\sin(\theta) + y.\cos(\theta)$.

The complex Gabor filter above may be split into real and imaginary portions:

$$gR = \exp\left(-\frac{X^2 + Y^2}{2\sigma^2}\right) \cdot \cos\left(\frac{2\pi X}{\lambda} + \varphi\right)$$

$$gI = \exp\left(-\frac{X^2 + Y^2}{2\sigma^2}\right) \cdot \sin\left(\frac{2\pi X}{\lambda} + \varphi\right)$$

The IPT function gabor is used to create a set of Gabor filters of specified wavelengths and angles. The filters can be visualized by displaying the real part of the spatial kernels. The IPT function imgaborfilt is then used to apply the set of Gabor filters to an image. The filtered outputs each has a magnitude and phase. In the following example, two values of wavelengths (5 and 10) and three values of angles (0°, 45°, and 90°) are specified which creates a set of six gabor filters *G*. This filter bank to image *I* and the magnitude *M* and phase *P* of the output images are displayed (Figure 1.61).

Example 1.54: Write a program to generate and apply a filter bank of 6 Gabor filters to an image and display the magnitude and phase of the output images.

```
clear; clc;
I = checkerboard(20);
wavelength = [5, 10];
angle = [0, 45, 90];
G = gabor(wavelength, angle);
[M,P] = imgaborfilt(I, G);

figure,
subplot(361), imshow(real(G(1).SpatialKernel),[]); title('G: 5,0');
subplot(362), imshow(real(G(2).SpatialKernel),[]); title('G: 10,0');
subplot(363), imshow(real(G(3).SpatialKernel),[]); title('G: 5,45');
subplot(364), imshow(real(G(4).SpatialKernel),[]); title('G: 10,45');
subplot(365), imshow(real(G(5).SpatialKernel),[]); title('G: 5,90');
subplot(366), imshow(real(G(6).SpatialKernel),[]); title('G: 10,90');

subplot(3,6,7),    imshow(M(:,:,1), []); title('M: 5,0');
subplot(3,6,8),    imshow(M(:,:,2), []); title('M: 10,0');
subplot(3,6,9),    imshow(M(:,:,3), []); title('M: 5,45');
subplot(3,6,10),   imshow(M(:,:,4), []); title('M: 10,45');
subplot(3,6,11),   imshow(M(:,:,5), []); title('M: 5,90');
subplot(3,6,12),   imshow(M(:,:,6), []); title('M: 10,90');

subplot(3,6,13),   imshow(P(:,:,1), []); title('P: 5,0');
subplot(3,6,14),   imshow(P(:,:,2), []); title('P: 10,0');
subplot(3,6,15),   imshow(P(:,:,3), []); title('P: 5,45');
subplot(3,6,16),   imshow(P(:,:,4), []); title('P: 10,45');
subplot(3,6,17),   imshow(P(:,:,5), []); title('P: 5,90');
subplot(3,6,18),   imshow(P(:,:,6), []); title('P: 10,90');
```

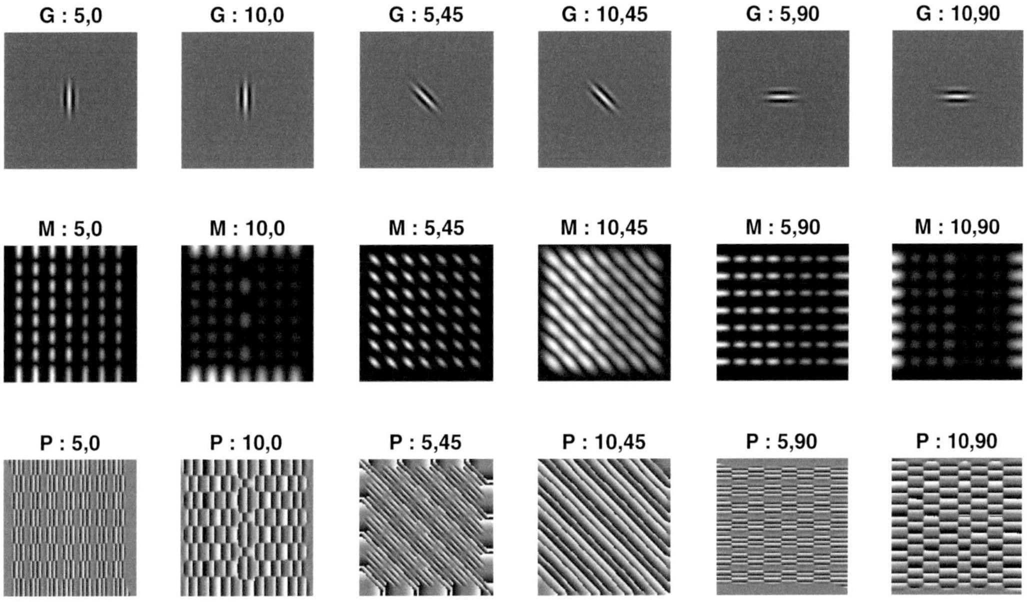

FIGURE 1.61
Output of Example 1.54.

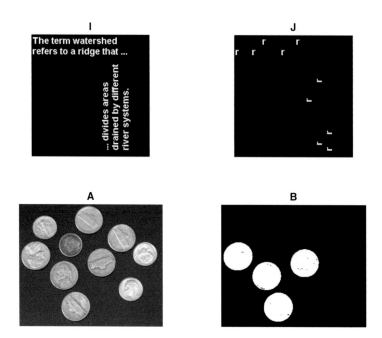

FIGURE 1.62
Output of Example 1.55.

Filters can also be applied to extract specific objects from an image based on their sizes. The sizes can be specified by their height and width, or alternatively by their areas. The IPT function **bwareafilt** is used to filter objects from a binary image. In the following example, all objects with height 35 and width 45 are extracted from an image in the first case, and the four largest objects based on area are extracted in the second case (Figure 1.62).

Example 1.55: Write a program to extract objects from an image based on their sizes and areas.

```
clear; clc;

% filter by size
I = imread('text.png');
J = bwareafilt(I,[35 45]);
figure,
subplot(221), imshow(I); title('I');
subplot(222), imshow(J); title('J');

%filter by area
A = imread('coins.png');
C = im2bw(A);
B = bwareafilt(C,4);
subplot(223), imshow(A); title('A');
subplot(224), imshow(B); title('B');
```

1.6.2 Edge Detection

Converse to image blurring which uses low-pass or integral filters, **image sharpening** is implemented using **high-pass** filters which are differential filters based on derivatives of the image. Their effect is to highlight the high frequency components like edges and corners. **Edge detection** is an important step in pattern recognition whereby foreground objects are separated from the background by highlighting their contours or edges. The kernels used for edge detection have a specialized property viz. the sum of their coefficients are zero. This is because edges are detected by a change in their intensities with respect to the surroundings i.e. there should a non-zero change in intensity as we move from the background to the foreground object. When the kernel is wholly placed above the background or above the foreground object where all the pixels under the kernel have similar values, the output produced is minimal since the sum of the coefficients is zero. However, when the kernel lies partly above the foreground and partly above the background, it produces a non-zero output since the pixels under the kernel have different intensities. This non-zero output is used to highlight the edge pixels. Three of the most frequently used edge detection kernels are the **Roberts** operator (Rx,Ry), **Sobel** operator (Sx,Sy) and **Prewitt** operator (Px,Py). Each of these actually consists of two kernels one for horizontal edges and one for vertical edges. The final result is a combination of both the outputs. Note that edge detection is actually performed on binary images, so a grayscale image needs to be first binarized using an appropriate threshold before the operators are applied.

$$Rx = \begin{bmatrix} -1 & 0 \\ 0 & +1 \end{bmatrix}; Ry = \begin{bmatrix} 0 & -1 \\ +1 & 0 \end{bmatrix}$$

$$Sx = \begin{bmatrix} -1 & 0 & +1 \\ -2 & 0 & +2 \\ -1 & 0 & +1 \end{bmatrix}; Sy = \begin{bmatrix} -1 & -2 & -1 \\ 0 & 0 & 0 \\ +1 & +2 & +1 \end{bmatrix}$$

$$Px = \begin{bmatrix} -1 & 0 & +1 \\ -1 & 0 & +1 \\ -1 & 0 & +1 \end{bmatrix}; Py = \begin{bmatrix} -1 & -1 & -1 \\ 0 & 0 & 0 \\ +1 & +1 & +1 \end{bmatrix}$$

An edge detector is essentially a **differential filter** i.e. it detects edges by calculating difference between pixel values along the same row and column. It calculates the derivative of a signal which for digital images is approximated by the difference.

$$sx = \frac{\partial f(x, y)}{\partial x} = \{f(x+1, y) - f(x-1, y)\}$$

$$sy = \frac{\partial f(x, y)}{\partial y} = \{f(x, y+1) - f(x, y-1)\}$$

These operations can be approximated by the following masks for detecting horizontal and vertical edges:

$$\begin{bmatrix} -1 \\ 0 \\ +1 \end{bmatrix} \text{and} \begin{bmatrix} -1, 0, +1 \end{bmatrix}$$

FIGURE 1.63
Output of Example 1.56.

The IPT function **edge** is used to detect edge pixels in an image using the specified kernels. The options *sobel, prewitt,* and *roberts* incorporate the corresponding masks. The IPT function **im2bw** converts an image into binary format using a specified threshold. The following example shows edges detected using the three operators (Figure 1.63).

> **Example 1.56: Write a program to perform edge detection in an image using sobel, prewitt and roberts kernels.**
>
> ```
> clear; clc;
> a = imread('cameraman.tif');
> b = im2bw(a, 0.5);
> c = edge(b, 'sobel');
> d = edge(b, 'prewitt');
> e = edge(b, 'roberts');
>
> subplot(221), imshow(a); title('image');
> subplot(222), imshow(c); title('sobel');
> subplot(223), imshow(d); title('prewitt');
> subplot(224), imshow(e); title('roberts');
> ```

Two slightly more complex edge detectors are the LoG and Canny operators. The **Canny** edge detector first smoothens the image using a Gaussian filter and then computes the local gradient magnitude and direction at each point of the resulting image. As mentioned before, the **image gradient** is the partial derivative of the image function along the *x*- and *y*- directions:

$$sx = \frac{\partial f}{\partial x}$$

$$sy = \frac{\partial f}{\partial y}$$

The gradient magnitude is $s = \sqrt{sx^2 + sy^2}$ and gradient direction is $\theta = \arctan(sy/sx)$. The third step involves a process called non-maximum suppression which retains those edges which have the maximum gradient magnitude along the maximum gradient direction. This results in keeping only the strong edges and discarding the weak edges. The fourth step is called hysteresis, where two threshold magnitudes are involved, a low and a high value. Edge pixels greater than the high threshold are considered strong edge pixels and are kept. Edge pixels smaller than the low threshold are discarded. Edge pixels between the two thresholds are considered weak edges and are only retained if they are connected to a strong edge otherwise discarded.

The **Laplacian of Gaussian** (LoG) edge detector first implements a Gaussian filter to blur the image for removing noise and then applies the Laplacian operator to the resulting image. The Laplacian operator is computed by adding the second derivatives of pixel intensities along the x- and y- directions i.e.

$$\nabla^2 f(x, y) = \frac{\partial^2 f(x, y)}{\partial x^2} + \frac{\partial^2 f(x, y)}{\partial y^2}$$

The following example uses a specified threshold *th* for binarization before the edge detectors are invoked, thereby over-riding the defaults calculated automatically from the images. The second argument for the LoG and Canny filters specify the standard deviation of the Gaussian filter for smoothing (Figure 1.64).

Example 1.57: Write a program to perform edge detection in an image using LoG and Canny edge detectors.

```
clear; clc;
a = imread('gantrycrane.png'); b = rgb2gray(a);
th = 0.1;
c = edge(b, 'sobel', th);
d = edge(b, 'prewitt', th);
e = edge(b, 'roberts', th);
```

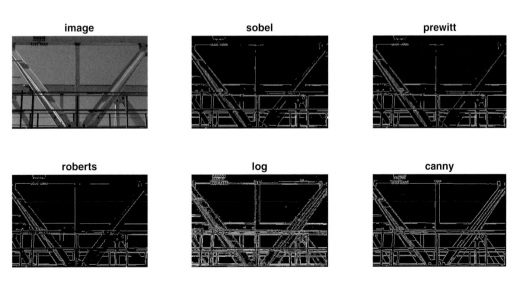

FIGURE 1.64
Output of Example 1.57.

```
f = edge(b, 'log', th, 0.5);
g = edge(b, 'canny', th, 0.5);

subplot(231), imshow(a); title('image');
subplot(232), imshow(c); title('sobel');
subplot(233), imshow(d); title('prewitt');
subplot(234), imshow(e); title('roberts');
subplot(235), imshow(f); title('log');
subplot(236), imshow(g); title('canny');
```

It has been mentioned before that directional image gradients can be derived by computing the partial derivatives of the pixel intensity values along two directions x and y. The magnitude of the gradient is the square root of the sum of the squares of the two directional gradients and the direction of the gradient is given by the tangent of their ratio. The IPT function **imgradientxy** computes the x-direction and y-direction gradients using the Sobel or Prewitt operators. The IPT function **imgradient** computes the gradient magnitude and direction. The following example displays the x-direction and y-direction gradients, the gradient magnitude and gradient direction of an image, using the Sobel operator (Figure 1.65).

Example 1.58: Write a program to calculate the horizontal and vertical image gradients, the gradient magnitude and direction.

```
clear; clc;
I = imread('coins.png');
[Gx, Gy] = imgradientxy(I,'sobel');
[Gmag, Gdir] = imgradient(I, 'sobel');

figure
subplot(221), imshow(Gx, []); title('Gx');
subplot(222), imshow(Gy, []); title('Gy');
subplot(223), imshow(Gmag, []); title('Gmag');
subplot(224), imshow(Gdir, []); title('Gdir');
```

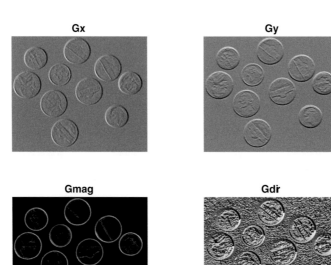

FIGURE 1.65
Output of Example 1.58.

It has already been mentioned before that the gradient of a digital image can be approximated by differences between pixel values. Based on a similar reasoning the second derivative of a digital image can be approximated by differences between pixel gradients,

$$\frac{\partial^2 f(x,y)}{\partial x^2} = \{f(x+1,y) - f(x,y)\} - \{f(x,y) - f(x-1,y)\} = f(x+1,y) + f(x-1,y) - 2f(x,y)$$

$$\frac{\partial^2 f(x,y)}{\partial y^2} = \{f(x,y+1) - f(x,y)\} - \{f(x,y) - f(x,y-1)\} = f(x,y+1) + f(x,y-1) - 2f(x,y)$$

The Laplacian operator being a sum of derivative of the image gradients can be written as follows:

$$\nabla^2 f(x,y) = f(x+1,y) + f(x-1,y) + f(x,y+1) + f(x,y-1) - 4f(x,y)$$

Expressed as a kernel the **Laplacian mask** becomes:

$$L = \begin{bmatrix} 0 & 1 & 0 \\ 1 & -4 & 1 \\ 0 & 1 & 0 \end{bmatrix}$$

A similar operation is carried out using an **unsharp mask** given below, where a blurred version of the image is subtracted from the original image to produce the sharpened image:

$$U = \begin{bmatrix} -1 & -4 & -1 \\ -4 & 26 & -4 \\ -1 & -4 & -1 \end{bmatrix}$$

In the following example, the Laplacian and unsharp masks are first created using the **fspecial** function and then convolved with the image to generate the sharpened images using the **imfilter** function. The masks and sharpened images are shown below. The Laplacian mask essentially detects edges in the image which is then subtracted from the original image to produce the sharpened image. The edge detected image is also shown (Figure 1.66).

Example 1.59: Write a program to perform image sharpening using Laplacian and unsharp masks.

```
clear; clc;
a = imread('rice.png');
k1 = fspecial('laplacian');
k2 = fspecial('unsharp');
b1 = imfilter(a,k1, 'conv');
b2 = imfilter(a,k2, 'conv');

subplot(231), imshow(a); title('original');
subplot(232), imshow(k1, []); title('laplacian mask');
subplot(233), imshow(b1, []); title('edges');
subplot(234), imshow(a-b1, []); title('filtered by Laplacian mask');
subplot(235), imshow(k2, []); title('unsharp mask');
subplot(236), imshow(b2, []); title('filtered by unsharp mask');
```

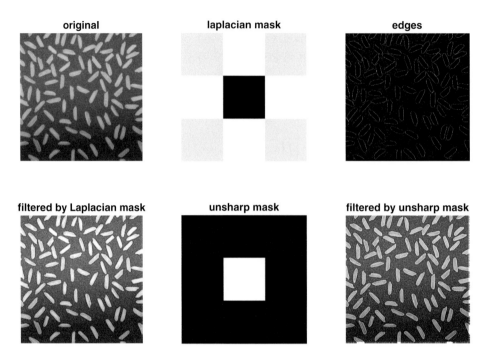

FIGURE 1.66
Output of Example 1.59.

1.6.3 Contrast Adjustment

Image enhancements use mathematical algorithms to manipulate pixel values of an image in order to enhance or improve its quality. Such techniques can broadly be classified into two categories viz. pixel-based and kernel-based. This section discusses pixel-based techniques while kernel-based techniques have been discussed later. The term *pixel-based* means that each pixel value is changed individually or in an isolated manner and is not influenced by the values of the neighboring pixels. Pixel-based techniques primarily use **gamma adjustments** which provides us with a procedure to change pixel values of an image using non-linear input–output transfer functions called **gamma curves**. If r denotes the original gray levels of an image, referred to as input levels, and s be the modified gray levels, referred to as output levels after applying a transfer function T, then the transformation relation can be written as

$$s = T(r)$$

For gamma adjustments the transfer function takes the form of power curves shown below, where c is constant and the index γ determines how the output changes with respect to the input i.e.

$$s = c.r^{\gamma}$$

A plot of the r vs. s curves for various range of γ values is shown below (Figure 1.67).It can be seen for $\gamma < 1$, a narrow range of input values are mapped to a broad range of output values. Since most input values are mapped to higher output values (above $\gamma = 1$ line) this

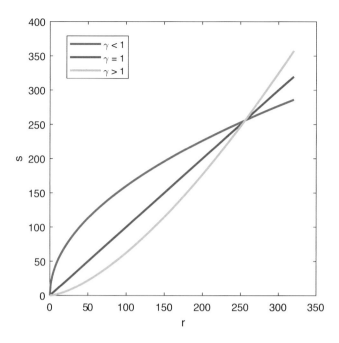

FIGURE 1.67
Gamma curves.

makes the image brighter. Conversely for $\gamma > 1$, a broad range of input values are mapped to a narrow range of output values. Since most input values are mapped to lower output values (below $\gamma = 1$ line) this makes the image darker.

In some cases, both of the gamma ranges are combined into a single S shaped curve to expand both the lower and upper gray levels (Figure 1.68). A piecewise version of the combined curve is often used for stretching the intensity of a small range of input values from r_1 to r_2 to a wider set of output values from s_1 to s_2 while keeping values below r_1 and those above r_2 to their original values [see figure (b)]. Such functionality is called **contrast stretching** as it increases the contrast of the image, which is the difference between the minimum and maximum intensities. A variant of contrast stretching forces the values below r_1 to 0 and those above r_2 to 255 [see figure (c)]. In a contrast stretching operation if $r_1 = r_2 = m$, then it becomes a binarization operation as all values below the threshold m becomes 0 and all values above m becomes 1 [see figure (d)]. In this case, the operation is referred to as **thresholding**. An image is called **inverted** if all gray levels are replaced by their opposite values i.e. $s = 255 - r$.

The IPT function **imadjust** is used to implement gamma adjustment by specifying three arguments viz. range of input values, range of output values, and the gamma index value. A set of empty braces [] implies all existing input values and all possible output values. To implement thresholding, a very small range of input values can be mapped to the entire set of output values. However, the dedicated IPT function **imbinarize** can also be used to binarize the image by specifying a threshold value. In this context it can be mentioned that this function can also be invoked without specifying any threshold value in which case an optimum threshold value is computed automatically by the system using Otsu's thresholding algorithm. The optimum value can be displayed by the function **graythresh**. In the following example, the original color image is converted to grayscale and stored as

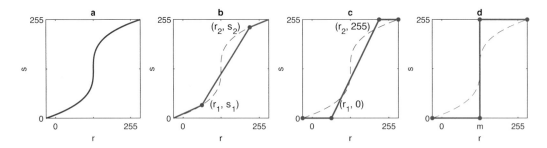

FIGURE 1.68
Contrast stretching and thresholding.

image *a*. For image *b*, all existing gray values of a are mapped to the range [0, 255] and the mapping takes place along the curve for gamma = 0.5. So for example if minimum value in a was 50 and the maximum value was 200, then in *b* value 50 will be converted to 0 and value 200 will be converted to 255 and all intermediate values between 50 and 200 will be converted to proportional values over the range [0,255] i.e. for any value *x* in *a*, the new value in *b* will be $\left\{ \dfrac{x-50}{200-50} \right\} \times 255$. For image *c*, a similar conversion takes place but now the mapping is done along the curve for gamma = 3 which makes it darker than the original image *a*. For image *d*, all values in range [0.4,0.6] are mapped to values in range [0.2,0.8] i.e. values between $0.4 \times 255 = 102$ to $0.6 \times 255 = 153$ are mapped to $0.2 \times 255 = 51$ to $0.8 \times 255 = 204$. In this case, since gamma is not mentioned the mapping takes place linearly i.e. along the curve gamma = 1 by default. Similar reasoning applies for images e and f. For image *g*, a very small range [0.3,0.301] is mapped to the entire range [0,1] which makes the value 0.3 act like a threshold for a binary image i.e. all values below 0.3 become black and all values above 0.301 become white. Image *h* illustrates an alternative method for binarization. Image *i* illustrates inversion of images i.e.0 is mapped to 1 and 1 is mapped to 0 which means that any value *x* in the input image gets converted to value $255 - x$ in the output image (Figure 1.69).

> **Example 1.60: Write a program to demonstrate brightening, darkening, contrast stretching and thresholding of an image by adjusting the gamma value. Also show how images can be inverted.**
>
> ```
> clear; clc;
> a = imread('peppers.png'); a = rgb2gray(a);
>
> % syntax: g = imadjust(f, [low_in high_in], [low_out high_out], gamma);
> b = imadjust(a, [ ], [ ], 0.5);
> c = imadjust(a, [ ], [ ], 3);
> d = imadjust(a, [0.4 0.6], [0.2 0.8]);
> e = imadjust(a, [0.4 0.6], [0 1]);
> f = imadjust(a, [0.2 0.8], [0.4 0.6]);
> g = imadjust(a, [0.3 0.301], [0 1]);
> h = imbinarize(a, 0.5);
> i = imadjust(a, [0 1], [1 0]);
>
> figure,
> subplot(331), imshow(a); title('a: Original');
> subplot(332), imshow(b); title('b: \gamma = 0.5');
> subplot(333), imshow(c); title('c: \gamma = 3');
> subplot(334), imshow(d); title('d: [0.4, 0.6] → [0.2, 0.8]');
> subplot(335), imshow(e); title('e: [0.4, 0.6] → [0, 1]');
> ```

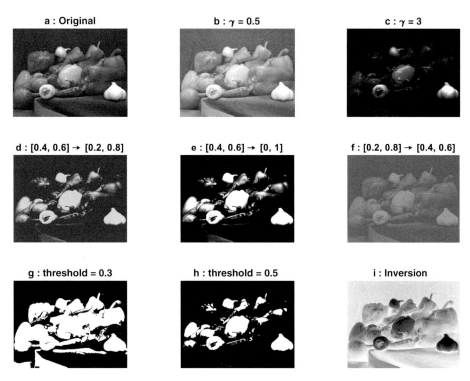

FIGURE 1.69
Output of Example 1.60.

```
subplot(336), imshow(f); title('f: [0.2, 0.8] → [0.4, 0.6]');
subplot(337), imshow(g); title('g: threshold = 0.3');
subplot(338), imshow(h); title('h: threshold = 0.5');
subplot(339), imshow(i); title('i: Inversion');
```

Sharpness of an image is enhanced if there is a quick transition from low intensity values like black to high intensity values like white. Sharpness is perceived as increase in contrast along the edges in an image. The **unsharp masking** technique involves subtracting a blurred (unsharp) version of the image from the original image. The IPT function **imsharpen** uses unsharp masking to improve the contrast of an image. For an RGB color image, the function converts the color image to L*a*b* color space, applies sharpening to the L* channel only, and then converts the image back to the RGB colorspace. The option *Radius* specifies the standard deviation of the Gaussian low-pass filter used for blurring the image and has a default value of 1. The option *Amount* denotes the strength of the sharpening effect and has a default value of 0.8. The following example illustrates sharpening of a color image (Figure 1.70). Unsharp masking can also be implemented using a convolution filter, as shown in Example 1.59 above.

Example 1.61: Write a program to improve contrast of an image by applying a sharpening technique

```
clear; clc;
a = imread('tape.png');
b = imsharpen(a, 'Radius', 2,'Amount',1);
figure, imshowpair(a, b, 'montage')
```

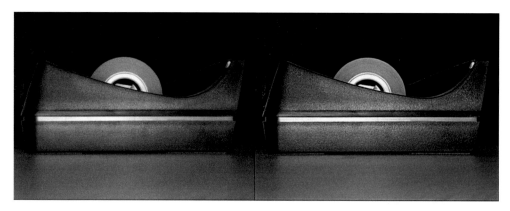

FIGURE 1.70
Output of Example 1.61.

An image **histogram** is a 2-D plot in which the horizontal axis specifies the intensity levels of the pixels and the vertical axis specifies the number of occurrence of pixels belonging to each intensity level. Histograms can be represented mathematically as vectors (collection of values) of pixel occurrences, sometimes called **pixel frequencies**. To make histograms independent of the size of the image, each pixel frequency is divided by the total number of pixels in the image. This is referred to as normalization. The pixel frequencies then become fractional numbers between 0 and 1 and the sum of all pixel frequencies should be equal to 1. Visually a histogram denotes the distribution of gray levels over the image. Ideally a histogram should be distributed over the entire range of gray levels from 0 to 255. If the image is excessively dark, then the histogram is seen to be clustered toward the left end signifying an absence of light shades. Conversely if an image is excessively light, then the histogram is seen to be clustered toward the right end signifying an absence of dark shades. If an image has low contrast, then the histogram is clustered toward the middle signifying absence of both pure black and pure white shades and the difference between the dark and light shades is relatively small. The IPT function **imhist** displays the histogram of an image, where the *x*-axis denotes the range of gray-level intensities 0–255, while the *y*-axis denotes the number of pixels belonging to each gray-level intensity. In the following example, images are displayed on the top row and their corresponding histograms are displayed in the bottom row. The histogram of a normal image *a* is seen to be spread across the entire range of intensity values 0–255. Image a is converted to a bright image *b* using a low gamma value, and its histogram is seen to be shifted to the right signifying an absence of dark shades. Image *c* is a darker version of image *a* made using a high value of gamma, and its histogram is seen to be shifted to the left signifying an absence of light shades. Image *d* is created by mapping the entire range of values in *a* to a small range [0.4, 0.6] resulting in a low-brightness low-contrast image, and its histogram is seen to be clustered in a narrow region toward the center signifying an absence of both light shades and dark shades (Figure 1.71).

**Example 1.62: Write a program to display histograms of an
excessively bright, excessively dark and a low contrast image.**

```
clear; clc;
a = imread('peppers.png'); a = rgb2gray(a);
b = imadjust(a, [ ], [ ], 0.1);
```

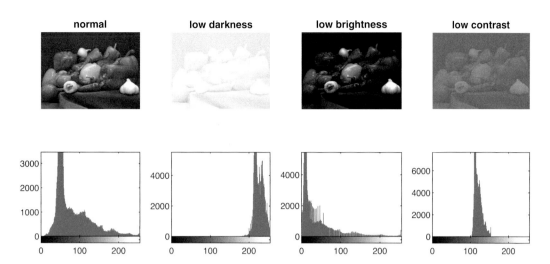

FIGURE 1.71
Output of Example 1.62.

```
c = imadjust(a, [ ], [ ], 2);
d = imadjust(a, [0 1], [0.4 0.6]);
subplot(241), imshow(a); title('normal');
subplot(242), imshow(b); title('low darkness');
subplot(243), imshow(c); title('low brightness');
subplot(244), imshow(d); title('low contrast');
subplot(245), imhist(a);
subplot(246), imhist(b);
subplot(247), imhist(c);
subplot(248), imhist(d);
```

Note: For MATLAB versions 2014 and later, an alternative function **histogram** has been introduced for displaying histograms. In the above code, the *imhist* terms are to be replaced by the term *histogram*.

Histogram equalization is a process of stretching the histogram so that it covers the entire range from 0 to 255. The deficit in gray tones as seen in the previous example could be corrected if the histogram can be stretched over the entire range of values, and this is referred to as **tonal correction** of images. If L is the total number of gray levels (typically 256) and k be a specific intensity level, then the histogram value $h(k)$ for level k i.e. the total number of pixel occurrences for level k after equalization, is given by the following:

$$h(k) = \text{round}\left[\frac{\text{CDF}(k) - \text{CDF}(\min)}{N - \text{CDF}(\min)}\right] \times (L-1)$$

Here, CDF is the cumulative distribution frequency and N is the total number of pixels. For example, consider an image segment divided into four equal sized quadrants having intensity values 0, 50, 100, 150. Since there are no pure white areas, the histogram of the image would be clustered toward the left end covering a range of 0–150. To stretch the histogram to cover the entire range from 0 to 255 we need to compute new values of intensities for each of the quadrants. The process is outlined below:

Index k:	0	1	2	3
Level r:	0,	50,	100,	150

Freq f: $N/4, N/4, N/4, N/4$

CDF: $N/4, N/2, 3N/4, N$

Here, $\text{CDF}(\min) = N/4$, $L = 256$

Using the above relation,

$$h(0) = \left(\frac{N}{4} - \frac{N}{4} \right) \times \frac{255}{N - \frac{N}{4}} = 0$$

$$h(1) = \left(\frac{N}{2} - \frac{N}{4} \right) \times \frac{255}{N - \frac{N}{4}} = \left(\frac{1}{3} \right) \times 255 = 85$$

$$h(2) = \left(\frac{3N}{4} - \frac{N}{4} \right) \times \frac{255}{N - \frac{N}{4}} = \left(\frac{2}{3} \right) \times 255 = 170$$

$$h(3) = \left(N - \frac{N}{4} \right) \times \frac{255}{N - \frac{N}{4}} = (1) \times 255 = 255$$

So the new target levels for the quadrants are 0, 85, 170, 255 which shows that the histogram covers the entire range of gray values. Also note that the target values are independent of the original values and depend on the CDF values for the area concerned. The IPT function **histeq** equalizes a histogram by stretching it so that it cover the entire intensity range of 0–255. The following example illustrates how the gray tones of an image change after tonal correction and how this is reflected by the stretching of its histogram due to the process of histogram equalization (Figure 1.72).

> **Example 1.63: Write a program to display histogram of low contrast image and enhance it using histogram equalization.**
>
> ```
> clear; clc;
> a = imread('peppers.png'); a = rgb2gray(a);
> b = imadjust(a, [0 1], [0.4 0.6]);
> c = histeq(b);
> subplot(221), imshow(b); title('before equalization');
> subplot(222), imshow(c); title('after equalization');
> subplot(223), imhist(b);
> subplot(224), imhist(c);
> ```

1.6.4 Morphological Operations

Morphological operations deal with the study of the geometrical structures and topology of an image. Mostly used with binary images, morphological operations can also be applied on grayscale images and involve manipulations of shape structures like edges, boundaries, and skeletons. A kernel called **structural element** (SE) is used for convolution with images producing effects like **dilation** which thickens foreground pixels, **erosion** which thins foreground pixels, **opening** which is erosion followed by dilation, **closing** which is dilation followed by erosion, **thinning** and **thickening** which selectively removes or adds foreground pixels, and **skeletonization** which exposes the skeletal structure of a shape.

before equalization

after equalization

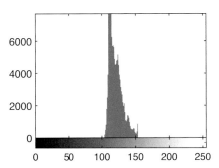

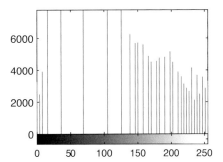

FIGURE 1.72
Output of Example 1.63.

The IPT function **bwmorph** is used to perform morphological operations using a SE which is a 3×3 matrix of all 1s. After the operation the resulting values are divided by the sum of the coefficient values of the kernel so that they do not exceed 255, hence such a kernel is called an **average kernel**. The number of times the operation is to be repeated can be specified. If the repetition value is set to *Inf* the operation is repeated until the image no longer changes. The *dilate* option performs dilation by adding pixels, during which if a 1 on the SE coincides with a 1 on the image, then the center pixel will be converted to 1 if not already so. The *erode* option performs erosion by removing pixels, during which if a 1 on the kernel coincides with 1 on the image, then the center pixel is set to 0 if not already so. The *close* option performs closing operation, which is dilation followed by erosion. The *open* option performs opening operation, which is erosion followed by dilation. The *bothat* option performs **bottom hat** operation which returns the image minus closing on the image. The *tophat* option performs **top hat** operation which returns the image minus opening on the image. The *remove* option removes interior pixels leaving only the boundary of objects. The *skel* option removes boundary of objects leaving the interior skeleton intact. The following example illustrates the effects of various morphological operators on a binary image using an average kernel (Figure 1.73).

Example 1.64: Write a program to demonstrate morphological operations on a binary image

```
clear; clc;
a = imread('circles.png');
h = bwmorph(a, 'thick', 10);
```

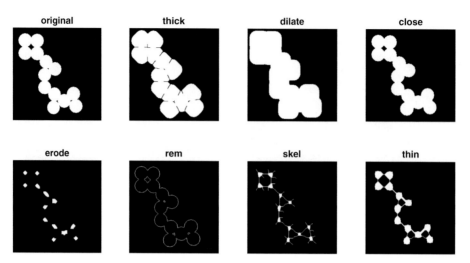

FIGURE 1.73
Output of Example 1.64.

```
c = bwmorph(a, 'close', Inf);
d = bwmorph(a, 'dilate', 10);
e = bwmorph(a, 'erode', 10);
n = bwmorph(a, 'thin', 10);
r = bwmorph(a,'remove');
s = bwmorph(a,'skel',15);

figure
subplot(241), imshow(a); title('original');
subplot(242), imshow(h); title('thick');
subplot(243), imshow(d); title('dilate');
subplot(244), imshow(c); title('close');
subplot(245), imshow(e); title('erode');
subplot(246), imshow(r); title('rem');
subplot(247), imshow(s); title('skel');
subplot(248), imshow(n); title('thin');
```

Morphological operations can also be performed using customized structural elements (SE) instead of the average kernel consisting of all 1s. The IPT function **strel** is used to create a structural element based on the specified shape. Valid shapes are disk, square, rectangle, line, diamond, octagon. The IPT functions **imclose, imopen, imdilate, imerode, imtophat, imbothat** performs the close, open, dilate, erode, top-hat, and bottom-hat operations, respectively, using a specified SE. The following example illustrates erosion operation using five different types of structuring elements which are displayed along the top row and the resulting images on the bottom row (Figure 1.74).

Example 1.65: Write a program to create customized structural elements for morphological operations on a binary image.

```
clear; clc;

a = imread('circles.png');
s1 = strel('disk',10);          % disk
s2 = strel('line', 10, 45);     % line
s3 = strel('square', 10);       % square
s4 = strel('diamond', 10);      % diamond
s5 = [0 1 0 ; 1 1 1 ; 0 1 0];   % custom

b = imerode(a,s1,5);
```

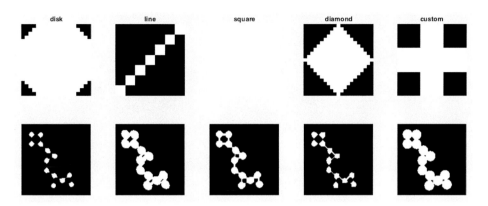

FIGURE 1.74
Output of Example 1.65.

```
c = imerode(a,s2,5);
d = imerode(a,s3,5);
e = imerode(a,s4,5);
f = imerode(a,s5,5);

figure
subplot(251), imshow(s1.Neighborhood, []); title('disk');
subplot(252), imshow(s2.Neighborhood, []); title('line');
subplot(253), imshow(s3.Neighborhood, []); title('square');
subplot(254), imshow(s4.Neighborhood, []); title('diamond');
subplot(255), imshow(s5, []); title('custom');
subplot(256), imshow(b, []);
subplot(257), imshow(c, []);
subplot(258), imshow(d, []);
subplot(259), imshow(e, []);
subplot(2,5,10), imshow(f, []);
```

1.6.5 ROI and Block Processing

While the above filters are each applied over the entire image, it is also possible to apply the filter over a specific **region-of-interest** (ROI) within the image. To achieve this, a binary mask is first created which is white (uncovered region), where the change is to take place and black (covered region) elsewhere. A filter is then applied on to the image through the mask, which applies the changes specified by the filter only over the region uncovered by the mask. In the following example, the IPT function **roifilt2** is used to apply a filter within an ROI specified by the coordinates of a mask over the image. The IPT function **roipoly** is used to create a polygonal mask having one-third the image dimensions, over the middle of the image. The IPT function **imrect** is used to specify the location and size of the mask. An edge detection Laplacian filter is applied over the image through the region allowed by the mask. The height of the mask is arbitrarily specified as one-third the height of the image and its width as one-third the width of the image (Figure 1.75).

Example 1.66: Write a program to specify a region of interest (ROI) within an image and apply a filter within the ROI.

```
clear; clc;
I = imread('coins.png');
h = size(I,1); w = size(I,2);
c = [w/3    2*w/3    2*w/3    w/3];
r = [h/3      h/3    2*h/3    2*h/3 ];
```

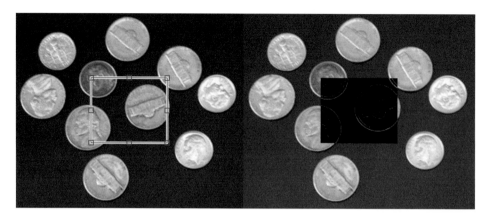

FIGURE 1.75
Output of Example 1.66.

```
BW = roipoly(I, c, r);      % binary mask
H = fspecial('laplacian');  % edge detection filter
J = roifilt2(H, I, BW);     % filter on image through mask
figure, imshowpair(I, J, 'montage')
imrect(imgca, [w/3, h/3, w/3, h/3]);
```

Block processing is a technique of dividing an image into a specified number of rectangular non-overlapping blocks, then applying a specified function to each distinct block and finally concatenating the results in the output image. The IPT function **blockproc** is used to process an image by applying a specified function to each distinct block and then concatenating the results into the output matrix. The IPT function **mean2** calculates the average of all values in a 2-D array. The IPT function **std2** calculates the standard deviation of all values in a 2-D array. In the following example, the input image is split into blocks of size 32×32, the mean and standard deviation of each block is computed, and finally the image is represented by the constant values for each block (Figure 1.76). The example also uses user-defined functions to compute the mean and standard deviation of pixel values in each block. User-defined functions are discussed in more detail toward the end of this section.

I

J

K

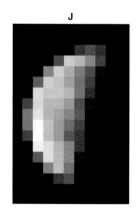
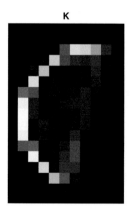

FIGURE 1.76
Output of Example 1.67.

Example 1.67: Write a program to split an image into non-overlapping blocks, calculate mean and standard deviation of each block, and finally represent the image by concatenating the block values.

```
clear; clc;
I = imread('moon.tif');
fun = @(b) mean2(b.data) * ones(size(b.data));   % function to calculate mean
J = blockproc(I,[32 32],fun);
fun = @(b) std2(b.data) * ones(size(b.data));     % function to calculate SD
K = blockproc(I,[32 32],fun);
figure,
subplot(131), imshow(I), title('I');
subplot(132), imshow(J, []), title('J');
subplot(133), imshow(K, []), title('K');
```

The IPT function **im2col** can be used to rearrange image blocks of a specified size into columns which returns the concatenated columns in the output image. For example, if the input image is split into four blocks, one block per quadrant i.e. $A = [A_{11}, A_{12}; A_{21}, A_{22}]$, then the output image $B = [A_{11} (:), A_{12} (:); A_{21} (:), A_{22} (:)]$, where the colon (:) represents concatenation of all values of a matrix into a single column. In the following example, the input image A of size 256×256 is split into non-overlapping distinct 16×16 blocks. Each block consist of 256 pixels which are arranged as a single column. Since there are 256 such blocks in the image, the output image B consists of 256 such columns, the order of block scanning being from left to right and top to bottom. The IPT function **col2im** takes in these columns from B and rearranges them back to the image i.e. it reads all 256 columns from the output image, converts them into 16×16 blocks, and arranges them back to form the image C which is expected to be identical to the original image A (Figure 1.77).

Example 1.68: Write a program to rearrange image blocks into columns and again convert the columns back to the image.

```
clear; clc;
A = imread('cameraman.tif');
[r, c] = size(A);
m = 16; n = 16;
B = im2col(A,[m n], 'distinct');
C = col2im(B,[m n],[r c],'distinct');
figure,
subplot(131), imshow(A); title('A');
subplot(132), imshow(B); title('B');
subplot(133), imshow(C); title('C');
```

A   B   C

 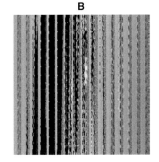 

FIGURE 1.77
Output of Example 1.68.

This section discusses how different types of **user-defined functions** are constructed and executed in MATLAB. Functions like scripts allow execution of command sequences. Functions however provide more flexibility since input values can be passed to them and output values can be returned from them. Functions are defined using the keyword function and are typically saved in a file. The following example shows a function which computes a selective sum of numbers between two specified limits. The function name is *ssum*, the input arguments are *a* and *b* and the output variable is *op* which contains the computed sum.

```
function op = ssum(a,b)
    op = sum(a:b)
end
```

When saving the function the file name should be same as the name of the function i.e. *ssum.m* and the function is executed from the command line by calling the function name and providing the necessary input arguments. Obviously the name should not be the same as that of an already existing function.

```
y = ssum(10, 15)
```

The above command produces *y* = 75. Starting with MATLAB ver 2016b, a function can be included at the end of a script file. For example, the following can be stored in a script file which when executed produces *op1* = 75, *op2* = 3603600. These compute sum and product of numbers between specified limits. The function definitions are included at the end of the script.

```
clear; clc;

x = 10;
y = 15;

op1 = ssum(x,y)
op2 = sprod(x,y)

function ss = ssum(a, b)
    ss = sum(a:b);
end

function sp = sprod(a, b)
    sp = prod(a:b);
end
```

If a function returns multiple outputs, then they can be included within square brackets. The following shows a function which generates two output values *ss* and *sp* containing the sum and product.

```
function [ss sp] = ssumprod(a,b)
    ss = sum(a:b)
    sp = prod(a:b)
end
```

The above function can be called using two input variables and produces two output values. The following produces *x* = 75, *y* = 3603600.

```
[x, y] = ssumprod(10, 15)
```

Functions can also be written as an anonymous function which means they are not stored in a file but defined on the fly in the command line statement. In the following code, variable *myfunction* has a data type *function_handle*, the @ operator creates the handle and the parentheses () immediately after the @ operator includes the function input arguments.

```
myfunction = @(x,y) (sum(x:y));
x = 10;
y = 15;
z = myfunction(x,y)
```

Each statement returns a single output which is assigned to the calling variable *z*. To handle multiple outputs, a cell array can be created, as shown below. The two output values are obtained as: $f\{1\}(10,15) = 75$ and $f\{2\}(10,15) = 3603600$.

```
f = {
    @(x,y) (sum(x:y));
    @(x,y) (prod(x:y));
    };

x = 10;
y = 15;

f{1}(x,y)
f{2}(x,y)
```

Variables defined within a function behave as local variables i.e. they exist only within a function and do not exist outside the function. In the above example, trying the print out the values of *ss* and *sp* will generate errors of undefined variables. Before executing a function, MATLAB checks whether a variable with the same name exists in the current workspace and if so executes the variable instead of the function. Use the BM function **path** to list the folders MATLAB uses to search for variables and functions. The BM function **addpath** adds the specified folders to the top of the search path. The BM function **userpath** denotes the first entry of the MATLAB search path.

1.6.6 Image Arithmetic

Arithmetic operations between images involve addition, subtraction, multiplication, and division operations of the pixel values of the images. The operations are performed on a pixel to pixel basis i.e. a pixel of one image and the corresponding pixel of another image. For this reason, the images involved should have the same number of pixels and same dimensions. Each pixel value can also be added or multiplied to a scalar quantity to increase its value if the scalar is positive and more than 1, else its value can be decreased. One point to be noted is that during increase if the pixel value exceeds 255, then it stays at 255, conversely during decrease if the pixel value becomes negative, then it stays at 0. This is because a grayscale image has an 8-bit representation whose values must range from 0 to 255. The complement of an image is obtained when each pixel is subtracted from 255. A linear combination of multiple images can be computed when images are multiplied by a positive or negative scaling factor and then added up. The IPT functions **imadd**, **imsubtract, imabsdiff, imcomplement, immultiply, imdivide, imlincomb** are used to

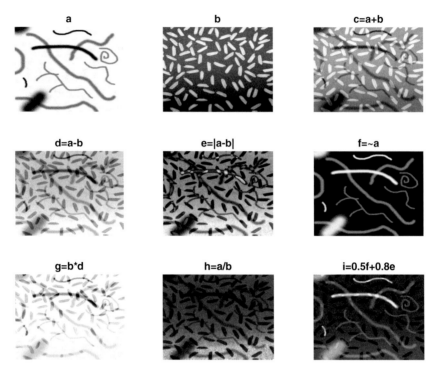

FIGURE 1.78
Output of Example 1.69.

implement the arithmetic operations addition, subtraction, absolute difference, complement, multiplication, division, and linear combination. The following example illustrates
these operations (Figure 1.78). Before the operations are performed the data types of the
images are converted to double precision from an unsigned integer 8-bit so that a larger
number of values can be represented without truncation. The BM function **double** converts a numerical value to 64-bit (8-byte) double-precision floating-point representation.
This is done according to the IEEE Standard 754 which specifies the range for a negative
number as $[-1.79769 \times 10^{308}$ to $-2.22507 - 10^{-308}]$ and the range for a positive number as
$[2.22507 \times 10^{-308}$ to $1.79769 \times 10^{308}]$.

**Example 1.69: Write a program to perform arithmetic
operations between two images.**

```
clear; clc;
a = imread('threads.png'); a = double(a);
b = imread('rice.png'); b = double(b);
a = imresize(a, [300, 400]);
b = imresize(b, [300, 400]);
c = imadd(a, b, 'uint16');
d = imsubtract(a, b);
e = imabsdiff(a, b);
f = imcomplement(a);
g = immultiply(b, d);
h = imdivide(a, b);
i = imlincomb(0.7,f,0.3,e);
subplot(331), imshow(a, []); title('a');
subplot(332), imshow(b, []); title('b');
subplot(333), imshow(c, []); title('c=a+b');
```

```
subplot(334), imshow(d, []); title('d=a-b');
subplot(335), imshow(e, []); title('e=|a-b|');
subplot(336), imshow(f, []); title('f=~a');
subplot(337), imshow(g, []); title('g=b*d');
subplot(338), imshow(h, []); title('h=a/b');
subplot(339), imshow(i, []); title('i=0.5f+0.8e');
```

Note: The BM functions **realmin** and **realmax** can be used to display the minimum and maximum positive values that can be represented in double precision e.g. `realmin('double')`.

Logical operations on images involve AND, OR, NOT operations between pixel values of the images. Like arithmetic operations, these are also performed on a pixel to pixel basis, because of which the images involved should have the same dimensions. The AND operation between two pixels is defined as the minimum of their values. The OR operation similarly is the maximum of their values. The NOT operation on a pixel returns the complement of its value.

$$p \text{ AND } q = \min(p, q) = p \cdot q$$

$$p \text{ OR } q = \max(p, q) = p + q$$

$$\text{NOT } p = 255 - p = p'$$

The basic logical operations can be combined to form extended operations, for example XOR operation:

$$p \text{ XOR } q = p \cdot q' + p' \cdot q$$

The following example illustrates results of logical operations between two grayscale images (Figure 1.79).

Example 1.70: Write a program to perform logical operations between two images.

```
clear; clc;
a = imread('toyobjects.png'); a = double(a);
b = imread('coins.png'); b = double(b);
a = imresize(a, [300, 400]);
b = imresize(b, [300, 400]);
c = min(a,b);
d = max(a,b);
na = 255-a; nb = 255-b;
c1 = min(a,nb); c2 = min(na,b);
d1 = max(a, nb); d2 = max(na, b);
subplot(331), imshow(a, []); title('a');
subplot(332), imshow(b, []); title('b');
subplot(333), imshow(na, []); title('a''');
subplot(334), imshow(c, []); title('a.b');
subplot(335), imshow(d, []); title('a + b');
subplot(336), imshow(min(na,nb), []); title('a''.b''');
subplot(337), imshow(max(na,nb), []); title('a''+b''');
subplot(338), imshow(max(c1, c2), []); title('a.b'' + a''.b');
subplot(339), imshow(min(d1, d2), []); title('(a+b'').(a''+b)');
```

1.6.7 De-blurring

Until now we have been trying to enhance an image by applying specific algorithms on an adhoc basis which we intuitively felt might address the problem at hand. In this section, we will use a mathematical model for image degradation and use the inverse of it for

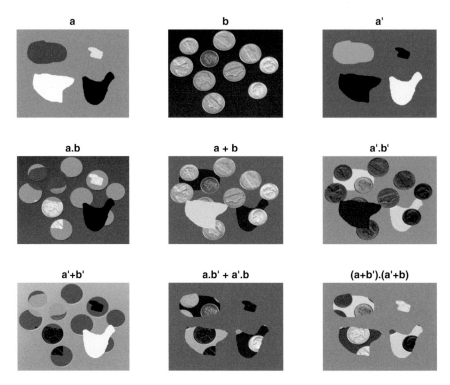

FIGURE 1.79
Output of Example 1.70.

image restoration. Accordingly, this model is called the **degradation-restoration model**. An original image $f(x, y)$ is assumed to be degraded in two steps: blurring and noise. In the first step, a **point spread function** (PSF) is thought to blur an image by causing lack of focus, usually arising from imperfections of the camera and/or motion of the object with respect to the camera. In the second step, a random noise from the environment, arising due to factors like electrical interference or low lighting conditions, is added to the blurred image (Figure 1.80).

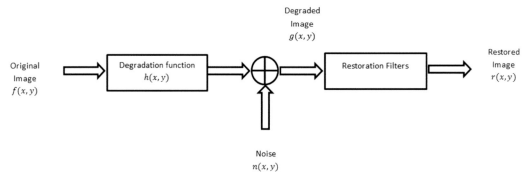

FIGURE 1.80
Image degradation-restoration model.

These two factors together lead to the degradation of the image. If $g(x,y)$ be the degraded image, then the original image $f(x,y)$ is converted to it as part of the degradation process. To create a mathematical model for degradation process we assume the PSF to be a function $h(x,y)$ which when convolved with the original image $f(x,y)$ causes a blurring effect, typically similar to a Gaussian effect which was shown to cause a blur in the image in previous sections. The second factor of additive noise is assumed to be caused by addition of a random noise function $n(x,y)$. Combining both the effects the degradation operation can be mathematically written as below, where $\otimes$ is the convolution operator:

$$g(x,y) = f(x,y) \otimes h(x,y) + n(x,y)$$

The purpose of image restoration is to reverse this process i.e. starting from the degraded image $g(x,y)$ get back the restored image $r(x,y)$. In most cases the restored image will not be identical to the original image $f(x,y)$ but a close estimate of it such that the squared error between them is minimized i.e. the error $(f - r)^2$ is as small as possible. This is because the actual degradation function may not exactly be known, and the noise being a random process cannot be accurately modeled either. Taking the Fourier Transform on either side and using the Convolution Theorem, we can rewrite the above as follows:

$$G(u,v) = F(u,v) \cdot H(u,v) + N(u,v)$$

The Convolution Theorem essentially states that a convolution operation in the spatial domain is equivalent to a multiplication operation in the frequency domain and vice versa. Fourier Transform (FT) and the frequency domain are discussed in Section 1.7. A restoration filter, as shown in the above figure, tries to achieve two goals: reduce or eliminate the noise, called **noise reduction** and reverse the convolution effect, called **deconvolution**. For noise reduction, techniques discussed in previous sections can be applied. In the absence of noise, a deconvolution operation reverses the effects of convolution. In this case, if $N(u, v) = 0$:

$$G(u, v) = F(u, v).H(u, v)$$

$$F(u, v) = \frac{G(u, v)}{H(u, v)}$$

The transfer function for restoration of the original image then becomes $T = \left\{ \dfrac{1}{H(u,v)} \right\}$. In the frequency domain, deconvolution is equivalent to the reciprocal of the PSF and hence it is also called **inverse filtering**. In the following example, a 2-D step function f is created by first filling a 30×30 matrix with 0s and then selectively filling a 10×10 matrix within it with 1s. A Gaussian filter h of size 5×5 is created which acts as the PSF. A convolution operation between the filter and image creates the blurred image g. Fourier transforms of the kernel and blurred images are computed and in the frequency domain, the size of the kernel is adjusted so as to be same as that of the image. A componentwise division operation is then used to create the FT of the restored image R. Subsequently an IFT generates the restored image r (Figure 1.81).

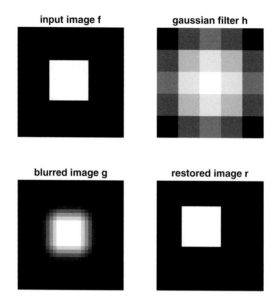

FIGURE 1.81
Output of Example 1.71.

Example 1.71: Write a program to implement inverse filtering on a 2-D step function in the frequency domain.

```
clear; clc;
f = zeros(30,30);
f(10:20,10:20) = 1;
h = fspecial('gaussian',5,5);
g = imfilter(f,h,'conv');
F = fft2(f);
[rows, cols] = size(f);
H = fft2(h, rows, cols);
G = fft2(g, rows, cols);
R = G./H;
r = ifft2(R);

subplot(221), imshow(f); title('input image f');
subplot(222), imshow(h, []); title('gaussian filter h');
subplot(223), imshow(g); title('blurred image g');
subplot(224), imshow(r, []); title('restored image r');
```

If both degradation and noise is present, then an inverse filtering is not sufficient. One of the filters used in such cases is the **Wiener filter,** proposed by Norbert Wiener (Wiener, 1949). As discussed before, the transfer function of an inverse filter is given by: $T = 1/H$. The inverse filter however makes no provision for handling noise. The objective of the Wiener filter is to find an estimate e of the uncorrupted image f such that the mean square error between them i.e. $(f - e)^2$, is minimized. It can be shown that the minimum error is given in the frequency domain by the following:

$$E(u,v) = \left[\frac{H^*(u,v)}{\left\{ |H(u,v)|^2 + \dfrac{S_n(u,v)}{S_f(u,v)} \right\}} \right] \cdot G(u,v)$$

This result is called the Wiener filter, also commonly referred to as the MMSQ (minimum mean square error) filter or the least square error filter. Here, $H(u,v)$ is the degradation function, $H^*(u,v)$ is the complex conjugate of $H(u,v)$, $S_n(u,v)$ is the power spectrum of noise and $S_f(u,v)$ the power spectrum of the undegraded image, both of which need to be estimated. This gives us the transfer function of the Wiener filter as below, where K is the ratio of the power spectrums of the noise and original signal:

$$T = \frac{H^*}{H^2 + K}$$

If the noise is zero, then the ratio K vanishes and the Wiener filter reduces to the inverse filter

$$T = \frac{1}{H}$$

The IPT function **wiener2** is used to filter a noisy image I by estimating the noise power using a local M by N neighborhood and produces the output image J. The function estimates the local mean (μ) and variance (σ) around each pixel of an image using a specified local neighborhood (Lim, 1990):

$$\mu = \left(\frac{1}{NM}\right)\sum I(x,y)$$

$$\sigma^2 = \left(\frac{1}{NM}\right)\sum \left\{ I^2(x,y) - \mu^2 \right\}$$

It then creates a pixelwise filter using these estimates as shown below, where v^2 is the estimated noise variance.

$$J(x,y) = \mu + \left\{ \frac{\sigma^2 - v^2}{\sigma^2} \right\}\left\{ I(x,y) - \mu \right\}$$

If the noise variance is not given, the filter uses the average of all the local estimated variances. In the following example, an image is first corrupted with Gaussian noise and then the Wiener filter is applied to restore it back. The original image (A), the noisy image (B), and the restored image (C) is displayed (Figure 1.82).

A	B	C
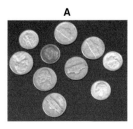	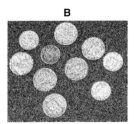	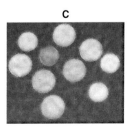

FIGURE 1.82
Output of Example 1.72.

Example 1.72: Write a program to implement Wiener filter for restoration of a noisy image.

```
clear; clc;
A = imread('coins.png');
m = 0.1;                          % noise mean
v = 0.05;                         % noise variance
B = imnoise(A,'gaussian', m, v);
M = 10; N = 10;                   % local neighborhood size
[C, V] = wiener2(B,[M N]);
subplot(131), imshow(A); title('A');
subplot(132), imshow(B); title('B');
subplot(133), imshow(C); title('C');
fprintf('Estimated noise variance: %f\n', V);
```

When both the PSF and noise is present, the IPT function **deconvwnr** is used to implement **Wiener deconvolution** using the known PSF and an estimated noise-to-signal ratio (NSR). In the following example, the function **im2double** converts an image from the unsigned 8-bit integer (*uint8*) to double precision data type (*double*). This step is necessary to avoid truncating the values to 255 when in the subsequent steps the image values are modified by the PSF and additive noise. The PSF is created using a 5 × 5 Gaussian kernel which is convolved with the image to generate a blurred version of the image *bl*. Gaussian noise of a specific mean and variance is added to create a blurred and noisy image *bn*. To restore the image in the first case, an inverse filtering is applied by assuming NSR = 0. In the second case, using an estimated NSR Wiener deconvolution is applied to generate an estimate of the original image (Figure 1.83).

Example 1.73: Write a program to implement Wiener deconvolution for restoration of a blurred and noisy image, using known PSF and noise parameters.

```
clear; clc;

a = im2double(imread('cameraman.tif'));
psf = fspecial('gaussian',5,5);        % gaussian blur
```

FIGURE 1.83
Output of Example 1.73.

```
bl = imfilter(a,psf,'conv');            % blurred image
n_mean = 0.5; n_var = 0.1;
bn = imnoise(bl, 'gaussian', n_mean, n_var);     % blurred & noisy image

% Assuming no noise
nsr = 0;
wnr1 = deconvwnr(bn, psf, nsr);         % same as inverse filtering

% Assuming estimate of the noise
nsr = n_var / var(a(:));                % noise to signal ratio
wnr2 = deconvwnr(bn, psf, nsr);

figure,
subplot(231), imshow(a, []); title('original image');
subplot(232), imshow(psf, []); title('PSF');
subplot(233), imshow(bl, []); title('blurred image');
subplot(234), imshow(bn, []); title('blurred noisy image');
subplot(235), imshow(wnr1, []); title('after inverse filtering');
subplot(236), imshow(wnr2, []); title('after wiener deconvolution');
```

Note: When an image is of type *double*, the function *imshow* expects the values to be in the range [0, 1] for it to be displayed correctly. Both the functions *double* and *im2double* converts an image to *double* datatype from *uint8*, but *im2double* scales the values to be in the range [0, 1] which can be displayed properly, while the function *double* keeps the values in the range [0, 255] which will not be properly displayed using *imshow*.

While Wiener deconvolution is a non-iterative frequency domain process and can be applied when both PSF and noise are known or can be estimated, an alternative method called **Lucy–Richardson deconvolution** (L-R), is an iterative process in the spatial domain which can be applied only when the PSF and not the noise is known. The method is named after Leon Lucy (Lucy, 1974) and William Richardson (Richardson, 1972) who described the method independently. As mentioned before, when an image is recorded on a sensor it is slightly blurred due to the PSF. It is assumed that the degraded image H can be represented in the form $H = W \otimes S$, where W is the original image, S is the PSF, and $\otimes$ denotes convolution operation. In order to estimate W given the observed H, an iterative procedure is followed in which the estimate of W for iteration number r is expressed as follows:

$$W_{i,r+1} = W_{i,r} \sum_k \frac{S_{i,k} \cdot H_k}{\sum_j S_{j,k} \cdot W_{j,r}}$$

Here, W_i, H_k, and S_j denote the *i*-th, *k*-th, and *j*-th element of W, H, and S, respectively. It can be shown that this equation converges to the maximum likelihood solution for W. In the following example, a Gaussian filter is used as PSF and noise of specified parameters is added. The IPT function **deconvlucy** is used to implement the deconvolution filter using only the known PSF but without any estimate of the noise (Figure 1.84).

> **Example 1.74: Write a program to implement Lucy-Richardson deconvolution for restoration of a blurred and noisy image, using a known PSF.**
>
> ```
> clear; clc;
>
> a = im2double(imread('cameraman.tif'));
> psf = fspecial('gaussian',5,5);          % gaussian blur
>
> bl = imfilter(a,psf,'conv');             % blurred
> n_mean = 0.1; n_var = 0.01;
> bn = imnoise(bl, 'gaussian', n_mean, n_var);    % blurred & noisy
> ```

Original Image

PSF

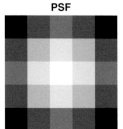

Blurred and Noisy Image

Restored Image

FIGURE 1.84
Output of Example 1.74.

```
luc = deconvlucy(bn, psf);

figure
subplot(221), imshow(a); title('Original Image')
subplot(222), imshow(psf, []); title('PSF')
subplot(223), imshow(bn); title('Blurred and Noisy Image')
subplot(224), imshow(luc); title('Restored Image')
```

When neither the PSF nor the noise is known, then a third type of deconvolution method known as **blind deconvolution** is employed which starts with an initial guess about the PSF, and then iteratively refines it to obtain the better results of the restored image (Lam, 2000). In the following example, the original image *a* is blurred using a motion blur PSF *p* to create the blurred image *b*. An initial guess *p*1 of the PSF is used as a parameter to the IPT function **deconvblind** for implementing blind deconvolution which returns an estimate of the restored image *a*2 and a restored PSF *p*2. The PSF is refined by removing some intermediate values and again used as an input parameter *p*3 for the next iteration of blind convolution. The returned deblurred image is now *a*3 and the restored PSF is *p*4. This process is repeated until the desired results are obtained. The BM function **find** finds out all values of *p*3 less than 0.01 which are then converted to 0 as a step for refining the initial guess of the PSF (Figure 1.85).

Example 1.75: Write a program to implement blind deconvolution for restoration of a blurred image, without a known PSF.

```
clear; clc;
a = imread('cameraman.tif');
p = fspecial('motion',13, 45);
b = imfilter(a, p, 'circ', 'conv');
p1 = ones(size(p));                    % initial guess
[a2, p2] = deconvblind(b, p1, 30);
p3 = p2;
```

a : Original image	p : Original PSF	b : Blurred image	a2 : Restored image

p2 : Restored PSF	p3 : Refined PSF	a3 : New Restored image	p4 : New Restored PSF

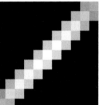

FIGURE 1.85
Output of Example 1.75.

```
p3(find(p3 < 0.01))= 0;              % refine the guess
[a3, p4] = deconvblind(b, p3, 50, []);

subplot(241), imshow(a, []);    title('a: Original image');
subplot(242), imshow(p, []);    title('p: Original PSF');
subplot(243), imshow(b);        title('b: Blurred image');
subplot(244), imshow(a2);       title('a2: Restored image');
subplot(245), imshow(p2, []);   title('p2: Restored PSF');
subplot(246), imshow(p3, []);   title('p3: Refined PSF');
subplot(247), imshow(a3);       title('a3: New Restored image');
subplot(248), imshow(p4, []);   title('p4: New Restored PSF');
```

1.7 Image Segmentation and Analysis

1.7.1 Image Segmentation

Image segmentation is the process of partitioning a digital image into characteristics regions usually for understanding the content of the image. It demarcates regions based on their properties like color, texture, and shape, such that pixels within each region share some common characteristics which are significantly different from those in other regions. Edge detection frequently forms the first step of segmentation to detect the edges and contours of these regions. Clustering algorithms and region-growing methods are used to group regions based on similarity criteria. Split and merge techniques based on quad-tree decomposition can be used to split larger regions into smaller parts, or merge multiple smaller parts into larger regions. The goal of the segmentation process is to assign a label to each pixel as to which region they should belong. Pixels with the same label therefore share certain characteristics. Image segmentation finds extensive use in applications like object detection, biometrics medical imaging, machine learning, computer vision and content-based retrieval (Belongie et al., 1998). The IPT function **superpixels** is used to demarcate regions of similar values around each pixel. Given the input image (A) and the number of regions required (N), the function returns a label matrix (L) and the actual number of regions computed (n). The label matrix specifies different integer values for each region computed, which is then used as the input to the IPT function **boundarymask** which highlights the boundary lines or borders of the regions. The IPT function **imoverlay** overlays or superimposes the border lines onto a 2-D image, typically the original image from which the regions were computed. The IPT function **labeloverlay** overlays or superimposes the label matrix onto a 2-D image. The IPT function **grayconnected** is used to generate a binary image from a grayscale regions by selecting contiguous image regions with respect to the intensity of a given point in the image, specified by coordinates (*seedrow*, *seedcol*) and a tolerance value. In the following example, five regions are computed from the input color image and these regions along with borders demarcating the region boundaries are subsequently overlaid onto the original image (Figure 1.86).

Example 1.76: Write a program to demarcate foreground objects of an image from the background by identifying boundary lines.

```
clear; clc;
A = imread('football.jpg');
N = 5;
[L, n] = superpixels(A,N);
BW = boundarymask(L);
B = imoverlay(A, BW, 'yellow');
C = labeloverlay(A, L);

D = rgb2gray(A);
seedrow = 100; seedcol = 200; tol = 50;
BB = grayconnected(D,seedrow,seedcol,tol);

figure,
subplot(241), imshow(A); title('A');
subplot(242), imshow(L, []); title('L');
subplot(243), imshow(BW); title('BW');
subplot(244), imshow(B); title('B');
subplot(245), imshow(C); title('C');
subplot(246), imshow(D); title('D'); hold on;
plot(seedrow, seedcol, 'yo'); hold off;
subplot(247), imshow(BB, []); title('BB');
```

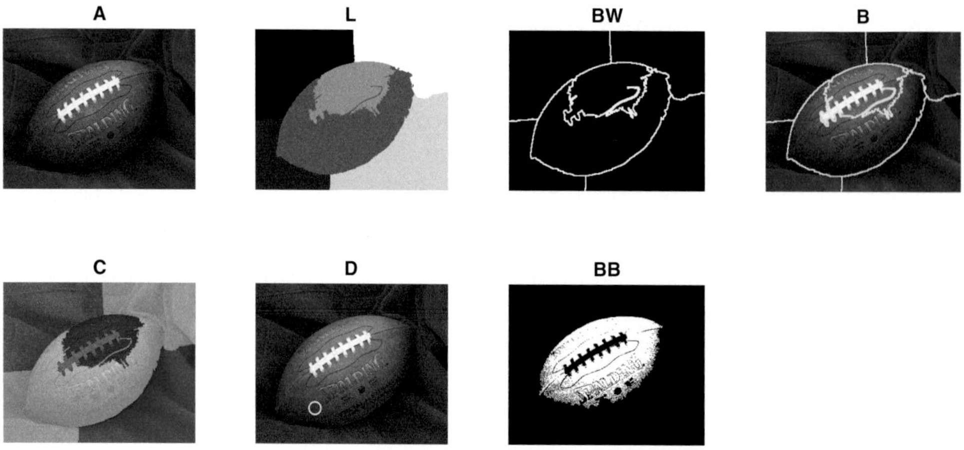

FIGURE 1.86
Output of Example 1.76.

1.7.2 Object Analysis

Object analysis is a set of processes, usually following image segmentation, which attempts to provide some useful information about the segmented regions in an image. As a first step, object segmentation tries to demarcate foreground objects from the background and extracts properties like size, shape, location, and number of objects detected. At a more advanced level, object analysis associates semantic meaning to the detected objects so that they can be labeled as some known entities like people, car, table, chair, road, houses, and so on. The IPT function **bwboundaries** traces region boundaries in a binary image. It returns a cell array of boundary pixel locations (B), a label matrix (L), and the number of objects found (N). In the following example, out of the ten objects found, the boundary of the first object from the left is traced out (Figure 1.87).

Example 1.77: Write a program to trace region boundaries in an image.

```
clear; clc;
D = imread('coins.png');
I = imbinarize(D, 0.4);
[B, L, N] = bwboundaries(I);
N              % return number of objects
b = B{1};      % consider the first object
y = b(:,1); x = b(:,2);      % boundary coordinates
figure,
subplot(121), imshow(D); title('D');
subplot(122), imshow(L); title('L');
hold on; plot(x,y, 'm','LineWidth', 2); hold off;
```

For more complex shapes, the IPT function **bwtraceboundary** traces the boundary of an object in a binary image given a point on the boundary and an initial direction to proceed forward. It returns the row and column coordinates of the boundary pixels for the region. It also needs to be specified the initial direction to find the next pixel connected to specified point that belongs to the same region. The direction is specified in terms of N, S, E, W. In the following example, P is a specified point and the initial direction to proceed forward is 'N' (north). The function returns the coordinates of the boundary points (B) which is used to plot the contour over the actual image. The initial point is shown as a red circle (Figure 1.88).

D

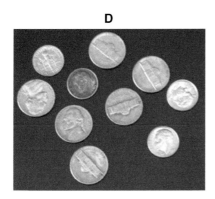

L

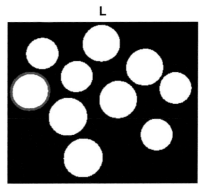

FIGURE 1.87
Output of Example 1.77.

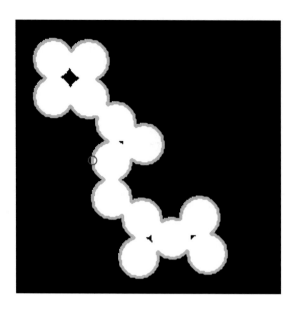

FIGURE 1.88
Output of Example 1.78.

Example 1.78: Write a program to trace boundaries of a complex shape in an image given a starting point and a direction.

```
clear; clc;
I = imread('circles.png');
x = 75; y = 132; P = [y x];
B = bwtraceboundary(I, P, 'N');
imshow(I); hold on;
plot(B(:,2),B(:,1),'g','LineWidth',2);
plot(x,y, 'ro'); hold off;
```

The IPT function **imfindcircles** finds circles in an image belonging to a specified range of radii lengths. The function returns the actual centers and radii found. In the following example, the function is used to find all circles with radius between 15 and 25 in the input image. The list of centers and radii are then used to draw a set of arrows with start points at the centers and length equal to the corresponding radii (Figure 1.89). The function **arrow** is used to draw arrows from specified centers having specified lengths and colors. This function is described in the following location: *https://in.mathworks.com/matlabcentral/fileexchange/278-arrow*

Example 1.79: Write a program to find circles in an image and identify their centers and radii.

```
clear; clc;
A = imread('coins.png');
[centers, radii] = imfindcircles(A,[15 25]);
figure, imshow(A); hold on;
plot(centers(:,1), centers(:,2), 'r*');
arrow([centers(:,1), centers(:,2)], ...
    [centers(:,1)+radii(:), centers(:,2)], 'EdgeColor','r','FaceColor','g')
```

The IPT function **viscircles** draws circles using specified center and radius. In the following example, the function *imfindcircles* is used find all circles with radius between 15 and

FIGURE 1.89
Output of Example 1.79.

FIGURE 1.90
Output of Example 1.80.

25 in the input image, which returns the relevant centers and radii. These are then used to draw circles around the located objects (Figure 1.90).

> **Example 1.80: Write a program to draw circles in an image given their centers and radii.**
>
> ```
> clear; clc;
> A = imread('coins.png');
> [centers, radii] = imfindcircles(A,[15 25]);
> imshow(A); hold on;
> viscircles(centers, radii,'EdgeColor','b');
> ```

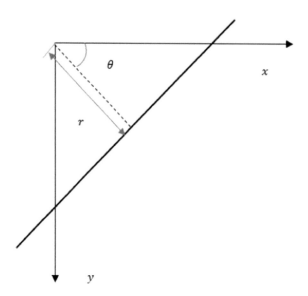

FIGURE 1.91
Relation between $x - y$ and $r - \theta$.

The **Hough Transform** detects lines in an image by using the parametric representation of a line $r = x \cdot \cos(\theta) + y \cdot \sin(\theta)$, where r is the length of the perpendicular projection from the origin on the line (also designated by Greek letter rho ρ), and θ is the angle the perpendicular projection makes with the positive x-axis. It is therefore possible to associate each line with a (r, θ) pair i.e. each line in the $x - y$ plane corresponds to a point in the r–θ space, also called the Hough space (Figure 1.91).

A point in the $x - y$ plane corresponds to a sinusoidal curve in the $r - \theta$ plane, since a large number of lines can be drawn through a single point. A value of q in the Hough space with coordinates (r,θ) means that q points in the $x - y$ plane lie on the line specified by r and θ. The IPT function **hough** computes the Hough transform from a binary image and returns the Hough space matrix. The IPT function **houghpeaks** identifies peaks in a Hough transform. Peaks signify the presence of lines in the $x - y$ plane. The IPT function **houghlines** extracts the line segments associated with the hough peaks. In the following example, the binary image I of size (100×100) containing two points, has the HI matrix below it which shows two sinusoids intersecting at $r = \left(100\sqrt{2}\right)/2 = 70.7$ (half diagonal length) and $\theta = 45°$. The next figure J to the right contains a straight line inclined at $-45°$ to the positive x-axis and the corresponding HJ matrix below it shows a collection of sinusoids, corresponding to all points on the line, intersecting at $\theta = -45°$. The third figure to the extreme right extracts the hough peaks from the HJ matrix and the corresponding extracted line is displayed below it (Figure 1.92).

Example 1.81: Write a program to detect lines in an image using Hough Transform

```
clear; clc;
I = zeros(100);
I (30,70) = 1; I(70,30) = 1;
figure, subplot(231),imshow(I); title('I');

subplot(234),
[HI,T,R] = hough(I,'RhoResolution',0.5,'Theta',-90:0.5:89);
```

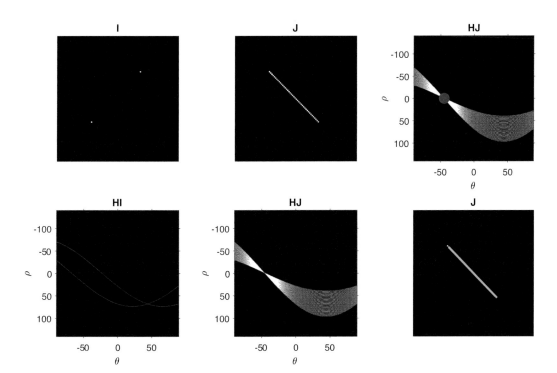

FIGURE 1.92
Output of Example 1.81.

```
imshow(imadjust(rescale(HI)),'XData',T,'YData',R);
title('HI'); xlabel('\theta'), ylabel('\rho');
axis on, axis square;

subplot(232),
J = zeros(100);
for i=30:70
    J(i,i) = 1;
end
imshow(J); title('J');

subplot(235),
[HJ,T,R] = hough(J,'RhoResolution',0.5,'Theta',-90:0.5:89);
imshow(imadjust(rescale(HJ)),'XData',T,'YData',R);
title('HJ'); xlabel('\theta'), ylabel('\rho');
axis on, axis square;

subplot(233)
P  = houghpeaks(HJ,5,'threshold',ceil(0.3*max(HJ(:))));
imshow(imadjust(rescale(HJ)),'XData',T,'YData',R)
xlabel('\theta'), ylabel('\rho'); title('HJ');
axis on, axis square, hold on;
p = plot(T(P(:,2)),R(P(:,1)),'o');

set(p,'MarkerSize', 10, 'MarkerEdgeColor','b','MarkerFaceColor', 'r')

subplot(236)
lines = houghlines(J,T,R,P,'FillGap',5,'MinLength',7);
imshow(J), hold on; title('J');
max_len = 0;
for k = 1:length(lines)
   xy = [lines(k).point1; lines(k).point2];
   plot(xy(:,1),xy(:,2),'LineWidth',2,'Color','green');
end
```

The **quad tree decomposition** (QT) is a data structure for representation of a grayscale or binary image in a compact form. It divides the image into four quadrants and checks to see whether pixels of each quadrant contain the same color (homogeneous) or different colors (heterogeneous). For homogeneous quadrants, the entire quadrant is represented by a single pixel value (since all other pixels in the quadrant have the same value), while heterogeneous quadrants are further divided into sub-quadrants until they become homogeneous. The IPT function **qtdecomp** is used to implement quad-tree decomposition and returns the QT structure S as a sparse matrix. If $S(x,y)$ is non-zero, then (x,y) is the upper left corner of a block of size $S(x,y)$. The BM function **disp** displays the quadtree structure in the sparse matrix S. The IPT function **qtgetblk** returns all blocks of the specified dimension and their row, column locations. Starting out with a blank matrix J, the IPT function **qtsetblk** arranges the blocks returned by the decomposition function and returns back the original matrix. The BM function **zeros** generates a blank matrix of the specified dimensions containing all zeros. The following example returns the QT decomposition structure of a 8×8 binary graphics region (I) and reconstructs back the graphics region (J) from the QT blocks (Figure 1.93). The detailed discussion about the decomposition and reconstruction is given after the example.

Example 1.82: Write a program to encode and decode an image section using quad-tree decomposition

```
clear; clc;

I = [
          1    1    1    1    1    1    1    1
          1    1    1    1    1    1    1    1
          1    1    1    1    0    0    0    1
          1    1    1    1    0    0    1    1
          0    0    0    0    0    0    1    1
          0    0    0    0    0    0    1    1
          0    0    0    0    1    1    1    1
          0    0    0    0    1    1    1    0
     ];
S = qtdecomp(I); disp(full(S));

dim1 = 4; [blocks1, r, c] = qtgetblk(I, S, dim1)
dim2 = 2; [blocks2, r, c] = qtgetblk(I, S, dim2)
```

I   J

FIGURE 1.93
Output of Example 1.82.

```
dim3 = 1; [blocks3, r, c] = qtgetblk(I, S, dim3)

J = zeros(8,8);
J = qtsetblk(J, S, dim1, blocks1);
J = qtsetblk(J, S, dim2, blocks2);
J = qtsetblk(J, S, dim3, blocks3);

subplot(121), imshow(I, []); title('I');
subplot(122), imshow(J, []); title('J');
```

The display function generates the quad-tree structure which shows the upper-left corner of a homogeneous block containing the block size:

4	0	0	0	2	0	2	0
0	0	0	0	0	0	0	0
0	0	0	0	2	0	1	1
0	0	0	0	0	0	1	1
4	0	0	0	2	0	2	0
0	0	0	0	0	0	0	0
0	0	0	0	2	0	1	1
0	0	0	0	0	0	1	1

In the first case blocks of dimension 4 are displayed along with row and column numbers indicating their locations in the original matrix *I*:

```
#    blocks           row     column
------------------------------------------------------------------------
1    1    1    1    1    1    1
     1    1    1    1
     1    1    1    1
     1    1    1    1


2    0    0    0    0    5    1
     0    0    0    0
     0    0    0    0
     0    0    0    0
```

In the second case blocks of dimension 2 are displayed along with row and column numbers indicating their locations in the original matrix *I*:

```
#    blocks    row   column
---------------------------------------------------------------------
1    1    1    1    5
     1    1


2    0    0    3    5
     0    0


3    0    0    5    5
     0    0


4    1    1    7    5
     1    1


5    1    1    1    7
     1    1
```

Similarly in the third case eight matrices of size 1 are returned along with their positions. Finally, starting out with a blank matrix *J*, all these blocks are inserted in their indicated locations to return back the original matrix *I*.

1.7.3 Region and Image Properties

Once object-based boundaries are detected in images, it might be required to extract shape-based features like centroid, radius, area, perimeter, and bounding box, to characterize the shape further. The bounding box is the smallest rectangle containing the region. The centroid is the coordinates of the center of mass of the region. The major axis length is the length in pixels of the major axis of the ellipse that has the same normalized second central moments as the region. The minor axis length is the length in pixels of the minor axis of the ellipse that has the same normalized second central moments as the region. The IPT function **regionprops** measures properties of objects in a binary image like area, centroid, major axis length, minor axis length, bounding box, eccentricity, orientation, convex hull etc. In the following example, objects with intensity less than 100 (dark objects) are first extracted and their boundaries are traced using their center and radius information. Secondly, objects with intensity more than 200 (bright objects) are likewise extracted and their boundaries traced out (Figure 1.94).

Example 1.83: Write a program to measure the centroid and mean radius from regions in a binary image.

```
clear; clc;
I = imread('circlesBrightDark.png');
figure,
subplot(131), imshow(I);

BW = I < 100;    % detect regions of low intensity
s = regionprops('table',BW, 'all');
c = s.Centroid;
d = mean([s.MajorAxisLength s.MinorAxisLength],2);
r = d/2;
subplot(132), imshow(BW); hold on;
viscircles(c,r, 'Color', [0.7 0.4 0]);
hold off;

BW = I > 200;    % detect regions of high intensity
s = regionprops('table',BW, 'all');
c = s.Centroid;
d = mean([s.MajorAxisLength s.MinorAxisLength],2);
r = d/2;
```

FIGURE 1.94
Output of Example 1.83.

```
subplot(133), imshow(BW); hold on;
viscircles(c,r, 'Color', [0 0.5 0]);
hold off;
```

The area of a region is the number of pixels contained within the region. The bounding box is the minimum rectangle containing the region. The area can be used to identify the largest and smallest regions in an image which can then be demarcated using bounding boxes. The following example illustrates how the largest and smallest regions in an image are identified using the area value and these are then subsequently highlighted using bounding boxes. The grayscale image (I) is binarized using a threshold of 120 (J), and the IPT function **imfill** is used to fill any holes in the regions using a flood-fill operation (K). The IPT function **bwlabel** is used to generate a label matrix L specifying each object and the total number of objects found n. Using the IPT function **regionprops**, the area of each region is computed, out of which the regions with the maximum and minimum areas are identified. The function also returns the location and size of the bounding boxes of the regions, The BM function **rectangle** is used to draw rectangular boundaries around these objects by using location and size of the bounding boxes returned earlier (Figure 1.95).

Example 1.84: Write a program to identify the largest and smallest of a group of image objects and demarcate them using bounding boxes.

```
clear; clc;
I = imread('coins.png');
J = I > 120;
K = imfill(J, 'holes');
[L, n] = bwlabel(K);
r = regionprops(L, 'all');
for i = 1:n, A(i) = r(i).Area; end
[xv, xi] = max(A);
[nv, ni] = min(A);

subplot(221), imshow(I); title('I');
subplot(222), imshow(J); title('J');
subplot(223), imshow(K); title('K');
subplot(224), imshow(I); hold on;

rectangle('Position', r(xi).BoundingBox, 'Edgecolor', 'g', 'Linewidth', 2);
rectangle('Position', r(ni).BoundingBox, 'Edgecolor', 'r', 'Linewidth', 2);
title(['\fontsize{13} {\color{red}smallest \color{green}largest}']);
```

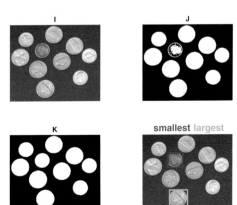

FIGURE 1.95
Output of Example 1.84.

The area is the actual number of pixels in the region. The perimeter is the distance around the boundary of the region measured in number of pixels. The Euler number is the total number of objects in the image minus the total number of holes in those objects. The IPT function **bwarea** estimates the area of objects in a binary image. The IPT function **bwperim** returns a binary image containing the perimeter of the objects in the image. The IPT function **bweuler** returns the Euler number of the binary image. In the following example, for the binary image $b1$, the white portion is considered as the object and the three black circles are considered as holes, and the Euler number is computed as $1 - 3 = -2$. For the second binary image $b2$, the three white areas are considered as objects which do not contain any holes, and so the Euler number is computed as $3 - 0 = 3$. The perimeter of the detected regions are finally traced out (Figure 1.96).

Example 1.85: Write a program to measure area, perimeter and Euler number from regions in a binary image.

```
clear; clc;
I = imread('circlesBrightDark.png');
b1 = im2bw(I, 0.5);
a1 = bwarea(b1);
p1 = bwperim(b1);
e1 = bweuler(b1);
b2 = im2bw(I, 0.7);
a2 = bwarea(b2);
p2 = bwperim(b2);
e2 = bweuler(b2);

subplot(231), imshow(I); title('I');
subplot(232), imshow(b1);  title(strcat('a=',num2str(a1), ', e=', num2str(e1)));
subplot(233), imshow(b2);  title(strcat('a=',num2str(a2), ', e=', num2str(e2)));
subplot(2,3,[4,6]),imshowpair(p1, p2, 'montage')
```

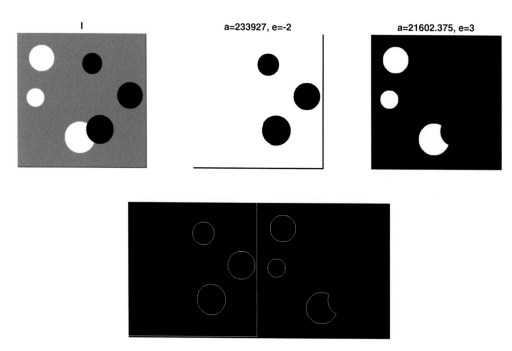

FIGURE 1.96
Output of Example 1.85.

Pixel connectivity is a scheme to identify which pixels in an image are connected to their neighbors. This is necessary to analyze image objects and patterns which consist of groups of connected pixels. There are mainly two types of connectivity schemes: 4-way and 8-way (Figure 1.97). In **4-way** connectivity, pixels share an edge with a neighbor. For a pixel P, its neighbors in a 4-way connectivity scheme are P2, P4, P6, and P8. In general, for a pixel P at coordinates (x,y) its 4-connected neighbors are at $(x+1,y)$, $(x-1,y)$, $(x,y+1)$, $(x,y-1)$. Each of them are at a unit distance from P. In a **8-way** connectivity pixels share either an edge or a corner with a neighbor. Thus, apart from its 4-connected neighbor, the pixel P also has four more neighbors at P1, P3, P5, and P7. In general, for a pixel P at coordinates (x,y), its 8-connected neighbors include the above four neighbors at $(x\pm1,y)$, $(x,y\pm1)$ and also the diagonal pixels at $(x+1, y+1)$, $(x+1, y-1)$, $(x-1, y+1)$, $(x\times1, y-1)$. The diagonal pixels are at an *Euclidean distance* of $\sqrt{2}$ from the central pixel. Note the *chessboard distance* metric assigns the same value of 1 to all 8 neighbors of a pixel.

The IPT function **bwconncomp** finds connected components in a binary image. It returns a matrix of all connected components found in the image estimated using a 4-way or 8-way connectivity, as specified. In the following example, the number of objects detected and their centroids are shown to be different depending on whether 4-way or 8-way connectivity is selected. When 4-way connectivity is specified, then pixels connected by an edge along the horizontal or vertical direction are considered part of a single object but pixels connected by a vertex point along diagonal directions are considered as parts of different objects. A total of six objects are detected, and their centroids are plotted over the corresponding objects. When 8-way connectivity is specified, then pixels connected by vertices along a diagonal are also considered part of the same object. A total of three objects are detected, and their centroids are plotted over the corresponding objects (Figure 1.98).

4-way connectivity		
P1	P2	P3
P8	P	P4
P7	P6	P5

8-way connectivity		
P1	P2	P3
P8	P	P4
P7	P6	P5

FIGURE 1.97
4-way and 8-way pixel connectivities.

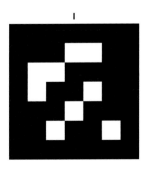
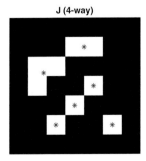
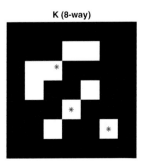

FIGURE 1.98
Output of Example 1.86.

Example 1.86: Write a program to demonstrate that number of objects detected and their centroids depend on 4-way and 8-way connectivities.

```
clear; clc;
I = [
    0 0 0 0 0 0 0 ;
    0 0 0 1 1 0 0 ;
    0 1 1 0 0 0 0 ;
    0 1 0 0 1 0 0 ;
    0 0 0 1 0 0 0 ;
    0 0 1 0 0 1 0 ;
    0 0 0 0 0 0 0 ;
    ];
subplot(131), imshow(I); title('I')

c1 = bwconncomp(I,4);
s1 = regionprops(c1,'Centroid');
n1 = cat(1, s1.Centroid);
subplot(132), imshow(I);  title('J (4-way)'); hold on;
plot(n1(:,1), n1(:,2), 'b*'); hold off;

c2 = bwconncomp(I,8);
s2 = regionprops(c2,'Centroid');
n2 = cat(1, s2.Centroid);
subplot(133), imshow(I);  title('K (8-way)'); hold on;
plot(n2(:,1), n2(:,2), 'r*'); hold off;
```

The objects detected in a binary image can be demarcated by plotting the centroid and bounding box of each over it. However, when a grayscale image is converted to binary, the number of objects detected can depend on the binarization threshold used. In the following example, a grayscale image *I* is converted to binary using two specific thresholds. In the first case, the white background is considered a single object whose centroid and bounding box is plotted over it. In the second case, there are three separate objects and three different centroids with their bounding boxes are plotted over them (Figure 1.99).

Example 1.87: Write a program to convert a grayscale image to binary using different thresholds and demarcate the different objects using bounding boxes

```
clear; clc;
I = imread('circlesBrightDark.png');
figure,
subplot(131)
imshow(I);

subplot(132)
```

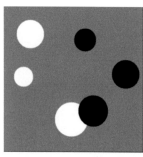
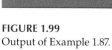
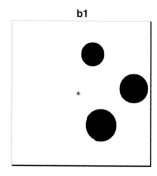
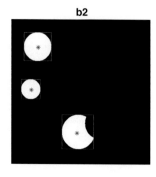

FIGURE 1.99
Output of Example 1.87.

```
b1 = im2bw(I, 0.5);
c1 = bwconncomp(b1);
s1 = regionprops(c1,'all');
n1 = cat(1, s1.Centroid);
bb = s1.BoundingBox;
c1 = insertShape(double(b1), 'Rectangle', bb, 'Color', 'red');
imshow(c1); title('b1'); hold on;
plot(n1(:,1),n1(:,2), 'b*');

subplot(133)
b2 = im2bw(I, 0.7);
c2 = bwconncomp(b2);
s2 = regionprops(c2,'all');
n2 = cat(1, s2.Centroid);
[bb1, bb2, bb3] = s2.BoundingBox;
bb = [bb1;bb2;bb3];
c2 = insertShape(double(b2), 'Rectangle', bb, 'Color', 'green');
imshow(c2);
title('b2'); hold on;
plot(n2(:,1),n2(:,2), 'r*');
```

The IPT function **bwdist** calculates distances for each pixel in a binary image using various metrics. For each pixel, the function assigns a number that is the distance between that pixel and the nearest nonzero pixel. It returns the matrix containing these distances. The following example shows Euclidean, chessboard, and cityblock distances from pixels being visualized as grayscale intensities i.e. larger the distance brighter is the spot (Figure 1.100). Note that the *chessboard* metric assigns the same value of 1 to all 8 neighbors of a pixel, the *Euclidean* metric assigns value 1 to the 4-connected neighbors and $\sqrt{2}$ to the diagonal pixels while the *cityblock* or *Manhattan* metric assigns value 1 to the 4-connected neighbors and 2 to the diagonal pixels. The actual distance values for the 8 neighbors of a pixel are:

$$\text{Euclidean:} \begin{bmatrix} \sqrt{2} & 1 & \sqrt{2} \\ 1 & 0 & 1 \\ \sqrt{2} & 1 & \sqrt{2} \end{bmatrix}$$

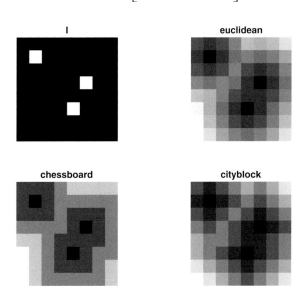

FIGURE 1.100
Output of Example 1.88.

$$\text{Chessboard:} \begin{bmatrix} 1 & 1 & 1 \\ 1 & 0 & 1 \\ 1 & 1 & 1 \end{bmatrix}$$

$$\text{Cityblock:} \begin{bmatrix} 2 & 1 & 2 \\ 1 & 0 & 1 \\ 2 & 1 & 2 \end{bmatrix}$$

Example 1.88: Write a program to depict distances from a pixel as grayscale intensities, using various metrics.

```
clear; clc;
a = zeros(8);
a(2,2)=1; a(4,6)=1; a(6,5)=1;
d1 = bwdist(a, 'euclidean');
d2 = bwdist(a, 'chessboard');
d3 = bwdist(a, 'cityblock');
figure,
subplot(221), imshow(a);  title('I');
subplot(222), imshow(d1, []);  title('euclidean');
subplot(223), imshow(d2, []);  title('chessboard');
subplot(224), imshow(d3, []);  title('cityblock');
```

The convex hull of a set of points X is the smallest convex region containing all of the points of X. The IPT function **bwconvhull** computes the convex hull of all objects in a binary image. The following example illustrates display of convex hull of different groups of binary objects. Image H1 shows the convex hull of the binary objects in image B1, and H2 shows the convex hull of the binary objects in B2. B1 is derived from I1 and B2 from I2 using a threshold value of 200. I2 is derived from I1 by inversion (Figure 1.101).

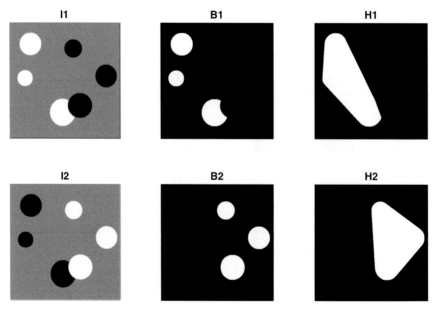

FIGURE 1.101
Output of Example 1.89.

Example 1.89: Write a program to display the convex hull of a group of binary objects.

```
clear; clc;
I = imread('circlesBrightDark.png');
figure,
subplot(231), imshow(I); title('I1');
B = I > 200;
subplot(232), imshow(B); title('B1');
H = bwconvhull(B);
subplot(233), imshow(H); title('H1')

I = 255 - I;
subplot(234), imshow(I); title('I2');
B = I > 200;
subplot(235), imshow(B); title('B2');
H = bwconvhull(B);
subplot(236), imshow(H); title('H2')
```

The IPT function **bwlabel** is used to generate a label i.e. a unique integer for each object found in a binary image using 4-way or 8-way connectivity. In the following example, it designates the three objects by labels 1, 2, 3 starting from left side of the image and the background by 0. The objects are then displayed by using gray intensities in the range [0,1] corresponding to their label index numbers i.e. index 0 with RGB 0, index 1 with RGB 0.33, index 2 with RGB 0.66 and index 3 with RGB 1 (Figure 1.102).

Example 1.90: Write a program to display image objects using intensity levels corresponding to their label index numbers.

```
clear; clc;
I = imread('circlesBrightDark.png');
B = im2bw(I, 0.7);
L = bwlabel(B);
subplot(121), imshow(I); title('I');
subplot(122), imshow(L, []); title('L');
```

1.7.4 Texture Analysis

Texture analysis refers to the characterization of regions in an image by their texture content. Texture refers to spatial patterns or arrangement of pixels that gives rise to visual perceptions like rough, smooth, coarse, fine, and so on. Several types of objects or surfaces

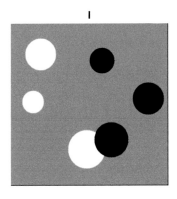
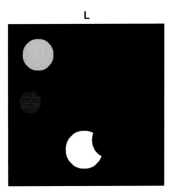

FIGURE 1.102
Output of Example 1.90.

can readily be identified using textures like grass, wood, stone, and cloth, due to which texture-based analysis forms an important part of pattern recognition and computer vision. One of the important methods in texture analysis is the **Gray Level Co-occurrence Matrix** (GLCM) proposed by Haralick et al. (Haralick, 1973). GLCM defines the probability of gray-level i occurring in the neighborhood of gray-level j at an offset distance d and angle θ i.e.

$$G = P(i, j \mid d, \theta)$$

GLCMs can be considered at various angles by varying θ: horizontal (0° and 180°), vertical (90° and 270°), right-diagonal (45° and 225°), left-diagonal (135° and 315°). Consider a 4×4 section I of an image having four gray-level intensities 1, 2, 3, 4. The image can be represented by the data matrix A:

$$A = \begin{bmatrix} 0 & 0 & 1 & 1 \\ 0 & 0 & 1 & 1 \\ 0 & 2 & 2 & 2 \\ 2 & 2 & 3 & 3 \end{bmatrix}$$

For simplicity if we consider offset $d = 1$, then the GLCM g can be computed along $\theta = 0°$ (i.e. along positive x-axis) as follows by noting 0 besides 0 occurs 2 times, 0 besides 1 occurs 2 times, 0 besides 2 occurs 1 times, 0 besides 3 occurs 0 times, and so on.

$$g = \begin{bmatrix} 2 & 2 & 1 & 0 \\ 0 & 2 & 0 & 0 \\ 0 & 0 & 3 & 1 \\ 0 & 0 & 0 & 1 \end{bmatrix}$$

For $\theta = 180°$ (i.e. along negative x-axis) the GLCM would be the transpose matrix g^T. The transpose is therefore added to the original matrix to make it symmetrical to take both the directions into account.

$$g + g^T = \begin{bmatrix} 2 & 2 & 1 & 0 \\ 0 & 2 & 0 & 0 \\ 0 & 0 & 3 & 1 \\ 0 & 0 & 0 & 1 \end{bmatrix} + \begin{bmatrix} 2 & 0 & 0 & 0 \\ 2 & 2 & 0 & 0 \\ 1 & 0 & 3 & 0 \\ 0 & 0 & 1 & 1 \end{bmatrix} = \begin{bmatrix} 4 & 2 & 1 & 0 \\ 2 & 4 & 0 & 0 \\ 1 & 0 & 6 & 1 \\ 0 & 0 & 1 & 2 \end{bmatrix} = S_0$$

To make the GLCM independent of the image size it is normalized by dividing by the sum of all elements: $G_0 = (1/24)S_0$. Following the same process for the other directional GLCMs we obtain a set of **symmetrical normalized directional** GLCMs as shown below:

$$G_0 = \frac{1}{24} \begin{bmatrix} 4 & 2 & 1 & 0 \\ 2 & 4 & 0 & 0 \\ 1 & 0 & 6 & 1 \\ 0 & 0 & 1 & 2 \end{bmatrix}$$

Image Processing   127

$$G_{45} = \frac{1}{18} \begin{bmatrix} 4 & 1 & 1 & 0 \\ 1 & 2 & 2 & 0 \\ 0 & 2 & 4 & 1 \\ 0 & 0 & 1 & 0 \end{bmatrix}$$

$$G_{90} = \frac{1}{24} \begin{bmatrix} 6 & 0 & 2 & 0 \\ 0 & 4 & 2 & 0 \\ 2 & 2 & 2 & 2 \\ 0 & 0 & 2 & 0 \end{bmatrix}$$

$$G_{135} = \frac{1}{18} \begin{bmatrix} 2 & 1 & 3 & 0 \\ 1 & 2 & 1 & 0 \\ 3 & 1 & 0 & 2 \\ 0 & 0 & 2 & 0 \end{bmatrix}$$

The IPT function **graycomatrix** is used to calculate GLCMs given the data matrix, off-set distance, number of gray levels, and the minimum and maximum gray intensities. The following example shows how the symmetrical GLCMs are computed along four directions. The BM functions **min** and **max** calculate the minimum value and maximum value of a vector. For a 2-D matrix, each function is repeated twice, the first occurrence calculates the value along each column and the second occurrence along the row (Figure 1.103).

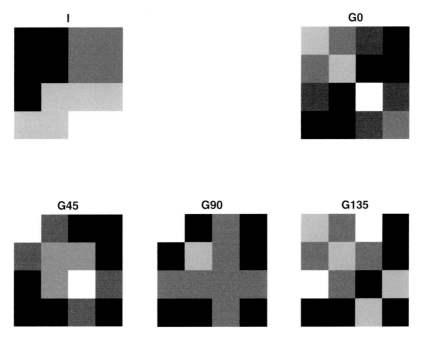

FIGURE 1.103
Output of Example 1.91.

Example 1.91: Write a program to compute symmetrical directional GLCMs from a grayscale image section.

```
clear; clc;
I = [0 0 1 1; 0 0 1 1; 0 2 2 2; 2 2 3 3];
d = 1; lev = 4;
minI = min(min(I));
maxI = max(max(I));
glcm0 = graycomatrix(I,      'Offset',[0 d],    'NumLevels', lev,   ...
        'Symmetric', true, 'GrayLimits', [minI, maxI]);
glcm45 = graycomatrix(I,     'Offset',[-d d],   'NumLevels', lev,   ...
        'Symmetric', true, 'GrayLimits', [minI, maxI]);
glcm90 = graycomatrix(I,     'Offset',[-d 0],   'NumLevels', lev,   ...
        'Symmetric', true, 'GrayLimits', [minI, maxI]);
glcm135 = graycomatrix(I,    'Offset',[-d -d],  'NumLevels', lev,   ...
        'Symmetric', true, 'GrayLimits', [minI, maxI]);

subplot(231), imshow(I,[]); title('I');
subplot(233), imshow(glcm0,[]); title('G0');
subplot(234), imshow(glcm45,[]); title('G45');
subplot(235), imshow(glcm90,[]); title('G90');
subplot(236), imshow(glcm135,[]); title('G135');
```

The dimensions of a GLCM depend on the total number of gray levels in the image. Since for a typical grayscale image total number of graylevels can be upto 256, the GLCM becomes a high dimensional matrix which cannot be directly incorporated within applications efficiently. It is therefore conventional to calculate some scalar statistical properties from a GLCM and incorporate these into calculations instead of the GLCM itself. Typical features include contrast, correlation, energy, homogeneity. If $G(i,j)$ represents the (i,j)-th element of a directional normalized symmetrical GLCM and N denotes the number of gray levels, then the features Contrast C, Homogeneity H, Energy E, and Correlation R are defined as follows:

$$C = \sum_{i=1}^{N} \sum_{j=1}^{N} G(i,j) \cdot (i-j)^2$$

$$H = \sum_{i=1}^{N} \sum_{j=1}^{N} \frac{G(i,j)}{1 + |i-j|}$$

$$E = \sum_{i=1}^{N} \sum_{j=1}^{N} \{G(i,j)\}^2$$

$$R = \sum_{i=1}^{N} \sum_{j=1}^{N} G(i,j) \cdot \frac{(i-\mu_i)(j-\mu_j)}{\sigma_i \cdot \sigma_j}$$

Here, the μ and the σ terms denote the mean and standard deviations, respectively, along specific directions indicated by the subscripts. The IPT function **graycoprops** calculates the above statistical properties from a GLCM. The following example illustrates calculation of features for the 0 degree GLCM of the previous example.

Example 1.92: Write a program to calculate statistical scalar values from a GLCM.

```
clear; clc;
G = glcm0;
s = graycoprops(G);
```

```
C = s.Contrast
E = s.Energy
H = s.Homogeneity
R = s.Correlation
```

Entropy is a statistical measure of randomness that can be used to characterize the texture of the input image. Entropy is defined as $E = -\sum p \cdot \log_2 p$, where p contains the normalized histogram counts returned from the function *imhist* (Gonzalez et al., 2003). The IPT function **entropy** is used to compute the entropy value from a grayscale image. The IPT function **entropyfilt** returns an array, where each output pixel contains the entropy value of a 9×9 neighborhood around the corresponding pixel in the input image. In the following example, each pixel in the original image is replaced by the entropy value of its 9×9 neighborhood (Figure 1.104).

> **Example 1.93: Write a program to represent each pixel by its local entropy value calculated from the pixel neighbourhood**

```
clear; clc;
I = imread('coins.png');
E = entropy(I);
J = entropyfilt(I);
imshowpair(I,J,'montage');
```

1.7.5 Image Quality

Image quality can degrade due to image acquisition and processing like noise handling and compression. To measure the amounts of degradation, the degraded image is compared with respect to a reference image without degradation, and the amount of change is computed using an index. Two of the most commonly used metrics are the **signal-to-noise ratio** (SNR) and **peak SNR** (pSNR). The SNR is a ratio of the average power p_I of the signal I to the average power p_N of the noise N, while the peak SNR is defined in terms of the ratio of the peak value of the signal P (255 for uint8 grayscale images) to the **mean square error** (MSE) between the signal and the noise, m and n being the dimensions of the images.

FIGURE 1.104
Output of Example 1.93.

The average power is also proportional to the square of the amplitude, where A denotes the RMS amplitude

$$\text{SNR} = \frac{p_I}{p_N} = \left(\frac{A_I}{A_N}\right)^2$$

$$p\text{SNR} = 10 \cdot \log_{10} \frac{P^2}{\text{MSE}}$$

$$\text{MSE} = \frac{1}{mn} \sum_{i=1}^{m} \sum_{j=1}^{n} \left\{ I(i,j) - N(i,j) \right\}^2$$

The IPT function **psnr** computes the pSNR for a given noisy image with respect to a reference image without noise and also returns the SNR value in decibels. In the following example, two types of noisy images are generated using *salt and pepper* noise and *Gaussian* noise. The SNR/pSNR values returned are 15.22/21.98 for the first image and 10.51/17.26 for the second image (Figure 1.105).

> **Example 1.94: Write a program to estimate degradation in image quality by computing SNR and pSNR values of noisy images.**
>
> ```
> clear; clc;
> I = imread('coins.png');
> A = imnoise(I,'salt & pepper', 0.02);
> B = imnoise(I,'gaussian', 0, 0.02);
>
> [peaksnr, snr] = psnr(A, I);
> fprintf('\n The Peak-SNR value is %0.4f', peaksnr);
> fprintf('\n The SNR value is %0.4f \n', snr);
> [peaksnr, snr] = psnr(B, I);
> fprintf('\n The Peak-SNR value is %0.4f', peaksnr);
> fprintf('\n The SNR value is %0.4f \n', snr);
>
> montage({I,A,B}, 'Size', [1 3]); title('I, A, B');
> ```

The SSIM index is used for measuring the similarity of two images x and y and has been designed to improve on pSNR and MSE. It is calculated as a product of three terms: luminance $L(x,y)$, contrast $C(x,y)$, and structure $S(x,y)$ as defined below (Zhou et al., 2004), where c_1, c_2, and c_3 are constants:

I, A, B

FIGURE 1.105
Output of Example 1.94.

$$L(x,y) = \frac{2\mu_x\mu_y + c_1}{\mu_x^2 + \mu_y^2 + c_1}$$

$$C(x,y) = \frac{2\sigma_x\sigma_y + c_2}{\sigma_x^2 + \sigma_y^2 + c_2}$$

$$S(x,y) = \frac{\sigma_{xy} + c_3}{\sigma_x\sigma_y + c_3}$$

Here, μ_x, μ_y, σ_x, σ_y, and σ_{xy} are the local means, standard deviations, and cross-covariance for images x,y. In the following example, the IPT function **ssim** calculates the SSIM of a noisy image with respect to the reference image. It returns a global value (sg) which is a scalar for the entire image and a local set of values (sl) calculated for each pixel. The IPT function **immse** calculates the MSE between a noisy image and the reference image. The global value returned is 0.8464, while the local values have been visualized as an intensity image. The MSE value returned is 0.0121. The images displayed are the original image I, the noisy image A, and the local SSIM value map sl (Figure 1.106).

> **Example 1.95: Write a program to estimate structural similarity of a noisy image with respect to the reference image.**

```
clear; clc;
I = checkerboard(8);
A = imnoise(I,'gaussian', 0, 0.02);
[sg, sl] = ssim(A,I);
montage({I,A, sl}, 'Size', [1 3]); title('I, A, sl');
m = immse(A, I);
fprintf('\n The SSIM value is %0.4f', sg);
fprintf('\n The MSE value is %0.4f\n', m);
```

1.7.6 Image Transforms

Until now all the filtering operations performed have been performed in the spatial domain i.e. the X-Y plane of the image. An alternative form of processing is called the frequency domain processing, where the image is transformed from the spatial domain to

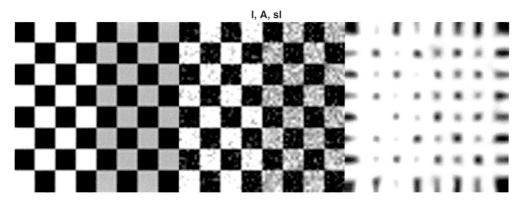

FIGURE 1.106
Output of Example 1.95.

the frequency domain using the **Fourier Transform** (FT). According to Fourier's theorem, any signal can be expressed as weighted sums of sine and cosine components. Since we are dealing with discrete signals, we use the **Discrete Fourier Transform** (DFT), where we use a summation sign instead of the integration sign in the standard FT meant for analog signals. Consider a 1-D time domain signal x with N samples. The i-th sample is denoted as $x(i)$. The signal is converted to a frequency domain signal X also having N values. The k-th frequency domain value is denoted by $X(k)$. The **forward DFT** expresses frequency domain values in terms of time domain values:

$$X(k) = \sum_{i=0}^{N-1} x(i) \cdot e^{-j \cdot \frac{2\pi ki}{N}}$$

Here, $j = \sqrt{-1}$, and e is exponential operator given by: $e = \lim\limits_{n \to \infty} \left(1 + \frac{1}{n}\right)^n \approx 2.718$.

We also know from Euler's formula: $e^{jx} = \cos(x) + j \cdot \sin(x)$
Combining the above two formulae we get:

$$X(k) = \sum_{i=0}^{N-1} x(i) \cdot \left(\cos \frac{2\pi ki}{N} - j \cdot \sin \frac{2\pi ki}{N} \right)$$

The **inverse DFT** expresses the $x(i)$ in terms of $X(k)$:

$$x(i) = \frac{1}{N} \sum_{k=0}^{N-1} X(k) . e^{j \cdot \frac{2\pi ki}{N}}$$

The values of the signal after the DFT is performed, are complex in nature consisting of real and imaginary portions. These values are called **coefficients** and they act as scaling factors for a series of unit magnitude sine and cosine components called **basis functions**. If $A(k)$ and $B(k)$ are the real and imaginary components of $X(k)$, then we have:

$$A(k) = Re\ \{X(k)\} = \sum_{i=0}^{N-1} x(i) \cdot \cos\left(\frac{2\pi ki}{N}\right)$$

$$B(k) = Im\ \{X(k)\} = \sum_{i=0}^{N-1} x(i) \cdot \sin\left(\frac{2\pi ki}{N}\right)$$

Collecting all the terms as k ranges from 0 to $(N-1)$ we get:

$$A = \{A(0), A(1), A(2), ..., A(N-1)\}$$

$$B = \{B(0), B(1), B(2), ..., B(N-1)\}$$

The original signal X can now be expressed as: $X = A - jB$. The absolute value provides the **magnitude** $|X(k)|$ and the **phase angle** $\varphi(k)$ provides the direction of the frequency domain signal, also called the **spectrum**:

$$|X(k)| = \sqrt{\{A(k)\}^2 + \{B(k)\}^2}$$

$$\varphi(k) = \tan^{-1}\left\{\frac{B(k)}{A(k)}\right\}$$

By default the zero-frequency component of the spectrum is displayed at the left corner of the plot. It is customary to shift this to the center of the plot for better viewing. In the following example, the BM function **fft** generates the DFT of a step function. The nomenclature FFT stands for **Fast Fourier Transform** which is a fast and computationally efficient algorithm for computing the DFT. The BM function **fftshift** is used to shift the zero-frequency component to the center of the plot. The BM functions **real** and **imag** calculate the real and imaginary portions of the complex values of the frequency domain signal. The following example shows the shifted spectrum of a 1-D step input signal and its phase (Figure 1.107).

Example 1.96: Write a program to plot the magnitude and phase of the spectrum of a 1-D discrete step function.

```
clear; clc;
a = zeros(1,1024);
a(1:10) = 1;
b = fft(a);
c = fftshift(b);
M = abs(b);
N = abs(c);
```

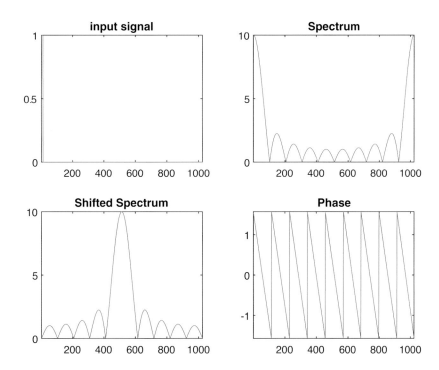

FIGURE 1.107
Output of Example 1.96.

```
R = real(c); I = imag(c); P = atan(I./R);
subplot(221), plot(a); axis tight; title('input signal')
subplot(222), plot(M); axis tight; title('Spectrum')
subplot(223), plot(N); axis tight; title('Shifted Spectrum')
subplot(224), plot(P); axis tight; title('Phase')
```

Since an image is a 2-D signal the 1-D version of DFT needs to be extended to the 2-D version to be applicable to images, in which case $x(i)$ is replaced by the image function $f(x,y)$, and $X(k)$ is replaced by $F(u,v)$, and the number of elements N is replaced by the two dimensions M (number of image columns) and N (number of image rows). Then, the **forward 2-D DFT** and the **inverse 2-D DFT** are given by:

$$F(u,v) = \sum_{x=0}^{M-1}\sum_{y=0}^{N-1} f(x,y) \cdot \exp\left\{-j \cdot 2\pi\left(\frac{ux}{M} + \frac{vy}{N}\right)\right\}$$

$$f(x,y) = \frac{1}{MN} \sum_{u=0}^{M-1}\sum_{v=0}^{N-1} F(u,v) \cdot \exp\left\{j \cdot 2\pi\left(\frac{ux}{M} + \frac{vy}{N}\right)\right\}$$

A shortcut, called the DFT operator, can be used to represent the forward and inverse operations as follows:

$$F(u,v) = \mathcal{F}\left\{f(x,y)\right\}$$

$$f(x,y) = \mathcal{F}^{-1}\left\{F(u,v)\right\}$$

To analyze the Fourier Transform of a 2-D image it is displayed as an image itself, since it is a 2-D matrix of coefficients. However, since the values are complex in nature we calculate its absolute value. The zero-frequency components are shifted to the center of the plot for better visualization, as for the 1-D case. The DFT coefficients usually span over a large range of values of the order of 10^6. To display these values in the form of an image they need to be compressed within the range 0–255. This is achieved by using the log function. To avoid possibility of a 0 value, 1 is added to it before calculating logarithm. The IPT function **fft2** is used to apply a 2-D Fourier Transform to the image and the absolute values of the coefficients are calculated to avoid the need for handling imaginary values. Since the original matrix is only of size 30 × 30, the size of the coefficients matrix is increased to 200 × 200 by padding with zeros for better visualization. The IPT function **ifft2** calculates the inverse DFT and returns the spatial domain image from its spectrum. The following example shows a 2-D step input signal and the logarithm value of its shifted spectrum. The spectrum is also transformed back to the spatial domain using the inverse DFT (Figure 1.108).

Example 1.97: Write a program to display the Fourier spectrum of a 2-D step function.

```
clear clc;
dim = 30; f = zeros(dim, dim);
h = size(f,1); w = size(f,2);
f(h/3:2*h/3, w/3:2*w/3) = 1;
F = fft2(f);
F1 = fft2(f, 200, 200);
F2 = fftshift(F1);
```

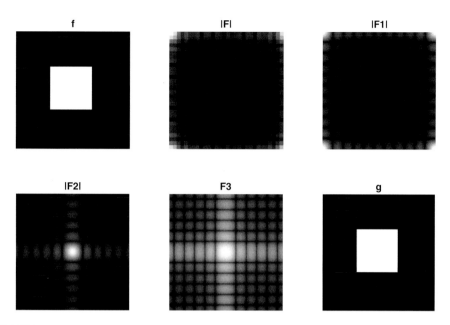

FIGURE 1.108
Output of Example 1.97.

```
F3 = log(1+abs(F2));
g = ifft2(F);
figure,
subplot(231), imshow(f); title('f');
subplot(232), imshow(abs(F), []); title('|F|');
subplot(233), imshow(abs(F1), []); title('|F1|');
subplot(234), imshow(abs(F2), []); title('|F2|');
subplot(235), imshow(F3, []); title('F3');
subplot(236), imshow(g, []); title('g');
```

The **Discrete Cosine Transform** (DCT) is an alternative to the DFT using which the signal can be transformed into the frequency domain using only real components and thereby avoid complex arithmetic. The basis functions consist of only cosine waveforms of unit amplitude. Consider a 1-D time domain signal x with N samples. The i-th sample is denoted as $x(i)$. The signal is converted to a frequency domain signal X also having N values. The k-th frequency domain value is denoted by $X(k)$. The **forward DCT** expresses frequency domain values in terms of time domain values:

$$X(k) = \alpha \cdot \sum_{i=0}^{N-1} x(i) \cdot \cos\left\{ \frac{(2i+1)\pi k}{2N} \right\}$$

$$\text{where, } \alpha = \sqrt{\frac{1}{N}} \text{ for } k = 0 \text{ and } \alpha = \sqrt{\frac{2}{N}} \text{ for } k \neq 0$$

The **inverse DCT** expresses the $x(i)$ in terms of $X(k)$:

$$x(i) = \sum_{k=0}^{N-1} \alpha \cdot X(k) \cdot \cos\left\{ \frac{(2i+1)\pi k}{2N} \right\}$$

$$\text{where, } \alpha = \sqrt{\frac{1}{N}} \text{ for } k = 0 \text{ and } \alpha = \sqrt{\frac{2}{N}} \text{ for } k \neq 0$$

The Signal Processing Toolbox (SPT) function **dct** computes the DCT from a 1-D signal and returns an output array of coefficients the same size as the input array. The SPT function **idct** computes the inverse DCT from the coefficients. In the following example, a square pulse is first transformed using DCT and then the program calculates how many DCT coefficients represents 99% of the energy in the signal. The signal is then reconstructed using those number of coefficients and compared with the case, where all coefficients are used for its reconstruction. The example shows that only 4.1992% of the sorted coefficients are sufficient to reconstruct 99% of the signal energy (Figure 1.109).

Example 1.98: Write a program to convert a 1-D signal to frequency domain using DCT and convert it back with a subset of the coefficients sufficient to represent 99% of the energy of the signal

```
clear; clc;
x = zeros(1,1024);
x(1:100) = 1;
X = dct(x);
[val, idx] = sort(abs(X),'descend');
i = 1;
while norm(X(idx(1:i)))/norm(X) < 0.99
    i = i + 1;
end
n = i;                  % no. of coefficients for 99% of energy
perc = n*100/numel(X)        % percentage of coefficients
val(idx(n+1:end)) = 0;
xx = idct(val);
y = idct(X);
figure,
subplot(131), plot(x); title('Original'); axis tight; axis square;
subplot(132), plot(xx); axis tight; axis square;
title('Reconstructed with 99% energy');
subplot(133), plot(y); axis tight; axis square;
title('Reconstructed with 100% energy');
```

The 2-D version of the DCT needs to be applied to images. If $f(x,y)$ is the (x,y)-th point on the image having dimensions M columns by N rows and $F(u,v)$ be the (u,v)-th frequency domain component, then the **forward 2-D DCT** is given by:

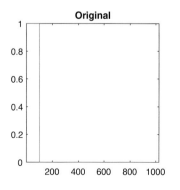 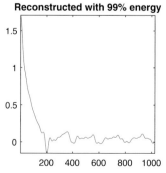 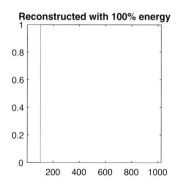

FIGURE 1.109
Output of Example 1.98.

$$F(u,v) = \alpha_u \cdot \alpha_v \cdot \sum_{x=0}^{M-1}\sum_{y=0}^{N-1} f(x,y) \cdot \cos\left\{\frac{(2x+1)\pi u}{2M}\right\} \cdot \cos\left\{\frac{(2y+1)\pi v}{2N}\right\}$$

$$\alpha_u = \sqrt{\frac{1}{M}} \text{ for } u = 0 \text{ and } \alpha_u = \sqrt{\frac{2}{N}} \text{ for } u \neq 0$$

$$\alpha_v = \sqrt{\frac{1}{N}} \text{ for } v = 0 \text{ and } \alpha_v = \sqrt{\frac{2}{N}} \text{ for } v \neq 0$$

The **inverse 2-D DCT** is given by:

$$f(x,y) = \sum_{u=0}^{M-1}\sum_{v=0}^{N-1} \alpha_u \cdot \alpha_v \cdot F(u,v) \cdot \cos\left\{\frac{(2x+1)\pi u}{2M}\right\} \cdot \cos\left\{\frac{(2y+1)\pi v}{2N}\right\}$$

$$\alpha_u = \sqrt{\frac{1}{M}} \text{ for } u = 0 \text{ and } \alpha_u = \sqrt{\frac{2}{N}} \text{ for } u \neq 0$$

$$\alpha_v = \sqrt{\frac{1}{N}} \text{ for } v = 0 \text{ and } \alpha_v = \sqrt{\frac{2}{N}} \text{ for } v \neq 0$$

The IPT function **dct2** is used to return the 2-D DCT of an input data matrix, and the IPT function **idct2** is used to return the inverse 2-D DCT. In the following example, a grayscale image is read and the 2-D DCT is computed (J). The DCT coefficients are displayed using a logarithmic scale to compress the large range of values within the display range of 0–255 (JL). In the first case, all coefficients with values less than 10 are removed, and the resulting DCT matrix is displayed (J1). These coefficients are converted back to the spatial domain, and the resulting image is displayed (K1). In the second case, all coefficients with values less than 50 are removed, and the resulting DCT matrix is displayed (J2). These coefficients are converted back to the spatial domain, and the resulting image is displayed (K2). In the second case, since a larger number of coefficients are removed, this results in more degradation of the image with corresponding reduced quality and more artifacts (Figure 1.110).

Example 1.99: Write a program to convert a grayscale image to frequency domain using DCT and convert it back with a subset of the coefficients to demonstrate the difference in image quality.

```
clear; clc;
I = imread('coins.png');
J = dct2(I);
JL = log(abs(J));
J(abs(J) < 10) = 0; % remove coefficients less than 10
J1 = log(abs(J));
K1 = idct2(J);
J(abs(J) < 50) = 0; % remove coefficients less than 50
J2 = log(abs(J));
K2 = idct2(J);

figure
subplot(231), imshow(I); title('I');
subplot(232), imshow(JL,[]); title('JL');
subplot(233), imshow(J1,[]); title('J1');
subplot(234), imshow(K1, []); title('K1');
subplot(235), imshow(J2, []); title('J2');
subplot(236), imshow(K2, []); title('K2');
```

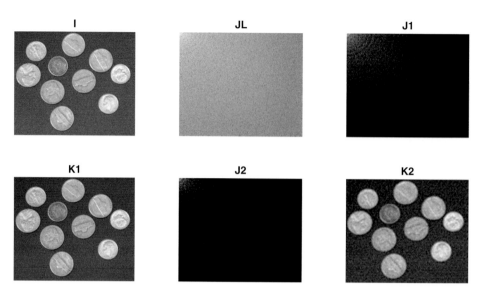

FIGURE 1.110
Output of Example 1.99.

Note: The IPT also includes a function **dctmtx** which returns a transformation matrix, that needs to be multiplied with the image to compute the 2-D DCT of the image matrix. The computation of the transformation matrix depends only on the size of the image I viz. $D = \text{dctmtx}(\text{size}(I, 1))$. The image DCT can then be computed using: dct = $D*I*D'$. This computation is sometimes faster than using the function dct2 especially when DCT of a large number of small equal sized image blocks are needed, since D needs to be computed only once, which will be faster than repeatedly calling dct2 for each individual block.

The **Discrete Wavelet Transform** (DWT) is a type of transform similar to DFT as both expresses the original signal as a combination of scaled basis functions, whereas the DFT basis functions are sinusoids of varying frequencies and infinite duration, DWT basis functions are small wave-like oscillations, called **wavelets**, of limited durations (Mallat, 1989). The original function is represented in terms of scaled and translated copies of a mother wavelet and are mostly used when the signal is non-periodic and has discontinuities. The DWT can be used to analyze a signal at different frequency bands by successive low-pass and high-pass filtering. The low-pass outputs are called **approximation coefficients**, while the high-pass outputs are called **detail coefficients**. When applied to a 2-D signal like an image I, a 2-D wavelet transform operation can be represented by the following scheme, which is called a 1-level or **single-level decomposition**, where W is the 2-D wavelet transform matrix:

$$W*I*W^T \rightarrow \begin{bmatrix} LL & LH \\ HL & HH \end{bmatrix} \rightarrow \begin{bmatrix} A & H \\ V & D \end{bmatrix}$$

Here, LL indicates low-pass filtering both along the columns and rows of an image, producing an approximation coefficient A which is a copy of the original image but scaled by half along both width and height. LH indicates low-pass filtering along the columns and high-pass filtering along the rows, producing the horizontal coefficient H which is an image produced by averaging along the width and differencing along the height and so able to identify horizontal edges. *HL* indicates high-pass filtering along the rows and low-pass filtering along the columns, producing vertical coefficient *V* which is an image produced by averaging along the height and differencing along the width and so able to identify vertical edges. *HH* indicates high-pass filtering along the rows and high-pass filtering along the columns, producing diagonal coefficient *D* which is an image produced by differencing along the height and differencing along the width and so able to identify diagonal edges. The above steps are collectively referred to as decomposition or analysis step. The approximation and detail components can be combined together to reconstruct the original image by again passing through low-pass and high-pass filters, which is called the reconstruction or synthesis step. The Wavelet Toolbox (WT) function **dwt2** computes a single-level 2-D wavelet transform using a specified wavelet name and returns the approximation, horizontal, vertical, and diagonal coefficients. In the following example, a single-level DWT is computed and the approximation and detail coefficients are displayed. The option *sym4* specifies the type of wavelet i.e. symlet order 4. The WT function **idwt2** performs the inverse wavelet transformation by reconstructing the original image from the coefficients (Figure 1.111).

Example 1.100: Write a program to perform 2-D Discrete Wavelet Transform and display the approximation, horizontal, vertical and diagonal components. Reconstruct the image back from the analysis components and compute its difference from the original image

```
clear; clc;
X = checkerboard(10);
wname = 'sym4';
```

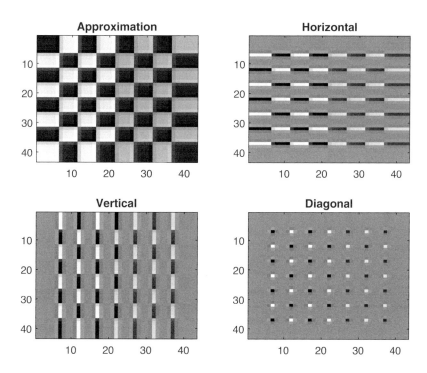

FIGURE 1.111
Output of Example 1.100.

```
[cA, cH, cV, cD] = dwt2(X, wname);
subplot(2,2,1)
imagesc(cA)
colormap gray
title('Approximation')
subplot(2,2,2)
imagesc(cH)
colormap gray
title('Horizontal')
subplot(2,2,3)
imagesc(cV)
colormap gray
title('Vertical')
subplot(2,2,4)
imagesc(cD)
colormap gray
title('Diagonal')

A0 = idwt2(cA, cH, cV, cD, wname, size(X));
figure
subplot(121), imshow(X); title('original');
subplot(122), imshow(A0); title('reconstructed');

max(max(abs(X-A0)))
```

Although the DFT uses cosine and sine basis function, the DWT uses a variety of basis functions depending on the waveform of the oscillating wavelet. The **waveletAnalyzer** app can be used to generate visual displays of different types of wavelets. The refinement menu allows one to choose the number of points that the wavelet function is computed over. The various wavelet types are organized using a number of family names e.g. haar, daubechies (db), symlet (sym), meyer (meyr), gaussian (gaus), coiflets (coif) etc. Figure 1.112 shows sym4 wavelets.

Each wavelet is associated with a set of low-pass and high-pass filters. Each type of filter can again be divided into a decomposition filter and a reconstruction filter. The above example illustrates a single-level decomposition. **Multi-level decomposition** can also be

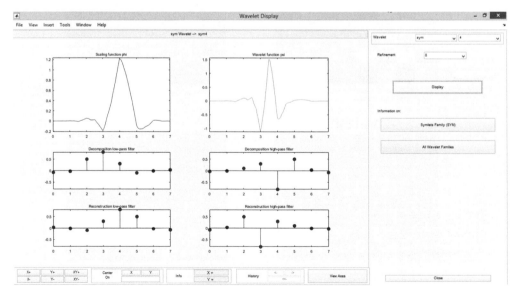

FIGURE 1.112
WaveletAnalyzer interface.

done by using the approximation image of the previous level as the starting image for the next level. The WT function **wavedec2** can be used to perform multi-level decomposition. The following example shows a 2-level decomposition and displays the approximation and detail components of both the levels. In the following example, X is an image of size 80×80, the first level components A1, H1, V1, D1 are of size 41×41, the second level components A2, H2, V2, D2 are of size 22×22. Outputs are the decomposition vector C and the corresponding bookkeeping matrix S. Vector C contains the vectorized components in the following order: [A2, H2, V2, D2, H1, V1, D1] and is therefore of the size ($22 \times 22 \times 4 + 41 \times 41 \times 3) = 6979$ consisting of ($3*N+1$) elements, where N denotes the decomposition level, which in this case is $3 \times 2 + 1 = 7$ elements. Matrix S contains the size of the detail coefficients in the order [size(A2,1), size(A2,2), size(H2,1), size(H2,2), size(H1,1), size(H1,2), size(X)] and in this case is therefore a 4×2 matrix [22, 22; 22, 22; 41, 41; 80, 80] i.e. of size ($N+2 \times 2$). The WT function **appcoef2** computes the approximation coefficients at the specified level, and the WT function **detcoef2** computes the detail coefficients at the specified level. The 8 coefficients of both the levels are subsequently displayed (Figure 1.113). All the matrices are arranged according to the following scheme.

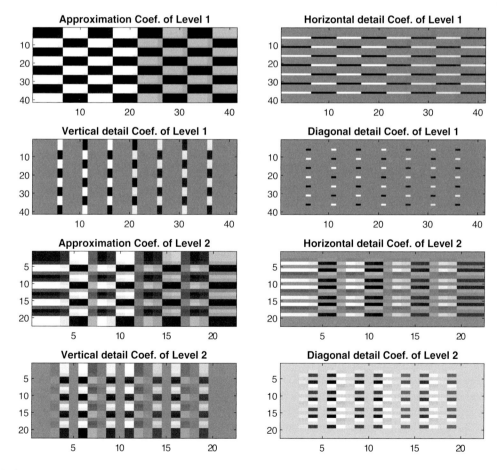

FIGURE 1.113
Output of Example 1.101.

```
C = [A2 | H2 | V2 | D2 | H1 | V1 | D1]
S = [A2(1), A2(2)]
[H2(1), H2(2)]
[H1(1), H1(2)]
[X(1),   X(2)]
```

Example 1.101: Write a program to perform 2-Level Discrete Wavelet Transform of an image and display all the approximation and detail coefficients of both the levels.

```
clear; clc;
X = checkerboard(10);

[C, S] = wavedec2(X,2,'db2');
A1 = appcoef2(C, S, 'db2', 1);
[H1, V1, D1] = detcoef2('all', C, S, 1);
A2 = appcoef2(C, S, 'db2', 2);
[H2, V2, D2] = detcoef2('all', C, S, 2);

subplot(4,2,1);
imagesc(A1); colormap gray;
title('Approximation Coef. of Level 1');

subplot(4,2,2);
imagesc(H1); colormap gray;
title('Horizontal detail Coef. of Level 1');

subplot(4,2,3);
imagesc(V1); colormap gray;
title('Vertical detail Coef. of Level 1');

subplot(4,2,4);
imagesc(D1); colormap gray;
title('Diagonal detail Coef. of Level 1');

subplot(4,2,5);
imagesc(A2); colormap gray;
title('Approximation Coef. of Level 2');

subplot(4,2,6);
imagesc(H2); colormap gray;
title('Horizontal detail Coef. of Level 2');

subplot(4,2,7);
imagesc(V2); colormap gray;
title('Vertical detail Coef. of Level 2');

subplot(4,2,8);
imagesc(D2); colormap gray;
title('Diagonal detail Coef. of Level 2');
```

Like DFT and DCT, the DWT is also able to decompose signals into their frequency components, due to which it has been applied in noise removal and compression. Since noise is frequently associated with high frequency components as also the fact that the human visual system is relatively insensitive to high frequency signals, DWT applications involve removal of high frequency components from the signal and reconstructing the signal back using only the low frequency components. The WT function **waverec2** is used to reconstruct the signal back from the decomposition components. By default the reconstruction takes place by using all the components. However, one can selectively make some of the components as zero in which case the reconstruction takes place using a subset of the decomposition components. In the following example, the structure of the vector C is as follows i.e. A2 ranges from elements 1 to 484, H2 from 485 to 968, and so on:

```
C = [ A2 (1:484) | H2 (485:968) | V2 (969:1452) | D2 (1453:1936) |
      H1 (1937:3617) | V1 (3618:5298) | D1 (5299:6979) ]
```

In order to selectively remove components, the corresponding values in the C are converted to zero. This also enables a system to generate multiple copies of the same image with different qualities. For X0, all components are used for reconstruction, for X1 the diagonal coefficient D1 is removed before reconstruction, for X2 all first level coefficients are removed, for X3 all second level coefficients are removed, for X4 both the diagonal coefficients are removed. The absolute difference between the original image and the reconstructed images are also computed to show the amount of degradation (Figure 1.114).

Example 1.102: Write a program to reconstruct an image after discrete wavelet decomposition by selectively removing some of the coefficients.

```
clear; clc;
X = checkerboard(10);
wname = 'db2';
[C, S] = wavedec2(X,2,wname);

X0 = waverec2(C,S,wname);
max(max(abs(X-X0)))

C1 = C; C1(5299:6979)=0;      % D1 removed
X1 = waverec2(C1,S,wname);
max(max(abs(X-X1)))

C2 = C; C2(1937:6979)=0;      % H1, V1, D1 removed
X2 = waverec2(C2,S,wname);
max(max(abs(X-X2)))
```

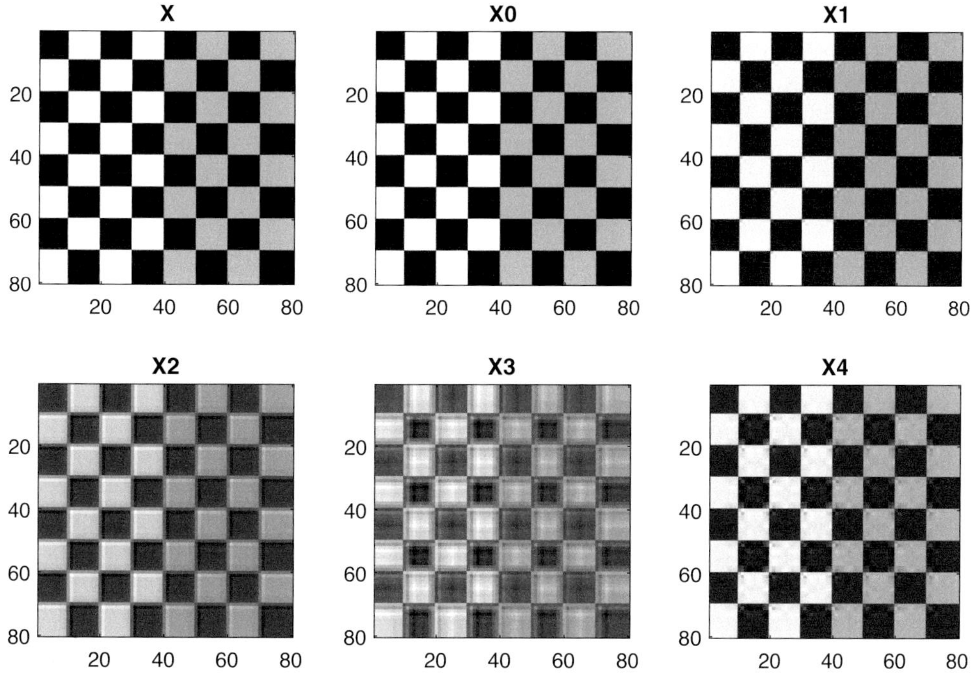

FIGURE 1.114
Output of Example 1.102.

```
C3 = C; C3(485:1936)=0;         % H2, V2, D2 removed
X3 = waverec2(C3,S,wname);
max(max(abs(X-X3)))

C4 = C; C4(1453:1936)=0; C4(5299:6979)=0;   % D1, D2 removed
X4 = waverec2(C4,S,wname);
max(max(abs(X-X4)))

subplot(231), imagesc(X);  colormap gray; title('X');
subplot(232), imagesc(X0); colormap gray; title('X0');
subplot(233), imagesc(X1); colormap gray; title('X1');
subplot(234), imagesc(X2); colormap gray; title('X2');
subplot(235), imagesc(X3); colormap gray; title('X3');
subplot(236), imagesc(X4); colormap gray; title('X4');
```

1.8 Working in Frequency Domain

Processing the image in the frequency domain is based upon a theorem called **Convolution Theorem** which establishes a relation between convolution operation and Fourier Transforms. According to the theorem, a convolution operation between an image $f(x,y)$ and a kernel $h(x,y)$ is equivalent to the product of their Fourier transforms. If $\otimes$ represents the convolution operator, and $F(u,v)$ and $H(u,v)$ be the Fourier Transforms of $f(x,y)$ and $h(x,y)$, respectively, then the following holds true:

$$h(x,y) \otimes f(x,y) \Leftrightarrow H(u,v) \cdot F(u,v)$$

If $\mathcal{F}$ represents the forward Fourier Transform operator and $\mathcal{F}^{-1}$ represents the inverse transform operator, then we can write:

$$F(u,v) = \mathcal{F}\{f(x,y)\}$$

$$H(u,v) = \mathcal{F}\{h(x,y)\}$$

$$f(x,y) = \mathcal{F}^{-1}\{F(u,v)\}$$

$$h(x,y) = \mathcal{F}^{-1}\{H(u,v)\}$$

The main advantage of this is to speed up computations. Convolutions involve shifting the kernel over the width and height of the image and computing a sum of products at each location, which is time consuming, especially for large images. The same results could be obtained in much less time if we simply compute the product of their Fourier transforms. Of course we will need to compute the inverse transform of the signal to switch back to spatial domain. In the following example, the image read and conversion to grayscale functions have been combined together. A Sobel filter is created and convolved with the image to generate the spatial domain result. The size of the image is used to adjust the size of the FT of the kernel, since these two matrices need to be multiplied together. Finally, the IFFT is used to bring the signal back to the spatial domain (Figure 1.115).

spatial domain frequency domain

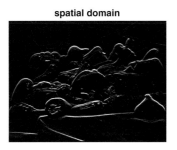 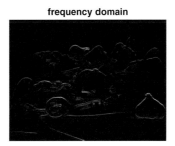

FIGURE 1.115
Output of Example 1.103.

Example 1.103: Write a program to perform edge detection using the Sobel operator using both the spatial domain and frequency domain.

```
clear; clc;

% Spatial domain
f = rgb2gray(imread('peppers.png'));
h = fspecial('sobel');
sd = imfilter(f, h,'conv');
subplot(121), imshow(sd);
title('spatial domain');

% Freq domain
h = fspecial('sobel');
[rows, cols] = size(f);
F = fft2(f);
H = fft2(h, rows, cols);
G = H.*F;
fd = ifft2(G);
subplot(122), imshow(abs(fd),[]);
title('frequency domain');
```

Low-pass filters (LPF) allow low frequencies in the image to remain intact and blocks the high frequencies. LPFs have a blurring effect on the image. An **ideal LPF** attenuates high frequencies larger than a fixed cut-off value D_0 from the origin of the (centered) transform. The transfer function of an ideal LPF is given by the following, where $D(u,v)$ is the distance of point (u,v) from the center of the frequency rectangle:

$$H(u,v) = 1, \text{ if } D(u,v) \leq D_0$$

$$H(u,v) = 0, \text{ if } D(u,v) > D_0$$

A 3-D plot and a 2-D image plot of an ideal LPF are shown in Figure 1.116 on the left as the function $H(u,v)$. The name *ideal* implies that all frequencies inside the circle of radius D_0

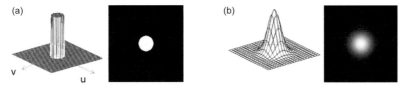

FIGURE 1.116
(a) Ideal LPF (b) Gaussian LPF.

are passed completely without any attenuation while all frequencies outside this circle are completely attenuated, which physically cannot be realized using electronic components.

A **Gaussian LPF** has a transfer function shown below obtained by putting $\sigma = D_0$ which is a measure of the spread of the Gaussian curve:

$$H(u,v) = e^{-\frac{D^2(u,v)}{2D_0^2}}$$

A 3-D plot and a 2-D image plot of a Gaussian LPF are shown in Figure 1.116 on the right.

In the following example, a color image is first converted to grayscale and then subjected to DFT after which its zero-frequency components are shifted to the center of the plot. To compress its dynamic range, the logarithm of the frequency components is calculated and 1 is added to avoid zero-error condition. This generates the spectrum of the original signal. The standard deviation of the Gaussian filter is chosen arbitrarily at 10% of the height of the image. The user-defined function **lpf** is used to create a Gaussian LPF directly in the frequency domain, of the specified size and cutoff frequency. Like the image, the filter components are centered and its dynamic range compressed using logarithm. The frequency domain representations of the image and filter are then multiplied componentwise to produce the frequency domain representation of the filtered image. Finally, the IDFT is used to bring it back to the spatial domain (Figure 1.117).

> **Example 1.104: Write a program to implement a Gaussian low-pass filter in the frequency domain.**
>
> ```
> clear; clc;
> a = rgb2gray(imread('football.jpg'));
> [M, N] = size(a);
> F = fft2(a);
> ```

original image

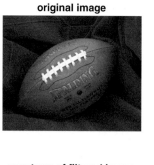

spectrum of original image

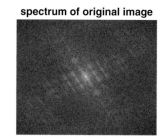

spectrum of filtered image

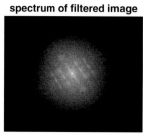

Filtered image

FIGURE 1.117
Output of Example 1.104.

```
F1 = fftshift(F);
S1 = log(1+abs(F1));
D0 = 0.1*min(M, N);
H = lpf(M, N, D0);
G = H.*F;
F2 = fftshift(G);
S2 = log(1+abs(F2));
g = real(ifft2(G));

subplot(221), imshow(a); title('original image');
subplot(222), imshow(S1, []); title('spectrum of original image');
subplot(223), imshow(S2, []); title('spectrum of filtered image');
subplot(224), imshow(g, []); title('Filtered image');

function H = lpf(M, N, D0)
    u = 0:(M-1);
    v = 0:(N-1);
    idx = find(u > M/2);
    u(idx) = u(idx) - M;
    idy = find(v > N/2);
    v(idy) = v(idy) - N;
    [V, U] = meshgrid(v, u);
    D = sqrt(U.^2 + V.^2);
    H = exp(-(D.^2)./(2*(D0^2)));
end
```

High-pass filters (HPF) allow high frequencies in the image to remain intact and blocks low frequencies. An **ideal HPF** allows high frequencies larger than a fixed cut-off value, while a Gaussian HPF uses a Gaussian curve as a transfer function. HPFs have a sharpening effect on the image. Because the HPFs perform the opposite function of LPFs, their transfer functions are related to the transfer functions of LPFs by the following relation:

$$H_{\mathrm{HPF}}(u,v) = 1 - H_{\mathrm{LPF}}(u,v)$$

An ideal HPF attenuates low frequencies smaller than a fixed cut-off value D_0 from the origin of the (centered) transform. The transfer function of an ideal LPF is given by the following, where $D(u,v)$ is the distance of point (u,v) from the center of the frequency rectangle:

$$H(u,v) = 0, \text{ if } D(u,v) \le D_0$$

$$H(u,v) = 1, \text{ if } D(u,v) > D_0$$

A 3-D plot and a 2-D image plot of an ideal HPF are shown in Figure 1.118 on the left as the function $H(u,v)$. The name *ideal* implies that all frequencies inside the circle of radius D_0 are completely attenuated while all frequencies outside this circle are passed completely without any attenuation, which physically cannot be realized using electronic components.

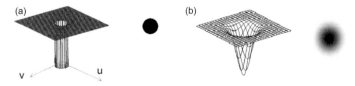

FIGURE 1.118
(a) Ideal HPF (b) Gaussian HPF.

A **Gaussian HPF** has a transfer function shown below obtained by putting $\sigma = D_0$ which is a measure of the spread of the Gaussian curve:

$$H(u, v) = 1 - e^{-\frac{D^2(u,v)}{2D_0^2}}$$

A 3-D plot and a 2-D image plot of a Gaussian HPF are shown in Figure 1.118 to the right.

In the following example, a color image is first converted to grayscale and then subjected to DFT after which its zero-frequency components are shifted to the center of the plot. To compress its dynamic range the logarithm of the frequency components are calculated and 1 is added to avoid zero-error condition. This generates the spectrum of the original signal. The cutoff frequency of the ideal HPF is chosen at 5% of the height of the image. The function **hpf** is used to create an HPF directly in the frequency domain, of the specified size and cutoff frequency. Like the image, the filter components are centered and its dynamic range compressed using the logarithm operator. The frequency domain representations of the image and filter are then multiplied componentwise to produce the frequency domain representation of the filtered image. Finally, the IDFT is used to bring it back to the spatial domain (Figure 1.119).

Example 1.105: Write a program to implement an ideal high-pass filter in the frequency domain.

```
clear; clc;
a = rgb2gray(imread('football.jpg'));
[M, N] = size(a);
F = fft2(a);
F1 = fftshift(F);
S1 = log(1+abs(F1));
D0 = 0.05*min(M, N);
```

original image

spectrum of original image

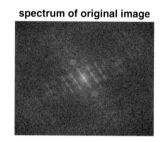

spectrum of filtered image

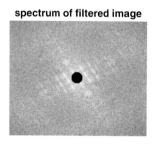

Filtered image

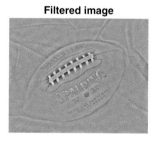

FIGURE 1.119
Output of Example 1.105.

```
H = hpf(M, N, D0);
G = H.*F;
F2 = fftshift(G);
S2 = log(1+abs(F2));
g = real(ifft2(G));

subplot(221), imshow(a); title('original image');
subplot(222), imshow(S1, []); title('spectrum of original image');
subplot(223), imshow(S2, []); title('spectrum of filtered image');
subplot(224), imshow(g, []); title('Filtered image');

function H = hpf(M, N, D0)
    u = 0:(M-1);
    v = 0:(N-1);
    idx = find(u > M/2);
    u(idx) = u(idx) - M;
    idy = find(v > N/2);
    v(idy) = v(idy) - N;
    [V, U] = meshgrid(v, u);
    D = sqrt(U.^2 + V.^2);
    Hlp = double(D <=D0);
    H = 1 - Hlp;
end
```

1.9 Image Processing Using Simulink

Simulink is a graphical environment included within MATLAB using which programming tasks can be performed without writing codes. It contains libraries of predefined blocks for modeling systems by defining dataflow between blocks of specific functionalities. It provides a graphical editor and viewer to see the final results of execution of algorithms.

To create a simulink model, type **simulink** at the command prompt or choose Simulink > Blank Model > Library Browser > Computer Vision System Toolbox (CVST). The video source files reside in the folder *(matlab-root)/toolbox/images/imdata/*. Within the CVST toolbox, the following libraries and blocks are included for image and video processing:

1. Sources: Image from file, Image from Workspace, From multimedia file
2. Sinks: Video viewer, To multimedia file
3. Analysis and Enhancement: Contrast adjustment, Histogram equalization, Edge detection, Median filter
4. Conversions: Autothreshold, Color space conversions, Image complement, Image data type conversion
5. Filtering: 2-D convolution, Median filter
6. Geometric transforms: Translate, Rotate, Scale, Shear, Affine, Warp
7. Morphological operations: Dilate, Erode, Open, Close, Top-hat, Bottom-hat
8. Statistics: 2-D Autocorrelation, 2-D correlation, 2-D mean, 2-D std, Blob analysis, PSNR
9. Text & Graphics: Compositing, Insert text
10. Transforms: 2-D DCT, 2-D IDCT, 2-D FFT, 2-D IFFT, Hough transform, Hough lines
11. Utilities: Block processing

The blocks are dragged from the Browser to the Editor and connected by dragging the mouse from the output port of one block to the input port of the other block. To branch a data line, place cursor on the line, press CTRL, and drag. To duplicate a block, press CTRL, and drag the block. After all the blocks are connected, click the Run button on the top of the Editor window to execute. The non-default parameters specified for each block is written in parenthesis beside the blocks in the examples below. Models created in Simulink can be saved using an .SLX file extension. If a Simulink model is saved as MODEL.SLX, then it can be printed out to a graphics file using the following command sequence:

```
print('-smodel', '-dpng', 'mymodel.png').
```

Example 1.106: Create a Simulink model to view an RGB image.

- CVST > Sources > Image from File ("peppers.png")
- CVST > Sinks > Video Viewer (Figure 1.120)

Example 1.107: Create a Simulink model for contrast adjustment of a grayscale image

- CVST > Sources > Image from File ("pout.tif")
- CVST > Sinks > Video Viewer
- CVST > Analysis & Enhancement > Contrast Adjustment (Figure 1.121)

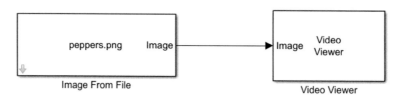

FIGURE 1.120
Output of Example 1.106.

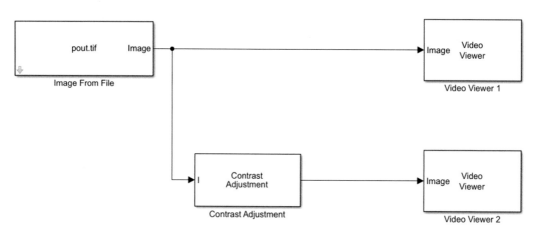

FIGURE 1.121
Output of Example 1.107.

Example 1.108: Create a Simulink model to convert RGB color images to grayscale and binary.

- CVST > Sources > Image from File ("peppers.png")
- CVST > Sinks > Video Viewer
- CVST > Conversions > Color Space Conversion (RGB to Intensity)
- CVST > Conversions > Autothreshold (Figure 1.122)

Example 1.109: Create a Simulink model for implementing an image filter

- CVST > Sources > Image from File ("coins.png")
- CVST > Sinks > Video Viewer
- CVST > Filtering > Median Filter (Neighbourhood size: [7 7]) (Figure 1.123)

Example 1.110: Create a Simulink model for geometrical transformation of an image.

- CVST > Sources > Image from File ("coins.png")
- CVST > Sinks > Video Viewer
- CVST > Geometric Transformation > Translate (Offset: [1.5 2.3])

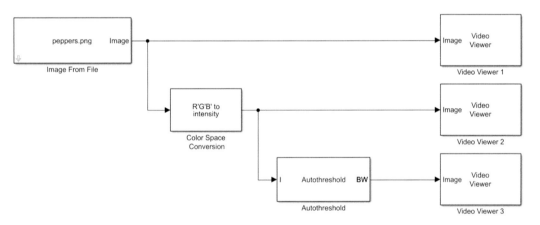

FIGURE 1.122
Output of Example 1.108.

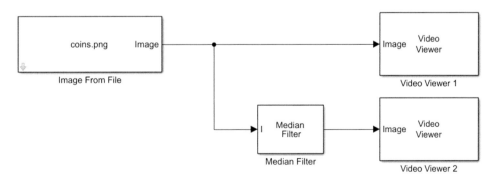

FIGURE 1.123
Output of Example 1.109.

- CVST > Geometric Transformation > Rotate (Angle: pi/6)
- CVST > Geometric Transformation > Resize (Resize factor: [200 150]) (Figure 1.124)

Example 1.111: Create a Simulink model for implementing morphological operations.

- CVST > Sources > Image from File ("circles.png")
- CVST > Sinks > Video Viewer
- CVST > Morphological Operations > Dilation (Structuring element: [1 1 ; 1 1])
- CVST > Morphological Operations > Erosion (Structuring element: strel('square', 10))
- CVST > Morphological Operations > Opening (Structuring element: strel('disk', 10)) (Figure 1.125)

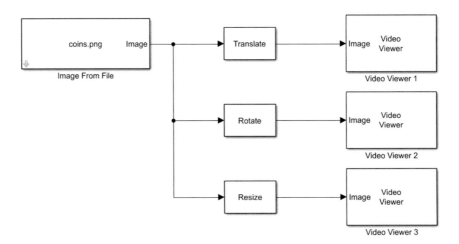

FIGURE 1.124
Output of Example 1.110.

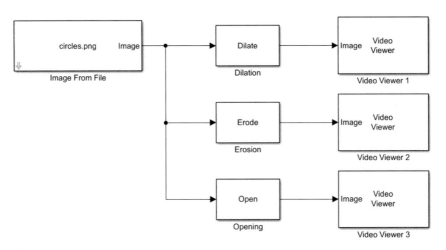

FIGURE 1.125
Output of Example 1.111.

Example 1.112: Create a Simulink model for performing edge detection and histogram equalization.

- CVST > Sources > Image from File (Filename: cameraman.tiff, Output data type: double)
- CVST > Sinks > Video Viewer
- CVST > Analysis & Enhancement > Edge Detection (Method: Canny)
- CVST > Analysis & Enhancement > Histogram Equalization (Figure 1.126)

Example 1.113: Create a Simulink model for performing color space conversion and text overlay on images.

- CVST > Sources > Image from File (Filename: peppers.png, Output Data Type: double)
- CVST > Sinks > Video Viewer
- CVST > Conversions > Color Space Conversion (Conversion: RGB to L*a*b*)
- CVST > Conversions > Color Space Conversion (Conversion: RGB to HSV)
- CVST > Text and Graphics > Insert Text (change parameters like text, color value, location, font etc.) (Figure 1.127)

Example 1.114: Create a Simulink model for displaying statistical values about an image.

- CVST > Sources > Image from File (Filename: eight.tif, Output data type: double)
- Simulink > Sinks > Display
- CVST > Statistics > 2-D Mean
- CVST > Statistics > 2-D Standard Deviation
- CVST > Statistics > 2-D Minimum
- CVST > Statistics > 2-D Maximum (Figure 1.128)

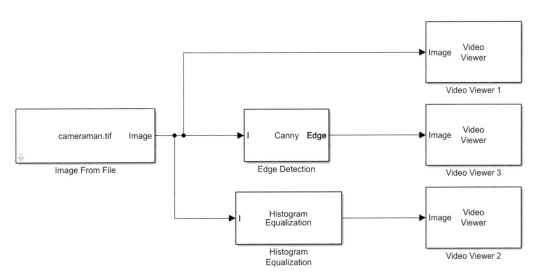

FIGURE 1.126
Output of Example 1.112.

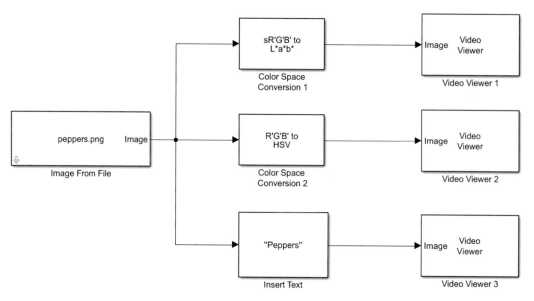

FIGURE 1.127
Output of Example 1.113.

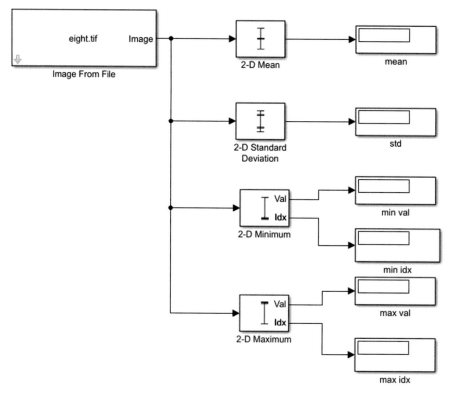

FIGURE 1.128
Output of Example 1.114.

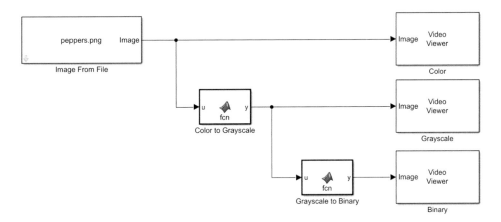

FIGURE 1.129
Output of Example 1.115.

Example 1.115: Create a Simulink model by implementing user-defined functions.

- CVST > Sources > Image from File > (Filename: circles.png)
- CVST > Sinks > Video Viewer
- Simulink > User-defined functions > MATLAB function ("function 1", "function 2") (Figure 1.129)

```
% Function 1: Color to grayscale:

function y = fcn(u)
    y = rgb2gray(u);
end
% Function 2: Grayscale to Binary:
function y = fcn(u)
    y = imbinarize(u, 0.5);
end
```

1.10 Notes on 2-D Plotting Functions

Since visualization is an important aspect of MATLAB, we dedicate a section on 2-D graphs and plotting functions. 2-D plots can be line plots, histograms, pie-charts, scatter plots, bar graphs, polar plots, contour plots, vector fields, and other similar types. The functions can also be used to specify properties of the figure window, axes appearance, titles, labels, legends, grids, colormaps, backgrounds, and so on. Each function is illustrated using one or more examples to indicate its options and parameters.

- **area**: Filled area 2-D plot
  This example plots the data in the variable Y as an area graph. Each subsequent column of Y is stacked on top of the previous data. The figure colormap controls the coloring of the individual areas (Figure 1.130).

```
Y = [   1, 5, 3;
    3, 2, 7;
    1, 5, 3;
    2, 6, 1];
```

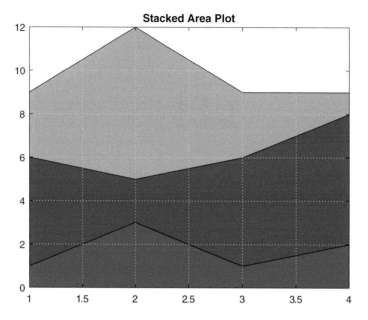

FIGURE 1.130
Plot using "area."

```
area(Y);
grid on;
colormap summer;
set(gca,'Layer','top');
title 'Stacked Area Plot'
```

- **axes**: Specify axes appearance and behavior
  The location is counted from the bottom-left of the figure window, not the parent axes (Fig. 1.131).

```
axes; grid;
rect = [0.3 0.3 0.5 0.5];
h = axes('position', rect);
% rect = [left, bottom, width, height]
```

- **axis**: Set axis limits and aspect ratios
  Use *axis tight* to set the axis limits to the range of the data, axis([xmin xmax ymin ymax]) to set minimum and maximum range of values, *axis equal* to set the aspect ratio so that equal tick mark increments on the $x-, y$ –axis are equal in size, *axis image* to fit the plot box tightly around the data, *axis square* to make the current axis box square in size (Figure 1.132).

```
x = -pi:0.01:pi;
subplot(231), plot(x, sin(x)); axis auto; title('auto');
subplot(232), plot(x, sin(x)); axis tight; title('tight');
subplot(233), plot(x, sin(x)); axis ([-5 5 -1.5 1.5]); title('range');
subplot(234), plot(x, sin(x)); axis equal; title('equal');
subplot(235), plot(x, sin(x)); axis square; title('square');
subplot(236), plot(x, sin(x)); axis image; title('image');
```

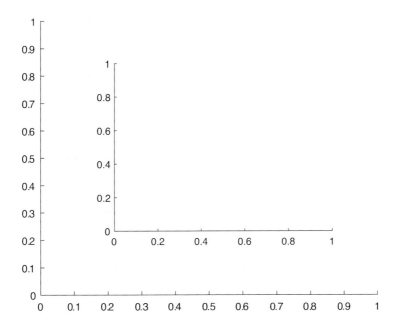

FIGURE 1.131
Plot using "axes."

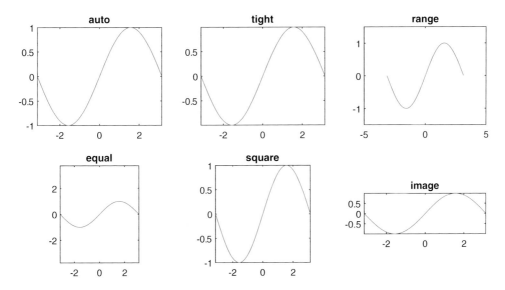

FIGURE 1.132
Plot using "axis."

- **bar**: Bar graph

Creates a bar graph. In the default orientation, a bar is displayed for each number in a vector. For a matrix, a group is created for each row of the matrix. In a stacked bar, display one bars for each row of the matrix, the height of each bar is the sum of the elements in the row (Figure 1.133).

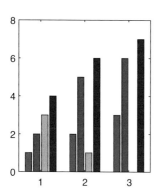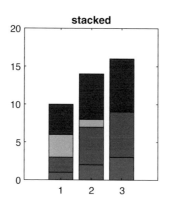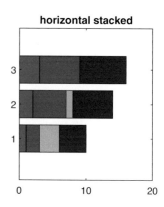

FIGURE 1.133
Plot using "bar" with various options.

```
a = [1 2 3 4; 2 5 1 6 ; 3 6 0 7]
subplot(131), bar(a, 0.8); title('width 0.8');
subplot(132), bar(a, 0.5); title('width 0.5');
subplot(133), bar(a, 1.6); title('width 1.6');
```

The default width of each bar is 0.8. If the width is reduced, the bars will be thinner. If the width is increased, then this would lead to overlapped bars (Figure 1.134).

```
a = [1 2 3 4; 2 5 1 6 ; 3 6 0 7]
subplot(131), bar(a, 0.8); title('width 0.8');
subplot(132), bar(a, 0.5); title('width 0.5');
subplot(133), bar(a, 1.6); title('width 1.6');
```

Converts baseline of bar graph to red dashed line and specifies a color for edges of the bars (Figure 1.135).

```
subplot(121),
a = bar(randn(10,1));
b = get(a,'BaseLine');
set(b,'LineStyle','--','Color','red');
subplot(122),
```

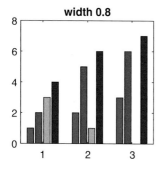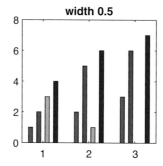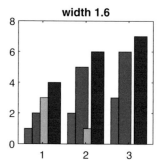

FIGURE 1.134
Plot using "bar" with various options.

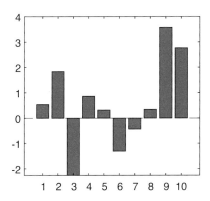 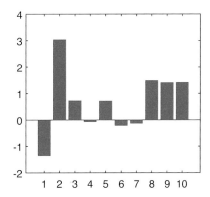

FIGURE 1.135
Plot using "bar" with various options.

```
a = bar(randn(10,1));
set(a,'EdgeColor','red');
```

- **colorbar**: Displays color scale in colormap

By default displays a vertical colorbar to the right of the current axis, but the location can be specified using options (Figure 1.136).

```
colormap jet
surf(peaks)
colorbar      % default eastoutside
colorbar('south')
colorbar('westoutside')
colorbar('northoutside')
```

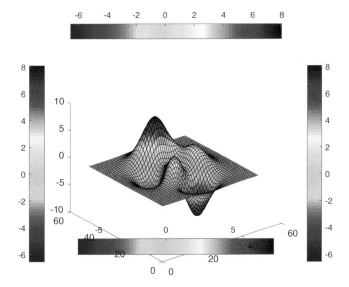

FIGURE 1.136
Plot using "colorbar."

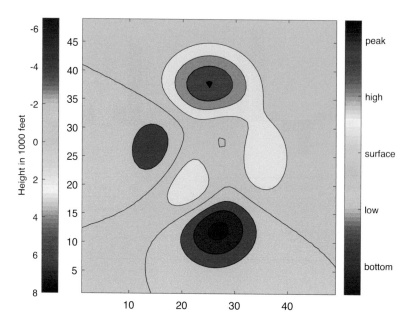

FIGURE 1.137
Plot using "colorbar" with customized options.

The direction of the colorbar can be reversed, and textual labels can be specified along it (Figure 1.137).

```
colormap jet;
contourf(peaks);
c = colorbar('westoutside','Direction','reverse');
c.Label.String = 'Height in 1000 feet';
colorbar('Ticks', [-5, -2, 1, 4, 7], ...
         'TickLabels', {'bottom','low','surface','high','peak'});
```

- **colormap**: Color look-up table

Built in colormaps: *parula, jet, hsv, hot, cool, spring, summer, autumn, winter, gray, bone, copper, pink, lines, colorcube, prism, flag, white*. The BM function **magic** returns a matrix with equal row and column sums (Figure 1.138).

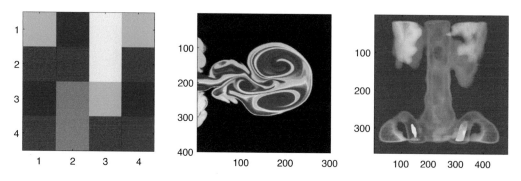

FIGURE 1.138
Plot using "colormap."

```
ax1 = subplot(131); map1 = prism;
I = magic(4); imagesc(I); colormap(ax1, map1)
ax2 = subplot(132); map2 = jet;
load flujet; imagesc(X); colormap(ax2, map2)
ax3 = subplot(133); map3 = autumn;
load spine ; imagesc(X); colormap(ax3, map3)
```

- **comet**: 2-D Comet plot
An animated graph in which a circle traces the data points on screen (Figure 1.139).

```
x = -pi:.1:pi;
y = tan(sin(x)) - sin(tan(x));
comet(x,y)
```

- **compass**: Plot arrows emanating from origin
A compass graph displays the vectors with components (x, y) as arrows emanating from the origin (Figure 1.140).

```
rng(0);
a = randn(1,20); A = fft(a);
rng(0);
b = randn(20,20); B = eig(b);
figure
subplot(121), compass(A)
subplot(122), compass(B)
```

- **contour, contourf**: Contour plot of a matrix
A contour plot of a matrix is displayed by joining lines of equal values. A filled version of the contour plot is generated by *contourf* (Figure 1.141).

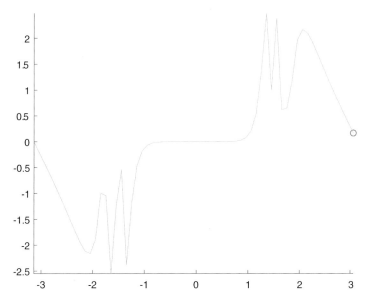

FIGURE 1.139
Plot using "comet."

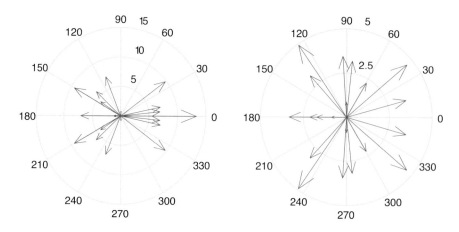

FIGURE 1.140
Plot using "compass."

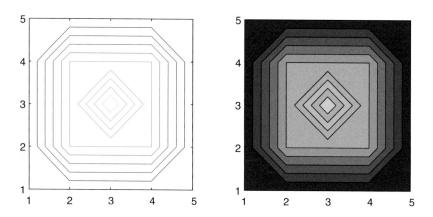

FIGURE 1.141
Plot using "contour" and "contourf."

```
a = [1 1 1 1 1 ; 1 2 2 2 1 ; 1 2 3 2 1 ; 1 2 2 2 1 ; 1 1 1 1 1];
subplot(121), contour(a); axis square;
subplot(122), contourf(a); axis square;
```

- **datetick**: Date formatted tick labels

Labels the tick lines of the axis using dates. The label format is based on the minimum and maximum limits of the specified axis. The BM function **datenum** creates a numeric array that represents each point in time as the number of days from January 0, 0000 (Figure 1.142).

```
clear;
rng(0);
t = [1:0.1:10]; x = rand(1,length(t));
subplot(121), plot(t,x); axis square;
datetick('x'); datetick('y');
```

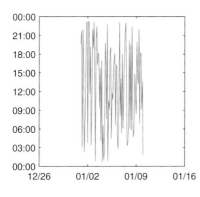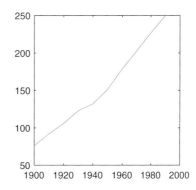

FIGURE 1.142
Plot using "datetick."

```
t = (1900:10:1990)'; % Time interval
p = [ 76 92 106 123 132 151 179 203 227 250]'; % Population
subplot(122), plot(datenum(t,1,1),p) % Convert years to date numbers
datetick('x','yyyy'); axis square;
```

- **errorbar**: Line plot with error bars

Creates a line plot of the data in y and draws a vertical error bar at each data point. The values in the error vector determine the lengths of each error bar above and below the data points, so the total error bar lengths are double the error values (Figure 1.143).

```
x = 1:10;
y = sin(x);
f = std(y)*ones(size(x));
e = abs(y - f);
subplot(121), plot(x, y, x, f); axis square; axis tight;
subplot(122), errorbar(y, e); axis square; axis tight;
```

- **ezplot**: Plots symbolic expressions

The symbolic expressions can also be expressions and equations in one or more variables, parametric equations and functions (Figure 1.144).

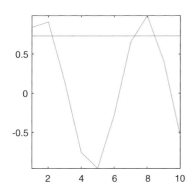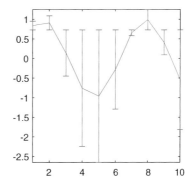

FIGURE 1.143
Plot using "errorbar."

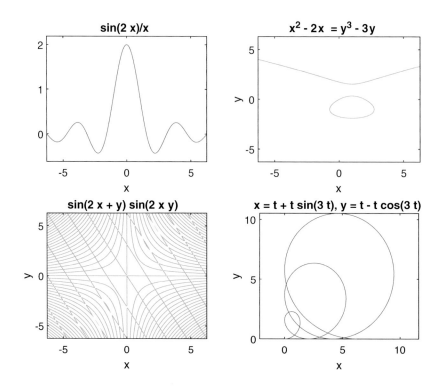

FIGURE 1.144
Plot using "ezplot."

```
clear;
syms x y t;
subplot(221), ezplot(sin(2*x)./x);
subplot(222), ezplot(x^2 - 2*x == y^3 - 3*y);
subplot(223), f(x, y) = sin(2*x + y)*sin(2*x*y); ezplot(f(x,y));
subplot(224), x = t + t*sin(3*t); y = t - t*cos(3*t); ezplot(x, y);
```

- **feather**: Plot velocity vectors
A feather plot displays vectors emanating from equally spaced points along a horizontal axis (Figure 1.145).

```
t = -2*pi:.2:2*pi; x = cos(t); y = sin(t); feather(x, y);
```

- **figure**: Create a figure window
The default properties of the figure window can be modified using name-value pairs like color, position, name etc. (Figure 1.146).

```
f = figure;
whitebg([0 0 0.5]);
f.Color = [0.5 0.5 0];
f.Position = [100 100 600 200];
f.Name = 'Sine Curve';
f.MenuBar = 'none'
t = 0:0.1:2*pi ;
```

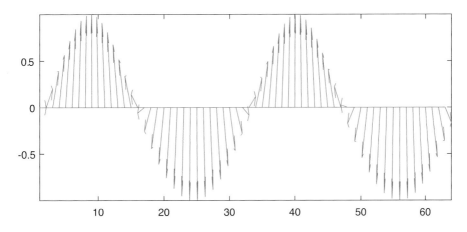

FIGURE 1.145
Plot using "feather."

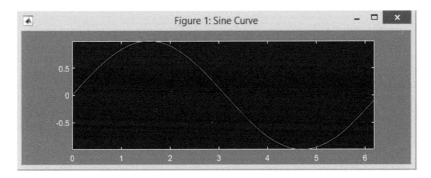

FIGURE 1.146
Plot using "figure."

```
x = sin(t);
plot(t,x); axis tight;
```

  • **gca**: Get current axis
  Used to modify properties of the axes (Figure 1.147).

```
x = -pi:0.1:pi;
y = sin(x);
plot(x,y); grid;
h = gca;
h.XTick = [-pi, -pi/2, 0, pi/2, pi];
h.XTickLabel = {'-\pi','-\pi/2','0','\pi/2','\pi'};
h.YTick = [-1, -0.8, -0.1, 0, 1/2, 0.9];
h.YTickLabel = {'-H','-0.8H', '-H/10', '0','H/2','0.9H'};
```

  • **gcf**: Current figure handle
  The current figure handle can be used to specify figure properties like background
color, menubar, toolbar etc. The BM function **whitebg** changes axes background color

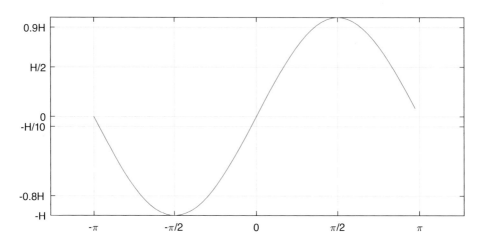

FIGURE 1.147
Plot using "gca."

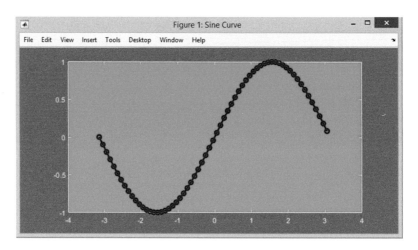

FIGURE 1.148
Plot using "gcf."

(default *white*), and the function **figure** is used to specify a figure title. To change the background back to white, use whitebg([1 1 1]) (Figure 1.148).

```
x = -pi:.1:pi;
y = sin(x);
figure('Name','Sine Curve');
p = plot(x, y, '-yo', 'LineWidth', 2);
whitebg([0.2 0.7 0.2])
set(gcf, 'Color', [1, 0.4, 0.6])
set(gcf, 'MenuBar', 'default')
set(gcf, 'ToolBar', 'none')
set(gcf, 'WindowState', 'maximized');
```

- **grid**: Display grid lines
Toggles display of major and minor grid lines (Figure 1.149).

```
clear;
t = -10:0.1:10;
x = sin(t)./t;
y = sec(x)./x;
subplot(121), plot(t, x); grid on;
subplot(122), plot(t, y); grid on; grid minor;
```

- **histogram**: Histogram plot
Displays a bar plot by grouping data into bins based on the frequency of occurrence. Optional parameters include face-color, edge-color, line-style etc. (Figure 1.150).

```
a = [3 3 1 4 0 0 1 1 0 1 2 4];
subplot(121)
h = histogram(a); axis square
```

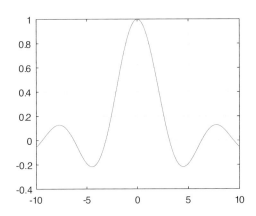
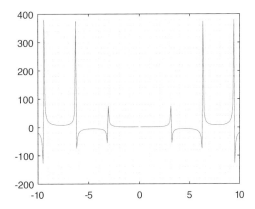

FIGURE 1.149
Plot using "grid."

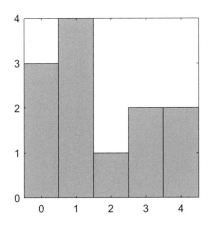
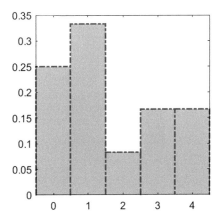

FIGURE 1.150
Plot using "histogram" with various options.

```
subplot(122)
h = histogram(a,'Normalization','probability'); axis square
h.FaceColor = [0.8 0.5 0.1]
h.EdgeColor = [1 0 0]
h.LineStyle = '-.'
h.LineWidth = 1.5
```

Once a histogram is generated, its parameters are listed within the structure h. If the histogram plot window is closed the structure h is deleted. The data from which the histogram is created can be listed using h.Data, the frequency values of the histogram are listed using h; Values, the number of bins using h.NumBins, the bin edge using h. BinEdges, the bin width using h.BinWidth, the bin limits using h.BinLimits. All these parameters can be changed to modify the histogram (Figure 1.151).

```
rng(0);
x = randn(1000,1);
subplot(221), h1 = histogram(x); title('h1');
subplot(222), h2 = histogram(x); h2.NumBins = 10; title('h2');
subplot(223), h3 = histogram(x); h3.BinEdges = [-2: 2];  title('h3');
subplot(224), h4 = histogram(x); h4.BinWidth = 0.5;  title('h4');
```

- **hold**: Hold on the current plot

Retains the current plot so that the subsequent commands can add to it. Works in a toggle mode (Figure 1.152).

```
ezplot('sin(x)');
```

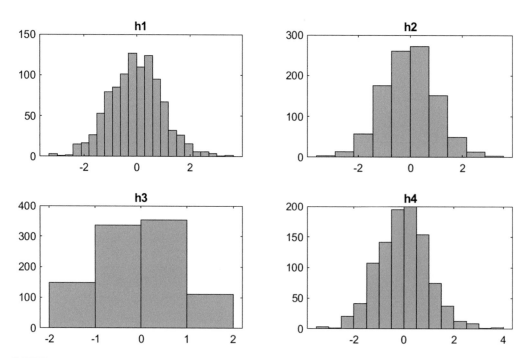

FIGURE 1.151
Plot using "histogram" with customized options.

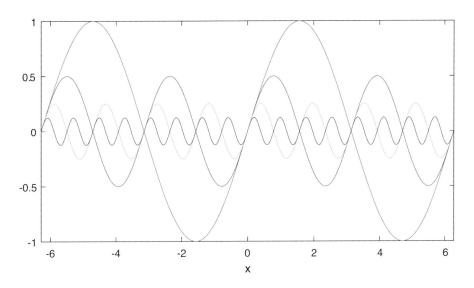

FIGURE 1.152
Plot using "hold."

```
hold on;
ezplot('0.5*sin(2*x)');
ezplot('0.25*sin(4*x)');
ezplot('0.125*sin(8*x)');
axis([-2*pi 2*pi -1 1])
title('')
hold off;
```

- **image, imagesc**: Display image from array

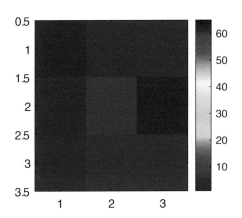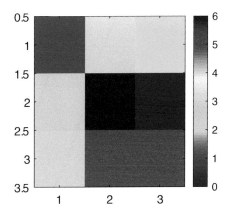

FIGURE 1.153
Plot using "image" and "imagesc."

Display matrix as an image. Elements of the matrix are used as indices into the current colormap. The second version scales the values so as to cover the entire range of values in the colormap (Figure 1.153).

```
a=[1 3 4 ; 2 6 0 ; 3 5 5];
colormap jet
subplot(121), image(a); colorbar;
subplot(122), imagesc(a); colorbar;
```

- **legend**: Add legend to axes

Creates descriptive labels for each plotted data series (Figure 1.154).

```
x = linspace(0,2*pi);
y1 = sin(x);
y2 = (1/2)*sin(2*x);
y3 = (1/3)*sin(3*x);
subplot(121), plot(x, y1, x, y2, x, y3, 'LineWidth', 2)
legend('sin(x)', '(1/2)sin(2x)', '(1/3)sin(3x)'); axis square;

subplot(122),
plot(x, y1+y2, x, y2+y3, x, y3+y1, 'LineWidth', 2)
g = legend('y1+y2', 'y2+y3', 'y3+y1', 'Location', 'southwest');
axis square;
title(g,'Sum of Sine Curves')
```

- **line**: Create line

Draws a line connecting specified data points using specified colors and styles (Figure 1.155).

```
x = [2 6 9];
y = [3 8 4];
figure
subplot(121)
plot(x,y)
```

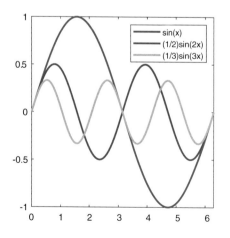
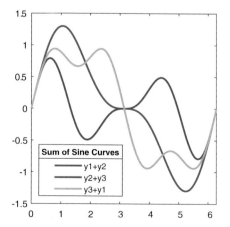

FIGURE 1.154
Plot using "legend."

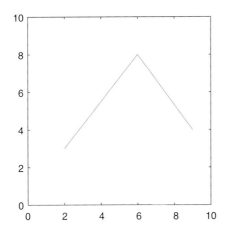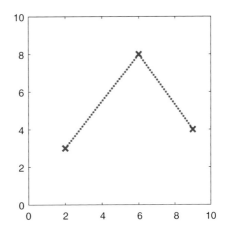

FIGURE 1.155
Plot using "line."

```
line(x,y)
axis([0 10 0 10])
axis square

subplot(122)
plot(x,y)
p = line(x,y)
p.Color = [0.6 0.3 0.1]
p.LineStyle = ':'
p.LineWidth = 2
p.Marker = 'x'
p.MarkerSize = 8
axis([0 10 0 10])
axis square
```

- **linspace**: Generate linearly spaced vector

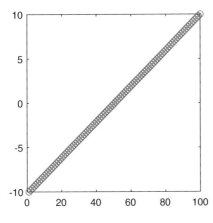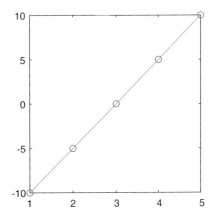

FIGURE 1.156
Plot using "linspace."

Returns a row vector of 100 evenly spaced points (default) between specified end points, or a specified number of points (Figure 1.156).

```
y1 = linspace(-10,10); subplot(121), plot(y1, 'o-');
y2 = linspace(-10,10,5); subplot(122), plot(y2, 'o-');
```

- **pie**: Pie chart

Draws a pie chart for a specified data such that slice of the pie chart represents an element of the data. The total area of the pie represents the sum of all the elements and the area of each slice is represented as a percentage of the total area (Figure 1.157).

```
x = [1 7 2 5 3];
subplot(121), pie(x);
y = 1:4;
ax = subplot(122), explode = [0 0 1 0];
labels = {'north','south','east','west'};
pie(y, explode, labels);
title('Directions'); colormap(ax, autumn);
```

- **plot**: 2-D line plot

Generates a 2-D line plot of vectors or matrices. Options can be specified for line width, type of markers, marker edge color, marker face color, marker size (Figure 1.158).

```
x = -pi:.1:pi;
y = tan(sin(x)) -sin(tan(x));
plot(x, y, '--rs', ...
    'LineWidth', 2, ...
    'MarkerEdgeColor', 'k', ...
    'MarkerFaceColor', 'g', ...
    'MarkerSize', 10);
```

- **plotyy**: Plot using two *y*-axis labelings

Plot using two different *y*-axis labels displayed on the left and right sides of the plot (Figure 1.159).

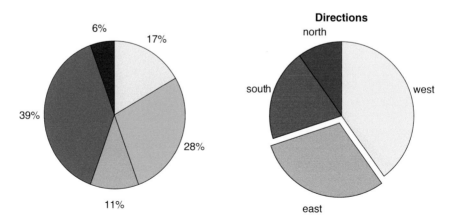

FIGURE 1.157
Plot using "pie."

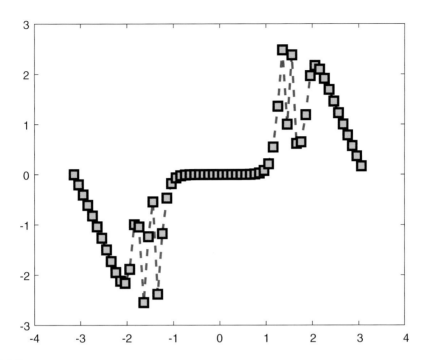

FIGURE 1.158
Plot using "plot."

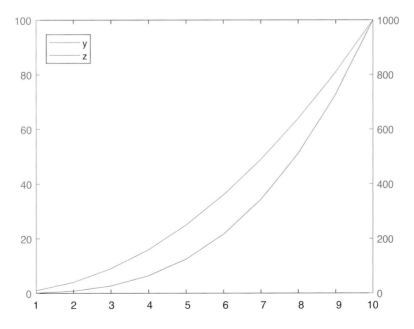

FIGURE 1.159
Plot using "plotyy."

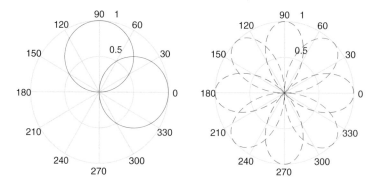

FIGURE 1.160
Plot using "polar."

```
x = 1:10;
y = x.^2;
z = x.^3;
plotyy(x, y, x, z)
legend('y', 'z')
```

- **polar**: Polar plot
Generates a polar plot using the specified angle and radius (Figure 1.160).

```
theta = 0:0.01:2*pi;
subplot(121),
polar(theta, sin(theta));
hold on;
polar(theta, cos(theta));
subplot(122),
polar(theta, sin(2*theta), '--r');
hold on;
polar(theta, cos(2*theta), '--b');
```

- **polarhistogram**: Polar histogram
Generates a polar histogram of a values in a vector between 1° and 360° into specified number of bins (Figure 1.161).

```
rng(0);
r = randi([1 360], 1, 50); % generate 50 random integers between 1 and 360
bins = 12;
polarhistogram(r, bins)
```

- **quiver**: Arrow plot
Displays vectors as arrows with specified starting point and specified width and height. The BM function **humps** uses the following in-bult function: $y = \dfrac{1}{(x-0.3)^2 + 0.01} + \dfrac{1}{(x-0.9)^2 + 0.04} - 6$ and x uses the range 0:0.05:1 by default (Figure 1.162).

```
quiver(1:length(humps), humps)
```

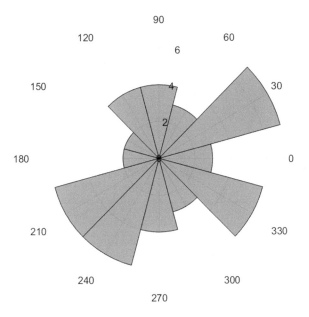

FIGURE 1.161
Plot using "polarhistogram."

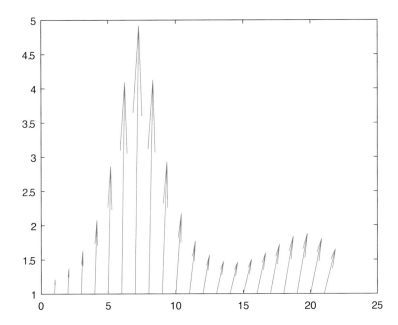

FIGURE 1.162
Plot using "quiver."

- **scatter**: Scatter plot
Displays circles at specified locations with specified sizes and colors (Figure 1.163).

```
clear;
```

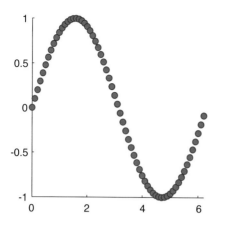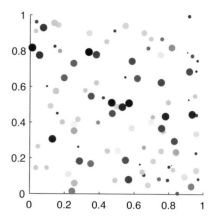

FIGURE 1.163
Plot using "scatter."

```
clear;

x = 0:0.1:2*pi;
y = sin(x);
sz = 30;
subplot(121),
s = scatter(x,y, sz, 'filled')
s.LineWidth = 0.6;
s.MarkerEdgeColor = 'b';
s.MarkerFaceColor = [0 0.5 0.5];
axis square

subplot(122),
r = rand(100,100);
x = r(:,1); y = r(:,2);
c = linspace(1,10,length(x));
sz = linspace(1,50,length(x));
scatter(x,y,sz,c, 'filled')
axis square
colormap jet
```

- **set**: Set graphics object properties

Used to specify value of the property name on the object identified by h using the format
set(h, name, value), as shown in (Figure 1.164).

```
x = -pi:.1:pi;
y = sin(x);
p = plot(x, y, '-o');
set(gca, 'XTick', -pi:pi/2:pi)
set(gca, 'XTickLabel', {'-pi', '-pi/2', '0', 'pi/2', 'pi'}) ;
set(p, 'Color', [1 1 0]);
set(p, 'LineWidth', 5);
set(p, 'MarkerSize', 10);
set(p, 'MarkerIndices', 1:3:length(y));
set(p, 'MarkerEdgeColor', 'b', 'MarkerFaceColor', [0.5, 0.5, 0.5])
```

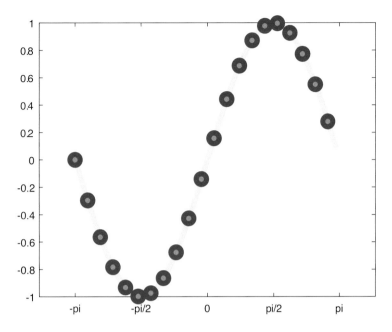

FIGURE 1.164
Plot using "set."

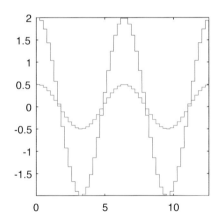
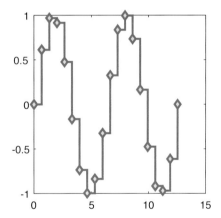

FIGURE 1.165
Plot using "stairs."

- **stairs**: Stairs plot
Draws a stairstep graph of the elements of a vector (Figure 1.165).

```
X = linspace(0,4*pi, 50)';
Y = [0.5*cos(X), 2*cos(X)];
subplot(121),
stairs(X, Y); axis square;
subplot(122),
X = linspace(0,4*pi, 20);
Y = sin(X);
```

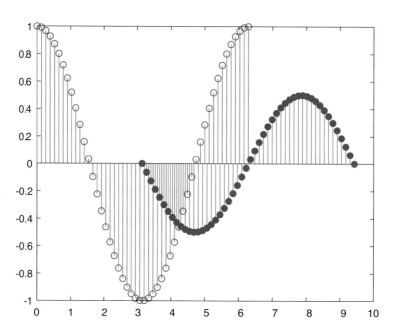

FIGURE 1.166
Plot using "stem."

```
stairs(X, Y, 'LineWidth', 2,'Marker', 'd', 'MarkerFaceColor', 'c');
axis square;
```

- **stem**: Stem plot
Plot data sequence as stems that extend from the baseline along the *x*-axis, terminating in circles (Figure 1.166).

```
x1 = linspace(0,2*pi, 50)';
x2 = linspace(pi, 3*pi, 50)';
y1 = cos(x1);
y2 = 0.5*sin(x2);
stem(x1, y1); hold on;
stem(x2, y2, 'filled');
```

- **subimage**: Multiple images in a single figure
Displays each image using a different specified colormap (Figure 1.167).

```
load trees;
subplot(2,2,1), subimage(X, map);
subplot(2,2,2), subimage(X, winter);
subplot(2,2,3), subimage(X, autumn);
subplot(2,2,4), subimage(X, summer);
```

- **text**: Text descriptions
Insert text descriptions in graphical plots at specified locations (Figure 1.168).

```
x = linspace(0,2*pi);
y = sin(x);
```

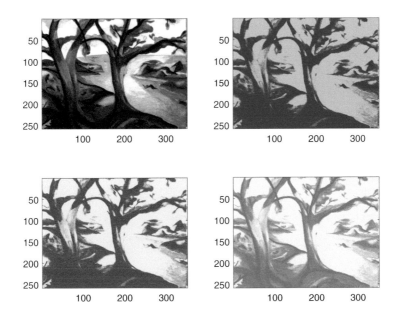

FIGURE 1.167
Plot using "subimage."

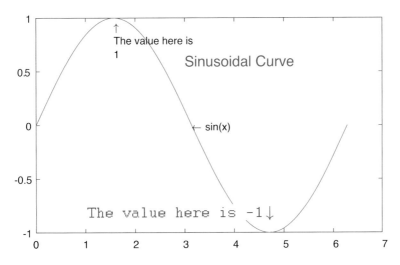

FIGURE 1.168
Plot using "text."

```
plot(x, y);
text(pi, 0,'\leftarrow sin(x)');
text(3,0.6,'Sinusoidal Curve', 'Color', 'red', 'FontSize', 14);
v1 = sin(pi/2);
str1 = {'\uparrow', 'The value here is ', num2str(v1)};
v2 = sin(3*pi/2);
str2 = ['The value here is -1', '\downarrow'];
text(pi/2, v1-0.2, str1);
```

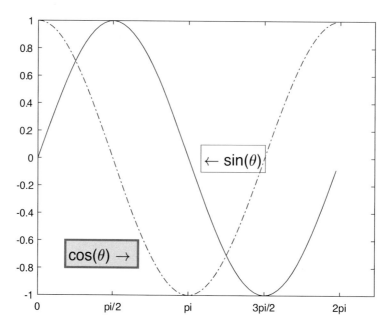

FIGURE 1.169
Plot using "text" with additional options.

```
t = text(3*pi/2, v2+0.2, str2);
t.Color = 'blue';
t.FontName = 'Cambria';
t.FontSize = 15;
t.Position = [2, -0.8];
```

Additional options can include inserting text boxes with specified edge color and background color (Figure 1.169).

```
x = 0: 0.1: 2*pi;
y1 = sin(x);
y2 = cos(x);
plot(x, y1, 'k-'); hold on;
plot(x, y2, 'k-.');
set(gca, 'XTick', 0:pi/2:2*pi)
set(gca, 'XTickLabel', {'0', 'pi/2', 'pi', '3pi/2', '2pi'});
text(1.1*pi, 0, '\leftarrow sin(\theta)', 'EdgeColor', 'red', ...
    'FontSize', 15, 'HorizontalAlignment', 'left');
text(0.65*pi, -0.7, 'cos(\theta) \rightarrow ', 'EdgeColor', [0 .5 0], ...
    'FontSize', 15, 'HorizontalAlignment', 'right', ...
    'BackgroundColor', [.7 .9 .7], 'LineWidth', 2);
```

- **title**: Plot title
  Add title to the current plot (Figure 1.170).

```
syms c f;
c = (f-32)/9;
ezplot(f, c); grid;
```

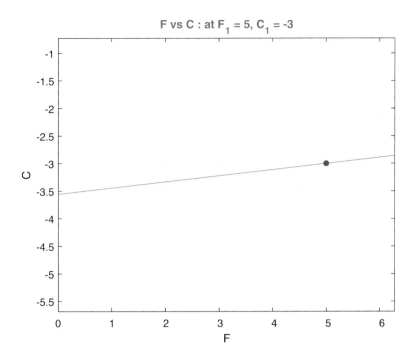

FIGURE 1.170
Plot using "title."

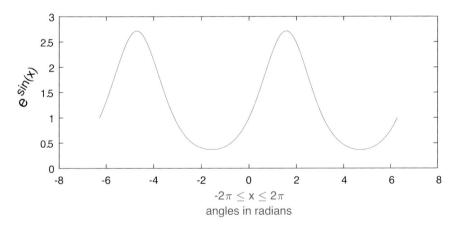

FIGURE 1.171
Plot using "xlabel" and "ylabel."

```
f1 = 5; c1 = double(subs(c, f, 5));
hold on;
scatter(f1, c1, 30, 'r', 'filled');
xlabel('F'); ylabel('C');
t = title(['F vs C: at F_1 = 5, C_1 = ', num2str(c1)]);
t.Color = [0 0.5 0];
```

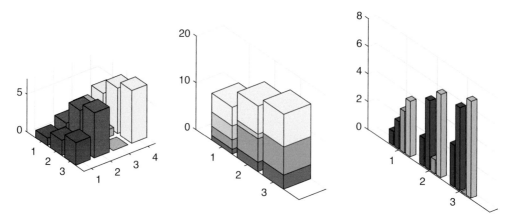

FIGURE 1.172
Plot using "bar3."

- **xlabel, ylabel**: Label *x*-axis and *y*-axis
  Adds descriptive textual labels to the *x*-axis and *y*-axis of the plot (Figure 1.171).

```
x = linspace(-2*pi, 2*pi);
y = exp(sin(x));
plot(x, y)
xl = xlabel({'-2\pi \leq x \leq 2\pi', 'angles in radians'});
yl = ylabel('e^{sin(x)}')
xl.Color = 'red';
yl.FontSize = 15;
yl.Rotation = 60;
```

1.11 Notes on 3-D Plotting Functions

This section deals with the 3-D plotting functions. 3-D plots can include surfaces, mesh-grids, 3-D line plots, 3-D pie charts, 3-D bar graphs, solid geometrical figures like spheres, cylinders, and so on. Each function is illustrated using one or more examples to indicate its options and parameters.

bar3: 3-D Bar graphs
Generates 3-D bar graphs in a specified format (Figure 1.172).

```
a = [1 2 3 4; 2 5 1 6 ; 3 6 0 7];
ax = subplot(131), bar3(a); colormap(ax, autumn);
bx = subplot(132), bar3(a, 'Stacked'); colormap(bx, summer);
cx = subplot(133), bar3(a, 'Grouped'); colormap(cx, winter);
```

- **cylinder**: 3-D cylinder

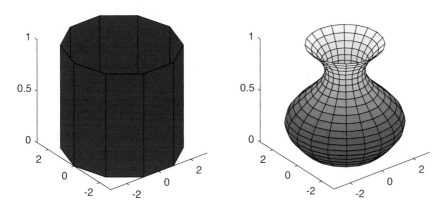

FIGURE 1.173
Plot using "cylinder."

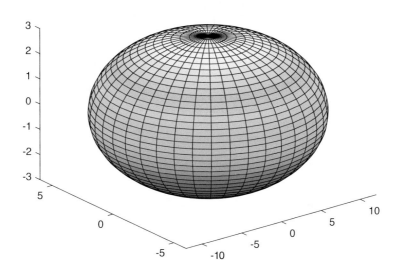

FIGURE 1.174
Plot using "ellipsoid."

Create a 3-D cylinder using the specified radius or defining function (Figure 1.173).

```
subplot(121), cylinder(3,10); % radius 3 with 10 points around circumference
ax = subplot(122);
t = 0:pi/10:2*pi;
cylinder(2 + sin(t));                  % defining function
colormap(ax, autumn)
```

• **ellipsoid**: 3-D ellipsoid

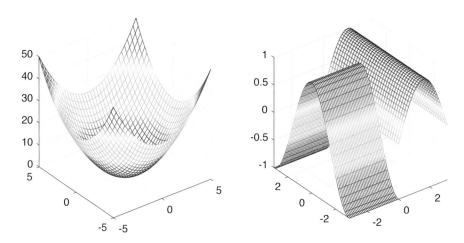

FIGURE 1.175
Plot using "fmesh."

Create a 3-D ellipsoid using the specified center coordinates and radii lengths along three axes and specified number of points along the circumference. The default number of points is 20 (Figure 1.174).

```
ellipsoid(0,0,0, 12,6,3, 40);
% center (0,0,0), radii (12,6,3), number of points 4
```

- **fmesh**: 3-D mesh

Creates a 3-D mesh by plotting $z = f(x,y)$ as specified functions of x and y over the default intervals [–5, 5] or over specified intervals. Number of points generated in each direction (mesh density) is 35 by default (Figure 1.175).

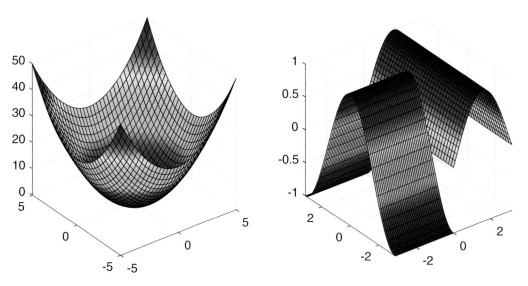

FIGURE 1.176
Plot using "fsurf."

```
subplot(121), fmesh(@(x,y) x^2 + y^2);  % uses default interval [-5,5]
axis square;

subplot(122),                      % uses interval [-pi, pi]
fmesh(@(x,y) cos(y), [-pi, 0, -pi, pi]);
% uses interval [-pi, 0] for first function
hold on;
fmesh(@(x,y) sin(x), [0, pi, -pi, pi]);
% uses interval [0, pi] for second function
hold off; axis square; colormap jet;
```

- **fsurf**: 3-D surface

Creates a 3-D surface by plotting $z = f(x,y)$ as specified functions of x and y over the default intervals [–5, 5] or over specified intervals. Number of points generated in each direction (mesh density) is 35 by default (Figure 1.176).

```
subplot(121), fsurf(@(x,y) x^2 + y^2); axis square;
% uses default interval [-5,5]
subplot(122),                      % uses interval [-pi, pi]
fsurf(@(x,y) cos(y), [-pi, 0, -pi, pi]);
% uses interval [-pi, 0] for first function
hold on;
fsurf(@(x,y) sin(x), [0, pi, -pi, pi]);
% uses interval [0, pi] for second function
hold off; axis square; colormap jet;
```

- **meshgrid**: Rectangular mesh grid

Generates a 2-D grid of points from two specified vectors x $(1 \times m)$ and y $(n \times 1)$. Returns two matrices X and Y both with $n \times m$ elements. Each row of X is a copy of x and there are n such rows. Each column of Y is a copy of y and there are m such columns. The values of X and Y are combined into a single $n \times m$ matrix using a function F. The matrix F can be plotted as a surface for visualization (Figure 1.177).

```
x = -1:3
y = 2:4; y = y'
```

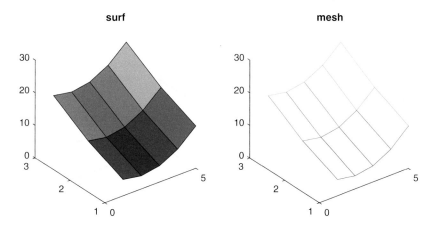

FIGURE 1.177
Plot using "meshgrid."

```
[X, Y] = meshgrid(x,y)
F = X.^2 + Y.^2
subplot(121), surf(F); title('surf');
subplot(122), mesh(F); title('mesh');


x =
     -1      0      1      2      3

y =
      2
      3
      4

X =
     -1      0      1      2      3
     -1      0      1      2      3
     -1      0      1      2      3

Y =
      2      2      2      2      2
      3      3      3      3      3
      4      4      4      4      4

F =

      5      4      5      8     13
     10      9     10     13     18
     17     16     17     20     25
```

- **mesh meshc**: Mesh plot

Creates a mesh plot to visualize a meshgrid. Mesh plot can be combined with a contour plot using 'meshc' (Figure 1.178).

```
[X, Y] = meshgrid(-3*pi:.5:3*pi);
```

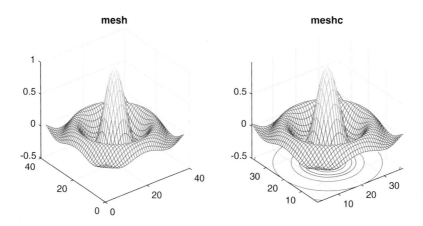

FIGURE 1.178
Plot using "mesh" and "meshc."

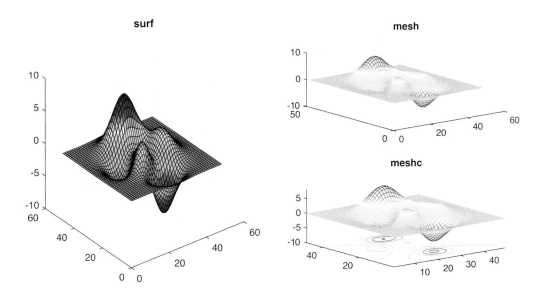

FIGURE 1.179
Plot using "peaks."

```
R = sqrt(X.^2 + Y.^2) + eps;
F = sin(R)./R;
subplot(121), mesh(F); title('mesh'); axis square;
subplot(122), meshc(F); title('meshc'); axis square;
```

- **peaks**: Sample function of two variables

The function of two variables x and y is plotted along the z-axis using a 49×49 grid by default (Figure 1.179):

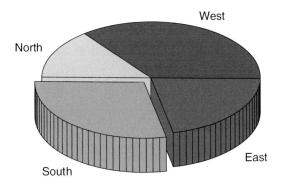

FIGURE 1.180
Plot using "pie3."

$$f = \left[3(1-x)^2 \cdot \exp\left\{ -\left(x^2\right) - \left(y+1\right)^2 \right\} \right] - \left[10\left(\frac{x}{5} - x^3 - y^5 \right) \cdot \exp\left(-x^2 - y^2\right) \right]$$

$$- \left[\left(\frac{1}{3} \right) \cdot \exp\left\{ -(x+1)^2 - y^2 \right\} \right]$$

```
f = peaks;
subplot(2,2,[1,3]), surf(f); title('surf');
subplot(222), mesh(f); title('mesh');
subplot(224), meshc(f); title('meshc');
colormap jet
```

- **pie3**: 3-D pie chart

Generates a 3-D pie chart for a vector. Each element of the vector is represented as a slice of the pie chart (Figure 1.180).

```
x = [2 4 3 5];
explode = [0 1 0 0];
pie3(x, explode, {'North', 'South', 'East', 'West'});
colormap cool
```

- **plot3**: 3-D line plot

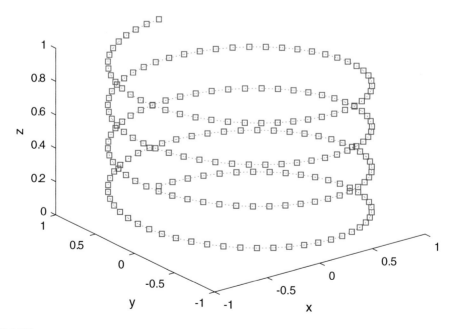

FIGURE 1.181
Plot using "plot3."

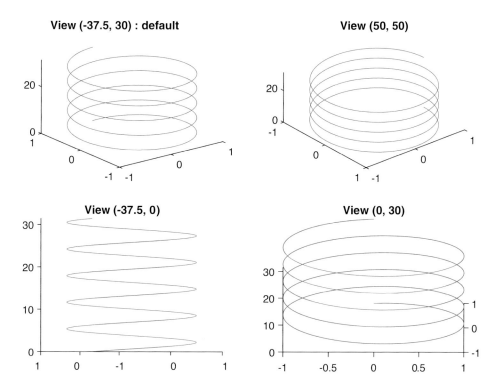

FIGURE 1.182
Plot using "view."

Generates a 3-D line plot for three vectors along *x*-, *y*-, and *z*- directions (Figure 1.181).

```
N = 200;
x = linspace(0,8*pi, N);
x = sin(x);
y = linspace(0,8*pi, N);
y = cos(y);
z = linspace(0,1,N);
plot3(x, y, z, 'rs:');
xlabel('x'); ylabel('y'); zlabel('z');
title('Spiral');
```

- **view**: Viewpoint specification
3-D graph viewpoint: view (AZ, EL), where AZ (azimuth) is the horizontal rotation (default – 37.5°) and EL (elevation) is the vertical rotation (default 30°) (Figure 1.182).

```
t = 0:pi/50:10*pi;
subplot(221), plot3(sin(t), cos(t), t); title('View (-37.5, 30): default');
subplot(222), plot3(sin(t), cos(t), t); view (50, 50); title('View (50, 50)');
subplot(223), plot3(sin(t), cos(t), t); view (-37.5, 0); title('View
(-37.5, 0)');
subplot(224), plot3(sin(t), cos(t), t); view (0, 30); title('View (0, 30)');
```

Review Questions

1. Differentiate between BM functions and IPT functions for image processing.
2. Differentiate between binary, grayscale, and color images and their internal matrix representations.
3. What is the need of a threshold for binarization? How can it be determined using Otsu's method?
4. How can an indexed image be displayed using color look up tables (CLUT)? What are inbuilt colormaps?
5. What are RGB, CMY, HSV, and L*a*b* color models? How does color conversion between them take place?
6. What are color tristimulus values and how are they used to generate the CIE chromaticity diagram?
7. How can noise be added to and removed from an image?
8. Explain different ways how multiple images can be read and displayed in a system.
9. What is geometric transformation of an image? When is such transformation affine and projective?
10. How can similarity between images can be measured? How can it be used to identify matching regions?
11. How can images be enhanced using gamma curves, kernels, and histograms?
12. What are the different types of image filters in spatial and frequency domains?
13. Explain different methods of edge detection. In this context define image gradients.
14. How can arithmetical, logical, and morphological operators be applied on images?
15. Explain the degradation-restoration model mentioning the significance of a PSF. What is image quality?
16. What is convolution and deconvolution? Differentiate between the various deconvolution techniques.
17. What is image segmentation and how is it implemented?
18. What are region of interests and region boundaries? How are they used in object analysis?
19. What is texture analysis? How can texture features be represented?
20. How is image transformed to frequency domain using DFT, DCT, and DWT?

2

Audio Processing

2.1 Introduction

Sound is a form of energy similar to heat and light which propagates from one location to another using a material medium. Sound is generated from vibrating objects when the kinetic energy of vibration flows through the layers of the medium until it reaches our ears. This sets our eardrums into a similar vibration and our brain recognizes this as the sensation of sound. **Acoustics** is the branch of science dealing with the generation, transmission, and reception of sound. **Psycho-acoustics** is a branch of acoustics which deals with how humans perceive sound and provide important guidelines for designing audio equipment. As sound propagates through air or any other gaseous medium, it sets up alternate regions of compression and rarefaction which is pictorially represented as waves. **Sound waves** are mathematically represented using sinusoidal equations of the form $y = a.\sin (bt - \theta)$, where a is the amplitude, b is the frequency, and θ is the phase difference. **Amplitude** of a wave is the maximum displacement of a particle in the path of a wave from its mean position and indicates energy content of the wave. For a sound wave, this corresponds to **loudness**. Loudness is measured in a unit called **decibels** (dB). **Frequency** of a wave is the number of vibrations of a particle in the path of a wave per unit time. The physical manifestation of this is the **pitch** of sound. A high pitched sound like a whistle has higher frequency than a low pitched sound like a drum. Frequency is measured in a unit called **hertz** (Hz).

To process sound, environmental sound waves need to be converted to electrical signals. There are three types of components that are used for this purpose viz. microphone, amplifier, and loudspeaker. A **microphone** converts environmental sound waves into electrical signals, while an **amplifier** enhances the amplitude of these electrical signals so that they can be utilized more conveniently. A **loudspeaker** converts the electrical energy back to sound energy which can then be heard by our ears. The human range of hearing is around 0–120 dB for loudness and 20 Hz–20 kHz for frequency. Audio recording and playback systems are designed keeping these range of values in mind. A single loudspeaker unit is often split into three sub-units for faithful reproduction of sounds with different frequency ranges viz. **woofer** for low-frequency values, **mid-range** for middle-frequency values, and **twitter** for high-frequency values. Typically when sound is recorded and played back using a recording system, multiple sound sources, captured using multiple microphones, are combined using an **audio-mixer** into one or two audio channels. Sounds having a single channel is called **mono** sound and can be played back using a single speaker, while sounds with two channels are called **stereo** and requires two speakers, left and right. **Surround sound** systems use multiple speakers typically five or seven, placed at different locations around the room to produce a 3-D audio effect. Surround sound systems often use a unit called a **sub-woofer** to reproduce very low-frequency sounds. Electrical audio signals can also be recorded onto magnetic audio tapes using local polarization of magnetic dipoles.

Analog audio is converted into the digital form through the processes of sampling, quantization, and code-word generation. The structural unit of a digital audio signal is called a **sample** similar to a pixel for a digital image. The sampling rate specifies the number of samples generated per second. An important theorem frequently used in digital signal processing to determine the sampling rate is the **Nyquist Sampling Theorem** named after Harry Nyquist. The theorem states that the sampling frequency needs to be twice the highest frequency of the input signal for faithful reproduction. This condition is often referred to as the Nyquist criterion. If the sampling frequency falls below this threshold, then the reconstruction of the signal exhibits imperfections or artifacts which is referred to as **aliasing** defects. If f is the highest frequency of the analog audio signal, then as per Nyquist's postulate, the sampling rate needs to be $F = 2f$. If b is the bit-depth of the samples and c is the number of audio channels then the data rate of the digital audio is $= F.b.c.$ If T is the duration of the audio in seconds, then the number of samples $N = F.T$ and the file size of the digital audio is $= D.T$. The **sound card** inside a computer is the device responsible for digitization of audio signals. It has input ports to allow analog audio from microphones or external cassette players to be fed to an analog-to-digital converter (ADC) in the card and the converted digital signal to be stored as a file on the hard disk via the bus connector of the card. During playback, the digital file is fed back to the card from the disk and gets converted to the analog form using a digital-to-analog converter (DAC) in the card, finally the analog signal is fed to a loudspeaker connected to the card via output ports for audio playback.

To encode the entire 20 kHz frequency range of human hearing, the sampling rate needs to be around 40 kHz. In practical systems, however, a slightly higher value of 44.1 kHz is used to produce high-quality digital audio e.g. audio-CD quality. The reason behind choosing this value is related to making it compatible to both the popular TV broadcasting standards NTSC and PAL and will be mentioned in the next chapter. Nowadays, **CD-quality** sound is synonymous with a sampling rate of 44,100 Hz, a bit-depth of 16 and 2 channels for stereo sound. This implies that the file size of 1 minute of stereo CD-quality sound is around 10 MB [(44,100×16×2×60) / (8×1024×1024)] although in practical applications this is reduced using compression algorithms. However, for digitization of sounds having smaller frequency ranges, a lower value of sampling rate and bit-depth can be used e.g. human speech can be encoded at 11 kHz, 8-bit, with a single channel.

Audio filtering forms an important step in audio processing applications, whereby specific frequencies in audio signals are attenuated or enhanced. Three common types of filters are low-pass, high-pass, and band-pass. **Low-pass** filters allow frequencies below a certain threshold to pass through but block higher frequencies, **high-pass** filters allow frequencies above a certain threshold to pass through but block lower frequencies, while **band-pass** filters allow frequencies within a range specified by a minimum value and maximum value to pass through but block all other frequencies. Filters are implemented by converting the audio signal from the time domain to frequency domain using **Fourier Transform**, named after the French mathematician Jean-Baptiste Joseph Fourier (Fourier, 1822). This transform enables the signal to be represented as a weighted sum of frequency components which can then be selectively processed to suit specific requirements. Another class of digital audio applications requiring frequency-based modulations is called MIDI. **MIDI**, which is a short form for Musical Instrument Digital Interface, is a protocol for connecting digital musical instruments like synthesizers to computers. Synthesizers are broadly of two types: **FM synthesizers** combine elementary tones to build up composite notes having a desired waveform, while **wavetable synthesizers** manipulate the amplitude and frequencies of stored digital recordings of actual instruments using filters.

Like images, compression schemes called CODEC (coder/decoder) can be used to reduce the size of digital audio files. Lossless compression algorithms manipulate sample values so that they can be stored using less space without making permanent changes to the file. Lossy compression algorithms delete information from the file to reduce its size and hence degrades its quality but offers much higher amounts of compression to be achieved. **File formats** in which the audio is saved depend on the compression scheme used. The Windows native audio file format is WAV which is typically uncompressed. Lossless compression algorithms can save digital audio in file formats like FLAC (Free Lossless Audio CODEC), ALAC (Apple Lossless Audio CODEC), and MPEG-4 SLS (Scalable Lossless), while lossy compression algorithms are associated with file formats like MP3 (MPEG Audio Layer 3), WMA (Window Media Audio), and RA (Real Audio).

2.2 Toolboxes and Functions

Audio processing functions in MATLAB® can be divided into two broad categories: basic MATLAB (BM) functions and Audio System Toolbox (AST) functions. The BM functions are a set of basic tools used for performing preliminary mathematical matrix operations and graphical plotting operations. The AST provides algorithms and tools for the design, simulation, and desktop prototyping of audio processing systems and includes a set of more advanced tools for specialized processing tasks like compressor, expander, noise gate, spectrum analyzer, and synthesizer. A small number of functions have been taken from two other toolboxes: DSP System Toolbox (DSPST) which provides algorithms, apps, and scopes for designing, simulating, and analyzing signal processing systems, and the *Signal Processing Toolbox* (SPT), which provides functions and apps to analyze, preprocess, and extract features from uniformly and nonuniformly sampled signals. Some functions like import/export and basic playback, however, are common to both BM set and toolbox collections. For solution of a specific task at hand, functions from both the BM set and the toolbox sets might be necessary. The source of these functions are accordingly mentioned as and when they are used to illustrate examples throughout this book. The MATLAB features for audio processing tasks are illustrated as solutions to specific examples throughout this chapter. This book has been written using MATLAB version 2018b; however, most of the functions can be used with versions 2015 and later with little or no change. The audio file formats supported by MATLAB includes WAVE (.wav), OGG (.ogg), FLAC (.flac), AIFF (.aif), MP3 (.mp3), and MPEG-4 AAC (.m4a, mp4). Most of the audio files used in the examples are included in the MATLAB package and do not need to be supplied by the user. The audio samples used in this book are available in the following folder: *(matlab-root)/toolbox/audio/samples/*.

2.2.1 Basic MATLAB® (BM) Functions

The BM functions used in this chapter fall into five different categories: Language Fundamentals, Mathematics, Graphics, Data Import and Analysis, and Programming Scripts and Functions. A list of these is provided below along with their hierarchical structure and a one-line description of each. The BM set consists of thousands of functions out of which a subset has been used in this chapter keeping in view the scope and coverage of this book.

1. Language Fundamentals:
1. cell: create cell array
2. clc: clear command window
3. floor: round toward negative infinity
4. length: length of largest array dimension
5. linspace: generate vector of evenly spaced values
6. numel: number of array elements
7. ones: create array of all ones
8. zeros: create array of all zeros

2. Mathematics
1. abs: absolute value
2. cos: cosine of argument in radians
3. fft: fast Fourier transform
4. filter: 1-D digital filter
5. imag: imaginary part of complex number
6. magic: magic square with equal row and column sums
7. real: real part of complex number
8. rng: control random number generation
9. sin: sine of argument in radians
10. sum: sum of elements

3. Graphics
1. axis: Set axis limits and aspect ratios
2. figure: Create a figure window
3. gca: Get current axis for modifying axes properties
4. grid: Display grid lines
5. legend: Add legend to axes
6. plot: 2-D line plot
7. subplot: Multiple plots in a single figure
8. title: Plot title
9. xlabel, ylabel: Label x-axis, y-axis

4. Data Import and Analysis
1. audioinfo: Information about audio files
2. audiodevinfo: Information about audio device
3. audioplayer: Create object for playing audio
4. audioread: Read audio files
5. audiorecorder: Create object for recording audio

6. audiowrite: Write audio files
7. clear: Remove items from workspace memory
8. disp: Display value of variable
9. getaudiodata: Store recorded audio in array
10. max: Maximum value
11. min: Minimum value
12. play: Play audio
13. recordblocking: Record audio holding control until recording completes
14. sound: Convert matrix to sound

5. *Programming Scripts and Functions*
   1. if ... end: Execute statements if condition is true
   2. continue: Pass control to next iteration of loop
   3. for ... end: for loop to repeat specified number of times
   4. pause: Stop MATLAB execution temporarily
   5. while ... end: loop while until condition is true

6. *Advanced Software Development*
   1. tic: Start stopwatch timer
   2. toc: Read elapsed time from stopwatch
   3. for ... end: for loop to repeat specified number of times
   4. pause: Stop MATLAB execution temporarily
   5. release: release resources
   6. while ... end: while loop until condition is true

2.2.2 Audio System Toolbox (AST) Functions

AST provides algorithms and tools for the design, simulation, and desktop prototyping of audio processing systems. AST includes libraries of audio processing algorithms for modifying digital audio signals. These are divided into a number of categories: (a) Audio I/O and Waveform Generation; (b) Audio Processing Algorithm Design; (c) Measurements and Feature Extraction; (d) Simulation, Tuning, and Visualization; and (e) MIDI. Some of the most commonly used functions from AST discussed in this chapter are listed below:

1. *Audio I/O and Waveform Generation:*
   1. getAudioDevices: List available audio devices
   2. audioPlayerRecorder: Simultaneously play and record using an audio device
   3. audioDeviceReader: Record from sound card
   4. audioDeviceWriter: Play to sound card
   5. audioOscillator: Generate sine, square, and sawtooth waveforms
   6. wavetableSynthesizer: Generate a periodic signal with tunable properties
   7. dsp.AudioFileReader: Stream from audio file
   8. dsp.AudioFileWriter: Stream to audio file

2. *Audio Processing Algorithm Design:*
   1. compressor: Dynamic range compressor
   2. expander: Dynamic range expander
   3. noiseGate: Dynamic range noise gate
   4. reverberator: Add reverberation to audio signal
   5. crossoverFilter: Audio crossover filter
   6. graphicEQ: Standard-based graphic equalizer

3. *Measurements and Feature Extraction:*
   1. integratedLoudness: Measure integrated loudness
   2. mfcc: Extract mel frequency cepstral coefficient (MFCC) from audio signal
   3. pitch: Estimate fundamental frequency of audio signal
   4. loudnessMeter: Standard-compliant loudness measurements
   5. voiceActivityDetector: Detect the presence of speech in audio signal
   6. cepstralFeatureExtractor: Extract cepstral features from audio segment

4. *Simulation, Tuning, and Visualization:*
   1. dsp.SpectrumAnalyzer: Display frequency spectrum of time-domain signals
   2. dsp.TimeScope: Time-domain signal display and measurement
   3. dsp.ArrayPlot: Display vectors or arrays

5. *Musical Instrument Digital Interface (MIDI):*
   1. getMIDIConnections: Get MIDI connections
   2. configureMIDI: Configure MIDI connections
   3. mididevice: Send and receive MIDI messages
   4. mididevinfo: MIDI device information
   5. midimsg: Create MIDI message
   6. midiread: Return most recent value of MIDI controls

2.2.3 DSP System Toolbox (DSPST) Functions

1. dsp.AudioFileReader: Stream from audio file
2. dsp.AudioFileWriter: Stream to audio file
3. dsp.TimeScope: Time-domain signal display and measurement
4. dsp.SineWave: Generate discrete sine wave
5. dsp.SpectrumAnalyzer: Display frequency spectrum of time-domain signals
6. dsp.ArrayPlot: Display vectors or arrays
7. fvtool: Visualize frequency response of DSP filters

2.2.4 Signal Processing Toolbox (SPT) Functions

1. designfilt: Design digital filters
2. spectrogram: Spectrogram using short-time Fourier transform
3. chirp: Swept-frequency cosine

4. dct: Discrete cosine transform

5. idct: Inverse discrete cosine transform

6. window: Create a window function of specified type

2.3 Sound Waves

As mentioned before, **sound waves** are mathematically represented using sinusoidal equations of the form $y = a.\sin(bt - \theta)$, where a is the amplitude, b is the frequency, and θ is the phase difference. A single frequency sound is called a **tone**, while composite sounds with multiple frequencies are called **notes**. Notes can be created by adding multiple tones having different amplitudes, frequencies, and phases. **Phase** is the angle ranging from 0 to 360 degrees or 0 to 2π radians over one full cycle of a periodic wave. Phase difference is the phase angle between the starting points of two waves and measures how much one wave is shifted with respect to the other. Although tones are represented by sine waves, notes can have different waveform shapes depending on the parameters of tonal components like amplitude, frequency, and phase shift. **Waveforms** are the pictorial representation of audio signals and physically indicate the timbre of sound, which enable us to distinguish sounds from different musical instruments. Different waveforms can be generated by combining sinusoidal tones in different proportions. In the following example, a sinusoidal tone of amplitude 1 and frequency 1 is added to another tone with amplitude 0.3 and frequency 3. The composite note does not have a sinusoidal waveform. The waveform is again seen to be different when the same tonal components are added using a phase difference of π (Figure 2.1).

Example 2.1: Write a program to display sinusoidal tones and composite notes

```
clear; clc;
t = 1 : 0.1 : 50;

subplot(411), plot(t, sin(t)); ylabel('sin(t)');
subplot(412), plot(t, 0.3*sin(3*t)); ylabel('0.3sin(3t)'); axis([0, 50, -1, 1]);
subplot(413), plot(t, sin(t) + 0.3*sin(3*t)); ylabel('sin(t) + 0.3sin(3t)');
subplot(414), plot(t, sin(t) + 0.3*sin(3*t - pi)); ylabel('sin(t) + 0.3sin(3t-\pi)');
```

An important theorem frequently used in digital signal processing is the **Nyquist Sampling Theorem** named after Harry Nyquist who proposed it while investigating the bandwidth requirements for transmitting information through a telegraph channel (Nyquist, 1928). **Sampling** is a step for converting a continuous analog signal to a discrete set of values by examining the magnitude of the signal at specific points in time or space. The number of times sampling is done per second is called the **sampling frequency**. The theorem states that "When converting an analog signal into digital form, the sampling frequency used must be at least twice the bandwidth of the input signal in order to be able to reconstruct the original signal accurately from the sampled version." For a baseband signal, the bandwidth is the same as the highest frequency component, and so the theorem is often stated as the sampling frequency needs to be twice the highest frequency on the input signal. This condition is often referred to as the Nyquist criterion. If the sampling frequency falls below this threshold, then the reconstruction of the signal exhibits imperfections or artifacts which is referred to as **aliasing** defects. The following example illustrates the

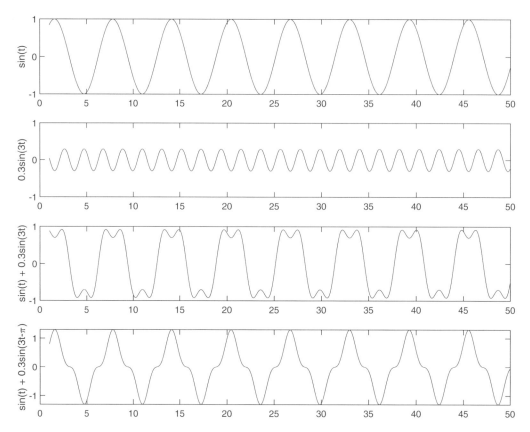

FIGURE 2.1
Output of Example 2.1.

Nyquist's theorem by progressively sampling a sinusoidal waveform at low frequencies to generate a staircase waveform. The first figure in the top row shows the original waveform, and the second row depicts the waveform after sampling more than ten times per cycle. The third row depicts the limiting condition when the original waveform is sampled at exactly twice the frequency i.e. two times per cycle at $\pi/2$, $3\pi/2$, $5\pi/2$, $7\pi/2$, $9\pi/2$, and $11\pi/2$ over a period of three cycles or 6π. This plot shows that since the sampling is done once in the positive half and once in the negative half of each cycle, the generated staircase waveform has at least some information to represent the two halves i.e. their approximate durations and approximate values. The fourth row depicts the case when sampling rate is equal to the frequency of the wave i.e. sampling is done once in each cycle. The resulting wave is a low-frequency flat wave which contains no information of the original wave. This distortion is the aliasing and occurs when the sampling rate falls below twice the wave frequency, which illustrates the Nyquist theorem. The BM function **stairs** is used to plot the staircase waveforms by joining the sampled points together (Figure 2.2).

Example 2.2: Write a program to demonstrate Nyquist Sampling Theorem.

```
clear; clc;
t0 = 0:0.1:6*pi;
t1 = 0:0.5:6*pi;
t2 = [pi/2, 3*pi/2, 5*pi/2, 7*pi/2, 9*pi/2, 11*pi/2];
```

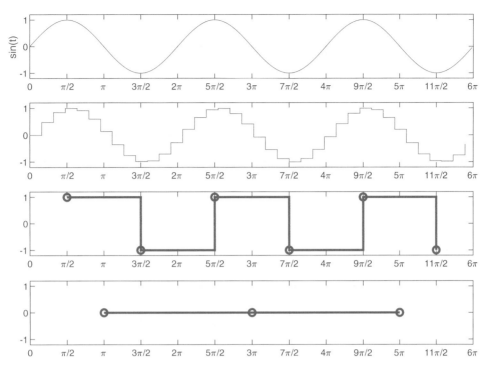

FIGURE 2.2
Output of Example 2.2.

```
t3 = [pi, 3*pi, 5*pi];
y1 = sin(t1);
y2 = sin(t2);
y3 = sin(t3);

subplot(411), plot(t0, sin(t0)); ylabel('sin(t)'); grid;
h = gca;
h.XTick = [0, pi/2, pi, 3*pi/2, 2*pi, 5*pi/2, 3*pi, ...
    7*pi/2, 4*pi, 9*pi/2, 5*pi, 11*pi/2, 6*pi];
h.XTickLabel = {'0', '\pi/2', '\pi', '3\pi/2', '2\pi', ...
    '5\pi/2', '3\pi', '7\pi/2', '4\pi', '9\pi/2', '5\pi', ...
    '11\pi/2', '6\pi'}
axis tight; axis ([0 6*pi -1.2 1.2]);

subplot(412), stairs(t1, y1); grid;
h = gca;
h.XTick = [0, pi/2, pi, 3*pi/2, 2*pi, 5*pi/2, 3*pi, ...
    7*pi/2, 4*pi, 9*pi/2, 5*pi, 11*pi/2, 6*pi];
h.XTickLabel = {'0', '\pi/2', '\pi', '3\pi/2', '2\pi', ...
    '5\pi/2', '3\pi', '7\pi/2', '4\pi', '9\pi/2', '5\pi', ...
    '11\pi/2', '6\pi'}
axis tight; axis ([0 6*pi -1.2 1.2]);

subplot(413), stairs(t2, y2, 'o-', 'LineWidth', 2); grid;
h = gca;
h.XTick = [0, pi/2, pi, 3*pi/2, 2*pi, 5*pi/2, 3*pi, ...
    7*pi/2, 4*pi, 9*pi/2, 5*pi, 11*pi/2, 6*pi];
h.XTickLabel = {'0', '\pi/2', '\pi', '3\pi/2', '2\pi', ...
    '5\pi/2', '3\pi', '7\pi/2', '4\pi', '9\pi/2', '5\pi', ...
    '11\pi/2', '6\pi'}
axis tight; axis ([0 6*pi -1.2 1.2]);

subplot(414), stairs(t3, y3, 'o-', 'LineWidth', 2); grid;
```

```
h = gca;
h.XTick = [0, pi/2, pi, 3*pi/2, 2*pi, 5*pi/2, 3*pi, ...
    7*pi/2, 4*pi, 9*pi/2, 5*pi, 11*pi/2, 6*pi];
h.XTickLabel = {'0', '\pi/2', '\pi', '3\pi/2', '2\pi', ...
    '5\pi/2', '3\pi', '7\pi/2', '4\pi', '9\pi/2', '5\pi', ...
    '11\pi/2', '6\pi'}
axis tight; axis ([0 6*pi -1.2 1.2]);
```

Another important theorem used frequently in signal processing is the **Fourier Theorem** named after French mathematician Jean-Baptiste Joseph (Fourier, 1822). The theorem proposes that any periodic signal can be represented as weighted sums of sinusoidal components. The summation of known weighted components to derive an arbitrary function is called Fourier *synthesis*, whereas the process of deriving the weights for describing a given function is called Fourier *analysis*. The collection of weighted sinusoidal terms is called the Fourier series. Some common periodic functions and their Fourier series are given below:

- Square wave: $\cos(t) - (1/3)\cos(3t) + (1/5)\cos(5t) - (1/7)\cos(7t) + \cdots$
- Triangular wave: $\cos(t) + (1/9)\cos(3t) + (1/25)\cos(5t) + \cdots$
- Sawtooth wave: $\sin(t) - (1/2)\sin(2t) + (1/3)\sin(3t) - (1/4)\sin(4t) + \cdots$
- Half-rectified wave: $1 + (\pi/2)\cos(t) + (2/3)\cos(2t) - (2/15)\cos(4t) + (2/35)\cos(6t)$;
- Full-rectified wave: $1 + (2/3)\cos(2t) - (2/15)\cos(4t) + (2/35)\cos(6t)$;
- Pulse wave: $1 + 2(f_1 \cdot \cos(t) + f_2 \cdot \cos(2t) + f_3 \cdot \cos(3t) + f_4 \cdot \cos(4t))$;

where $k < 1$, and $f_1 = (\sin(\pi k))/(\pi k)$, $f_2 = (\sin(2\pi k))/(2\pi k)$, $f_3 = (\sin(3\pi k))/(3\pi k)$, $f_4 = (\sin(4\pi k))/(4\pi k)$

The following example illustrates Fourier synthesis of some common waveforms from weighted sinusoidal components using the relations given above (Figure 2.3).

Example 2.3: Write a program to generate some common waveforms using Fourier series.

```
clear; clc;
t = 0:0.1:30;
% part a
figure,
a = cos(t) - (1/3)*cos(3*t) + (1/5)*cos(5*t) - (1/7)*cos(7*t) ;
b = cos(t) + (1/9)*cos(3*t) + (1/25)*cos(5*t) + (1/49)*cos(7*t);
c = sin(t) - (1/2)*sin(2*t) + (1/3)*sin(3*t) - (1/4)*sin(4*t);
subplot(311), plot(t, a); title('Square wave'); grid;
subplot(312), plot(t, b); title('Triangular wave'); grid;
subplot(313), plot(t, c); title('Sawtooth wave'); grid;
% part b
figure,
k = 0.3; f1 = (sin(pi*k))/(pi*k); f2 = (sin(2*pi*k))/(2*pi*k);
f3 = (sin(3*pi*k))/(3*pi*k); f4 = (sin(4*pi*k))/(4*pi*k);
d = 1 + (pi/2)*cos(t) + (2/3)*cos(2*t) - (2/15)*cos(4*t) + (2/35)*cos(6*t);
e = 1 + (2/3)*cos(2*t) - (2/15)*cos(4*t) + (2/35)*cos(6*t);
f = 1 + 2*(f1*cos(t) + f2*cos(2*t) + f3*cos(3*t) + f4*cos(4*t));
subplot(311), plot(t, d); title('Half rectified wave'); grid;
subplot(312), plot(t, e); title('Full-rectified wave'); grid;
subplot(313), plot(t, f); title('Pulse wave'); grid;
```

The accuracy of the signals improve incrementally as more terms are added to the series. So a loop can be designed to include an arbitrary number of terms, as illustrated in the following example for the sawtooth waveform. The waveforms are plotted by increasing the

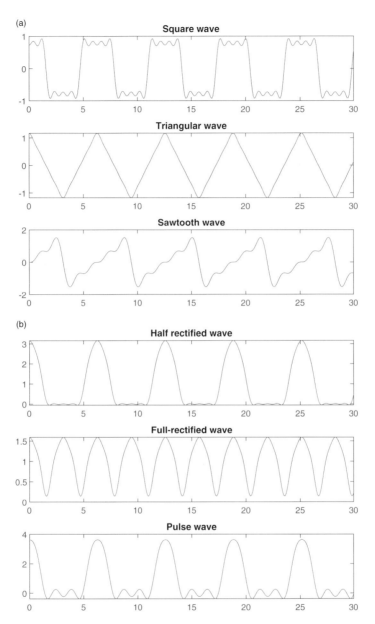

FIGURE 2.3
Output of Example 2.3 (part a and part b).

number of components in the series from 3 to 5 to 10 and finally 50. The inclusion of the negative sign for alternate terms serves to reverse the direction of the sawtooth (Figure 2.4).

Example 2.4: Write a program to illustrate the accuracy of the waveforms generated depends on the number of terms included in the Fourier series.

```
clear; clc;
t = 0:0.1:30;
```

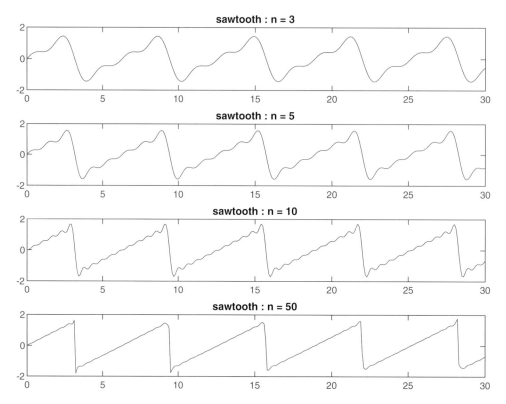

FIGURE 2.4
Output of Example 2.4.

```
sum = 0; n = 3;
for i=1:n, sum = sum - ((-1)^i)*(1/i)*sin(i*t); end;
subplot(411), plot(t, sum); title('sawtooth : n = 3');

sum = 0; n = 5;
for i=1:n, sum = sum - ((-1)^i)*(1/i)*sin(i*t); end;
subplot(412), plot(t, sum); title('sawtooth : n = 5');

sum = 0; n = 10;
for i=1:n, sum = sum - ((-1)^i)*(1/i)*sin(i*t); end;
subplot(413), plot(t, sum); title('sawtooth : n = 10');

sum = 0; n = 50;
for i=1:n, sum = sum - ((-1)^i)*(1/i)*sin(i*t); end;
subplot(414), plot(t, sum); title('sawtooth : n = 50');
```

It has already been discussed in Chapter 1 that a time-domain digital signal x is transformed to the frequency domain form X using Discrete Fourier Transform (DFT), the output represents a set of complex **DFT coefficients** which are scaling factors to a set of unit magnitude cosine and sine waves called basis functions. If $A(k)$ and $B(k)$ represent the real and imaginary parts of the k-th coefficient, then their magnitudes are given by the following expressions, where N is the total number of samples in:

$$A(k) = Re\{X(k)\} = \sum_{i=0}^{N-1} x(i) \cdot \cos\left(\frac{2\pi ki}{N}\right)$$

$$B(k) = Im\{X(k)\} = \sum_{i=0}^{N-1} x(i) \cdot \sin\left(\frac{2\pi ki}{N}\right)$$

Collecting all the terms as k ranges from 0 to $(N-1)$, we get:

$$A = \{A(0), A(1), A(2), \ldots, A(N-1)\}$$

$$B = \{B(0),\ B(1), B(2), \ldots, B(N-1)\}$$

The **DFT basis functions** are sinusoidal waves having pre-defined frequencies. The k-th cosine wave $C(k)$ and the k-th sine wave $S(k)$ are collection of terms represented by the following expressions:

$$C(k) = \bigcup_{i=0}^{N-1} \cos\left(\frac{2\pi ki}{N}\right)$$

$$S(k) = \bigcup_{i=0}^{N-1} \sin\left(\frac{2\pi ki}{N}\right)$$

Collecting all the terms as k ranges from 0 to $(N-1)$ we get:

$$C = \{C(0), C(1), C(2), \ldots, C(N-1)\}$$

$$S = \{S(0), S(1), S(2), \ldots, S(N-1)\}$$

It is to be noted that both the cosine and sine terms have amplitudes ranging from +1 to –1, and their frequencies depend on the value of N. So the basis functions are essentially independent of the original time-domain signal x and depend only on the total number of samples N in x (but not on the data values of x). The figure below shows the basis functions for $N = 8$, the C terms depicting the cosine functions and S terms depicting the sine functions. Note that $C(0)$ has a constant value of 1 and $S(0)$ has a constant value of 0. Also, $C(k)$ and $S(k)$ complete k cycles of the wave within 0 to $N–1$ samples (Figure 2.5).

In the following example, a square wave is digitized using 200 samples and it is transformed to the frequency domain using DFT. The real and imaginary parts of the coefficients A and B are extracted and plotted along with the signal waveform. The cosine and sine basis functions are then computed, each of which has 200 components since there are 200 samples, out of which eight components are displayed: 1, 2, 3, 4, 5, 10, 50, 100 (Figure 2.6).

Example 2.5: Write a program to decompose a signal into its coefficients and basis functions.

```
clear; clc;

t = 0 : 0.1 : 20;
x = sin(t) + (1/3)*sin(3*t) + (1/5)*sin(5*t) + (1/7)*sin(7*t);
% square wave
N = numel(x);
X = fft(x);
```

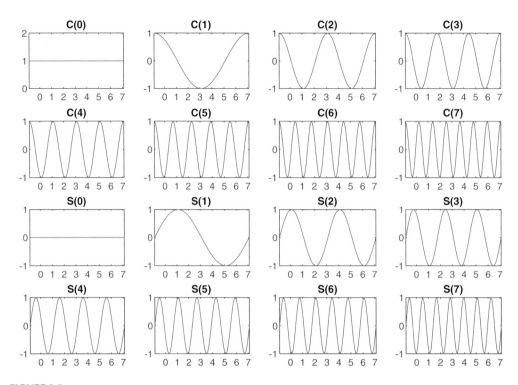

FIGURE 2.5
Basis functions.

```
A = real(X);
B = -imag(X);

figure,
subplot(311), plot(x, 'LineWidth', 2); axis tight; title('x');
subplot(312), plot(A); axis tight; title('A');
subplot(313), plot(B); axis tight; title('B');

C=cell(1,N);
for k=0:N-1
    for i=0:N-1
        C{k+1} = [C{k+1}, cos(2*pi*i*k/N)];
    end
end

S=cell(1,N);
for k=0:N-1
    for i=0:N-1
        S{k+1} = [S{k+1}, sin(2*pi*i*k/N)];
    end
end

figure,
subplot(441), plot(C{1}); axis tight; title('C (1)');
subplot(442), plot(C{2}); axis tight; title('C (2)');
subplot(443), plot(C{3}); axis tight; title('C (3)');
subplot(444), plot(C{4}); axis tight; title('C (4)');
subplot(445), plot(C{5}); axis tight; title('C (5)');
subplot(446), plot(C{10}); axis tight; title('C (10)');
subplot(447), plot(C{50}); axis tight; title('C (50)');
subplot(448), plot(C{100}); axis tight; title('C (100)');

subplot(449), plot(S{1}); axis tight; title('S (1)');
subplot(4,4,10), plot(S{2}); axis tight; title('S (2)');
```

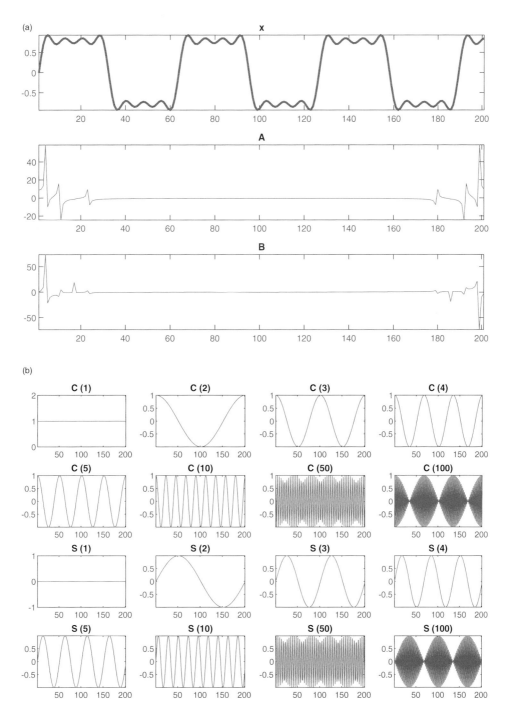

FIGURE 2.6
Output of Example 2.5 (a) Weighted coefficients (b) Basis functions.

```
subplot(4,4,11), plot(S{3}); axis tight; title('S (3)');
subplot(4,4,12), plot(S{4}); axis tight; title('S (4)');
subplot(4,4,13), plot(S{5}); axis tight; title('S (5)');
subplot(4,4,14), plot(S{10}); axis tight; title('S (10)');
subplot(4,4,15), plot(S{50}); axis tight; title('S (50)');
subplot(4,4,16), plot(S{100}); axis tight; title('S (100)');
```

The Fourier theorem states that the original time-domain signal can be represented as a weighted sum of the sinusoidal terms. It has already been mentioned above that the coefficients act as scaling factors for the basis functions. This means that if the real part of the coefficients (i.e. A terms) be multiplied with the cosine basis functions (i.e. C terms) while the imaginary part of the coefficients (i.e. B terms) be multiplied with the sine functions (i.e S terms) and these two products be added up together and normalized by dividing by the total number of terms N, then the resulting values should be the same as the data values of the original signal x. Mathematically:

$$xr = \left(\frac{1}{N}\right) \cdot A \cdot C^T = \left(\frac{1}{N}\right) \cdot \left[A(0) \cdot C(0) + \cdots + A(N-1) \cdot C(N-1)\right]$$

$$xi = \left(\frac{1}{N}\right) \cdot B \cdot S^T = \left(\frac{1}{N}\right) \cdot \left[B(0) \cdot S(0) + \cdots + B(N-1) \cdot S(N-1)\right]$$

$$x = xr + xi$$

In the following example, the coefficients are multiplied with the basis functions and added up and normalized to generate the estimated reconstructed wave. Since each sinusoidal wave consists of multiple values, a cell structure is used to store them. An error value is computed by subtracting the estimated reconstructed signal from the original signal to show the difference between them. The red line shows that the error is effectively zero (Figure 2.7).

Example 2.6: Write a program to illustrate that for the previous example, the signal can be reconstructed by adding up its weighted Fourier components.

```
clear; clc;

t = 0 : 0.1 : 20;
x = sin(t) + (1/3)*sin(3*t) + (1/5)*sin(5*t) + (1/7)*sin(7*t);
% square wave
N = numel(x);
X = fft(x);          % Fourier transform
A = real(X);         % Real coefficients
B = -imag(X);        % Imaginary coefficients

% cosine basis functions
C = cell(1,N);
for k = 0:N-1
    for i = 0:N-1
        C{k+1} = [C{k+1}, cos(2*pi*i*k/N)];
    end
end

% sine basis functions
S = cell(1,N);
for k = 0:N-1
    for i = 0:N-1
        S{k+1} = [S{k+1}, sin(2*pi*i*k/N)];
    end
end
```

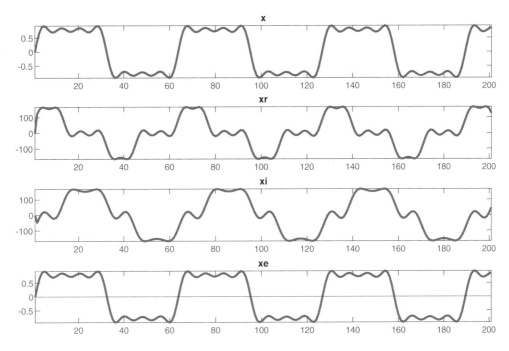

FIGURE 2.7
Output of Example 2.6.

```
% sum of weighted real components
xr = cell(1,N); sumxr = 0;
for i = 0:N-1
    xr{i+1} = A(i+1)*C{i+1};
    sumxr = sumxr + xr{i+1};
end

% sum of weighted imaginary components
xi = cell(1,N); sumxi = 0;
for i = 0:N-1
    xi{i+1} = B(i+1)*S{i+1};
    sumxi = sumxi + xi{i+1};
end

xe = (1/N)*(sumxr + sumxi);      % estimated reconstruction
err = (x - xe);                  % error

figure,
subplot(411), plot(x, 'LineWidth', 2); title('x'); axis tight;
subplot(412), plot(sumxr, 'LineWidth', 2); title('xr'); axis tight;
subplot(413), plot(sumxi, 'LineWidth', 2); title('xi'); axis tight;
subplot(414), plot(xe, 'LineWidth', 2);
hold on; plot(err); title('xe'); axis tight; hold off;
```

In the previous example, the error between the original signal and the reconstructed signal is seen to be zero. This is true since all the coefficients and basis functions have been utilized for reconstruction. However, if the number of reconstruction components are reduced, then the difference between the signals are increased which is reflected in the waveforms as well as in the computed sum square error. In the following example, the waveform of the previous example is reconstructed using variable number of components 3, 10, 50, 100, 201 and the plots of the waveforms are displayed along with the sum square error printed on top of each (Figure 2.8).

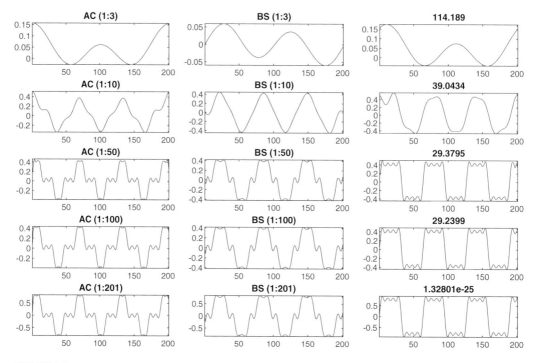

FIGURE 2.8
Output of Example 2.7.

Example 2.7: Write a program to illustrate that for the previous example, the signal can be reconstructed by using a variable number of components and compute the error in each case.

```
clear; clc;

t = 0 : 0.1 : 20;
x = sin(t) + (1/3)*sin(3*t) + (1/5)*sin(5*t) + (1/7)*sin(7*t); % square wave
N = numel(x);
X = fft(x);      % Fourier transform
A = real(X);     % Real coefficients
B = -imag(X);    % Imaginary coefficients

% cosine basis functions
C = cell(1,N);
for k = 0:N-1
    for i = 0:N-1
        C{k+1} = [C{k+1}, cos(2*pi*i*k/N)];
    end
end

% sine basis functions
S = cell(1,N);
for k = 0:N-1
    for i = 0:N-1
        S{k+1} = [S{k+1}, sin(2*pi*i*k/N)];
    end
end

% sum of weighted real components
xr = cell(1,N); sumxr = 0;
for i = 0:N-1
    xr{i+1} = A(i+1)*C{i+1};
    sumxr = sumxr + xr{i+1};
end
```

```
% sum of weighted imaginary components
xi = cell(1,N); sumxi = 0;
for i = 0:N-1
    xi{i+1} = B(i+1)*S{i+1};
    sumxi = sumxi + xi{i+1};
end

figure,
ac = 0; bs = 0; nc = 3; % number of components
for i = 1:nc
    ac = ac + xr{i};
    bs = bs + xi{i};
end

xe = (1/N)*(ac + bs);      % estimated reconstruction
se = sumsqr(x - xe);       % error

subplot(531), plot(ac/N); axis tight; title('AC (1:3)');
subplot(532), plot(bs/N); axis tight; title('BS (1:3)');
subplot(533), plot(xe);   axis tight; title(se);

ac = 0; bs = 0; nc = 10; % number of components
for i = 1:nc
    ac = ac + xr{i};
    bs = bs + xi{i};
end

xe = (1/N)*(ac + bs);      % estimated reconstruction
se = sumsqr(x - xe);       % error

subplot(534), plot(ac/N); axis tight; title('AC (1:10)');
subplot(535), plot(bs/N); axis tight; title('BS (1:10)');
subplot(536), plot(xe);   axis tight; title(se);

ac  =0; bs = 0; nc = 50; % number of components
for i = 1:nc
    ac = ac + xr{i};
    bs = bs + xi{i};
end

xe = (1/N)*(ac + bs);      % estimated reconstruction
se = sumsqr(x - xe);       % error

subplot(537), plot(ac/N); axis tight; title('AC (1:50)');
subplot(538), plot(bs/N); axis tight; title('BS (1:50)');
subplot(539), plot(xe);   axis tight; title(se);

ac = 0; bs = 0; nc = 100; % number of components
for i = 1:nc
    ac = ac + xr{i};
    bs = bs + xi{i};
end

xe = (1/N)*(ac + bs);      % estimated reconstruction
se = sumsqr(x - xe);       % error

subplot(5,3,10), plot(ac/N); axis tight; title('AC (1:100)');
subplot(5,3,11), plot(bs/N); axis tight; title('BS (1:100)');
subplot(5,3,12), plot(xe);   axis tight; title(se);

ac = 0; bs = 0; nc = 201; % number of components
for i = 1:nc
    ac = ac + xr{i};
    bs = bs + xi{i};
end

xe = (1/N)*(ac + bs);      % estimated reconstruction
se = sumsqr(x - xe);       % error

subplot(5,3,13), plot(ac/N); axis tight; title('AC (1:201)');
subplot(5,3,14), plot(bs/N); axis tight; title('BS (1:201)');
subplot(5,3,15), plot(xe);   axis tight; title(se);
```

2.4 Audio I/O and Waveform Generation

The first step in digital audio processing is to read the audio file and play it back. The BM function **audioread** is used to read audio from a specified file and returns the data matrix and the sample rate of the audio. The BM function **sound** is used to play the audio by specifying the data matrix and the sample rate. The BM function **audioinfo** displays information about the audio file like sample rate, total number of samples, duration in seconds, bits per sample, number of channels, and compression method. To plot the audio waveform, the BM function **plot** can be used. The function plots the data against the number of samples. The following example reads an audio file into the system, displays information about it, plays it back, and plots the audio waveform (Figure 2.9).

Example 2.8: Write a program to read an audio file, display information about it, play it and plot its waveform.

```
clear; clc;
f = 'Counting-16-44p1-mono-15secs.wav';
[x, fs] = audioread(f);
sound(x, fs);
audioinfo (f)
figure, plot(x);
xlabel('Samples'); ylabel('Amplitude'); axis tight;
```

By default, the waveform is plotted against the number of samples. To plot the waveform against time, the number of samples is divided by the sampling rate. The amplitude of the audio samples can be normalized within the range [0,1] by subtracting the minimum value from the samples and dividing the values by the difference between the maximum and minimum value (Figure 2.10).

Example 2.9: Write a program to normalize values of audio samples within the range [0, 1] and plot the audio waveform against the time duration in seconds

```
clear clc;
f = 'JetAirplane-16-11p025-mono-16secs.wav';
[x, fs] = audioread(f);
subplot(211), plot(x);
xlabel('Samples'); ylabel('Amplitude'); axis tight;

t = [1:length(x)]/fs;    % converting samples to secs
xn = (x - min(x))/abs(max(x) - min(x)); % normalize sample values
```

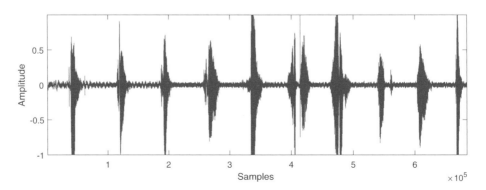

FIGURE 2.9
Output of Example 2.8.

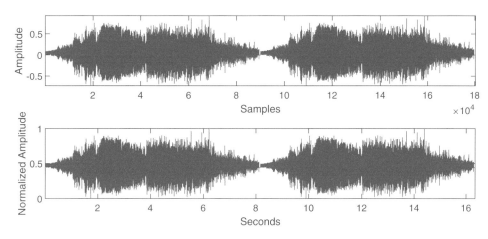

FIGURE 2.10
Output of Example 2.9.

```
subplot(212), plot(t,xn);
xlabel('Seconds'); ylabel('Normalized Amplitude'); axis tight;
```

The input and output devices available for reading and writing audio can be listed out using the BM function **audiodevinfo**

```
clear clc;
d = audiodevinfo;
fprintf('\n Input Devices : \n');
d.input(1)
d.input(2)
fprintf('\n Output Devices : \n');
d.output(1)
d.output(2)
```

The BM function **audioplayer** creates a player object for playing audio, given the specified data and sample rate. The BM function **play** plays the data from the player object. In the following example, a player object is used to playback an audio file.

> **Example 2.10: Write a program to create a player object and use it to play audio from a file.**
>
> ```
> clear clc;
> f = 'Counting-16-44p1-mono-15secs.wav';
> [x, fs] = audioread(f);
> player = audioplayer(x, fs);
> play(player);
> ```

The BM function **audiorecorder** creates an audio object for recording sound using the specified sample rate, bit-depth, and number of channels. The BM function **recordblocking** holds control for the audiorecorder object to record audio for the specified duration in seconds. The BM function **getaudiodata** stores the recorded audio signal in a numeric array. By default, the data type of the numeric array is of datatype *double* and the values of the audio samples range from –1 to +1. The BM function **audiowrite** is used to write the audio data to a specified filename using a specified sample rate. The file formats supported are mentioned at the beginning of this chapter. In the following example, an audiorecorder object is created to record 5 seconds of speech from the microphone, plays back the sound, and plots its waveform.

Example 2.11: Write a program to record audio from an input device and store it in a file.

```
clear; clc;
fs = 44100; bs = 16; ch = 1;
ar = audiorecorder(fs, bs, ch);
get(ar);

% Record your voice for 5 seconds.
disp('Start speaking.')
recordblocking(ar, 5);
disp('End of Recording.');
pause(10);

% Play back the recording.
play(ar);

% Plot the waveform.
r = getaudiodata(ar);
plot(r);

% Write audio to file
audiowrite('myrec.mp4', r, fs);
```

In addition to the BM functions, the AST also provides functions to read and write digital audio files. The AST function **audioDeviceReader** invokes a system object to read audio samples from an audio input device. The AST function **audioDeviceWriter** invokes a system object to write audio samples to an audio output device. The AST function **getAudioDevices** lists all available audio devices.

```
r = audioDeviceReader;
rdevices = getAudioDevices(r)
w = audioDeviceWriter;
wdevices = getAudioDevices(w)
```

The output from the above functions should be similar to that shown below:

```
rdevices =
    {'Default'}
    {'Primary Sound Capture Driver'}
    {'Microphone (High Definition Audio Device)'}
wdevices =
    {'Default'}
    {'Primary Sound Driver'}
    {'Headphones (High Definition Audio Device)'}
```

The *Digital Signal Processing System Toolbox* (DSPST) function **dsp.AudioFileReader**, which is also included in the AST, reads audio samples from an audio file. The DSPST function **dsp.AudioFileWriter** writes audio samples to an audio file. In the following example, audio is recorded from a default input device (microphone) for 10 seconds and written to a file. The file is then read, and the audio is written to the default output device (speakers).

Example 2.12: Write a program to use a DeviceReader to read audio from microphone, a FileWriter to write it to a file, a FileReader to read it from the file and a DeviceWriter to send it to the speakers for playback.

```
clear; clc;
dr = audioDeviceReader;        % create device reader
fn = 'myaudio.wav';
fw = dsp.AudioFileWriter(fn, 'FileFormat', 'WAV'); % create file writer
```

```
setup(dr);                    % initialize reader
disp('Speak into microphone now');
tic;
while toc < 10
    fw(record(dr));           % record from mic and write to file
end
release(dr);                  % release device reader
release(fw);                  % release file writer
disp('Recording complete');

fr = dsp.AudioFileReader(fn);    % create file reader
f = audioinfo(fn);
dw = audioDeviceWriter('SampleRate', f.SampleRate); % create device writer
setup(dw, zeros(fr.SamplesPerFrame, f.NumChannels)); % initialize writer
while ~isDone(fr)
    audioData = fr();         % read audio from file
    dw(audioData);            % write audio to speakers
end
release(fr);                  % release file reader
release(dw);                  % release device writer
```

An **oscillator** is an electronic device that generates periodic oscillating signals of specified amplitude and frequency. An audio oscillator produces signals with frequencies in the audio range i.e. 20 Hz–20kHz. Common waveforms generated are often sine waves or square waves. The AST function **audioOscillator** is used to display a specified signal in a time scope display. In the following example, a square wave of frequency 1000 and a sine wave of frequency 100 are added up to produce a resultant, whose waveform is displayed (Figure 2.11).

Example 2.13: Write a program to create an audioOscillator object to display a sine wave and a square wave along with their resultant.

```
clear; clc;
osc1 = audioOscillator('square',1000);
w1 = osc1();
osc2 = audioOscillator('sine',100);
```

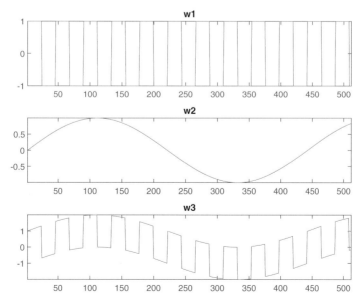

FIGURE 2.11
Output of Example 2.13.

```
w2 = osc2();
w3 = w1 + w2;

figure,
subplot(311), plot(w1); title('w1'); axis tight;
subplot(312), plot(w2); title('w2'); axis tight;
subplot(313), plot(w3); title('w3'); axis tight;
```

A **synthesizer** is an electronic musical instrument used to generate a variety of sounds by modulating sound waves with filters, envelopes, and oscillators. A synthesizer can broadly be of two types: the **FM synthesizer** selectively adds tonal components to produce complex waveforms while the **wavetable synthesizer** uses a storehouse of pre-recorded sounds whose amplitudes and frequencies can be controlled using oscillators. Moreover, digital interpolation techniques can be used between different waveforms to produce smooth and dynamic changes in pitch and timbre of the generated sounds. The AST function **wavetableSynthesizer** can be used to modify the frequency of an audio signal. In the following example, *syn* is a *wavetableSynthesizer* object which is used to generate an audio waveform whose frequency is at first increased and then decreased (Figure 2.12).

Example 2.14: Write a program to create a wavetableSynthesizer object for modifying frequency of a sound.

```
clear; clc;
fn = 'RockGuitar-16-44p1-stereo-72secs.wav';
[x,fs] = audioread(fn);

y = x(1:441000);

t = numel(y)/fs;
freq = 1/t;
```

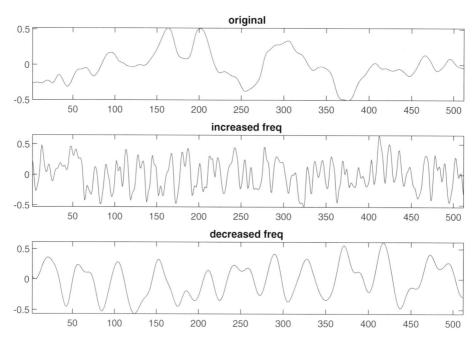

FIGURE 2.12
Output of Example 2.14.

```
syn = wavetableSynthesizer('Wavetable',y,'SampleRate',fs, 'Frequency',freq);
w = audioDeviceWriter('SampleRate',fs);

for i = 1:.01:10
    audioWave = syn();
    w(audioWave);
end
figure,
subplot(311),plot(audioWave); title('original'); axis tight;

freq1 = freq + 0.1;
syn = wavetableSynthesizer('Wavetable',y,'SampleRate',fs, 'Frequency',freq1);
for i = 1:.01:1/freq1
    audioWave = syn();
    w(audioWave);
end
subplot(312),plot(audioWave); title('increased freq'); axis tight;

freq2 = freq - 0.05;
zsyn = wavetableSynthesizer('Wavetable',y,'SampleRate',fs, 'Frequency',freq2);
for i = 1:.01:1/freq2
    audioWave = syn();
    w(audioWave);
end
subplot(313),plot(audioWave); title('decreased freq'); axis tight;
```

2.5 Audio Processing Algorithm Design

Since the characteristics of an audio signal change over time, the first step of audio processing involves segmenting the entire audio signal into audio frames and then processing each frame individually. The following example splits an audio file into non-overlapping frames of 0.5 second duration each and displays the first frame, last frame, and middle frame. The number of samples is divided by the sampling rate (fs) to return the duration in seconds ($t1$) which is then rounded to an integer number ($t2$). The newly computed number of samples (len_x) is used to isolate the relevant samples (new_x) from the original file. The number of samples in each audio frame (e) is used to calculate the number of audio frames (f) (Figure 2.13).

> **Example 2.15: Write a program to partition an audio signal into a number of frames each of 0.5 second duration. Display the first frame, middle frame and last frame.**

```
clear; clc;
fn = 'Counting-16-44p1-mono-15secs.wav';
[x, fs] = audioread(fn);
t1 = length(x)/fs;
t2 = floor(t1);
len_x = t2*fs;
new_x = x(1:len_x);

e = 0.5*fs;                    % samples per audio frame
f = floor(len_x/e);  % Total no. of audio frames/file
t = [1:len_x]/fs;

for c = 0 : f-1                     % do for all frames
    seg{c+1} = new_x(c*e+1:(c+1)*e);
% x represents collection of samples in each frame
end

subplot(2,3,[1:3]), plot(t,new_x);
```

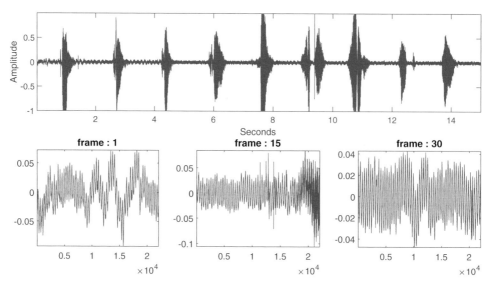

FIGURE 2.13
Output of Example 2.15.

```
xlabel('Seconds'); ylabel('Amplitude'); axis tight;
subplot(234), plot(seg{1}), axis tight; title('frame : 1');
subplot(235), plot(seg{round(f/2)}), axis tight;
title(['frame : ', num2str(round(f/2))]);
subplot(236), plot(seg{f}), axis tight;
title(['frame : ', num2str(f)]);
```

Reverberation is an acoustic phenomenon generated by multiple reflections or echoes of the original sound. It is created when a sound or signal is reflected multiple times which then gradually decay over a period of time. In music generation, the sound of reverberation is electronically added to the vocals of singers and original sounds of musical instruments to create effects of depth and space. The decay factor in reverberation decides how quickly the reflections die down or become zero. The AST function **reverberator** adds reverberation to mono or stereo audio signals. In the following example, reverberation of a specified decay factor is added to an audio signal and its waveform is plotted both before and after the modification (Figure 2.14).

Example 2.16: Write a program to generate reverberation of an audio signal and plot their waveforms

```
clear; clc;
fn = 'Counting-16-44p1-mono-15secs.wav';
r = dsp.AudioFileReader(fn);
fs = r.SampleRate;
w = audioDeviceWriter('SampleRate',fs);

while ~isDone(r)
    audio = r();
    w(audio);
end
release(r)

reverb = reverberator('DecayFactor',0.1);
while ~isDone(r)
    audio = r();
```

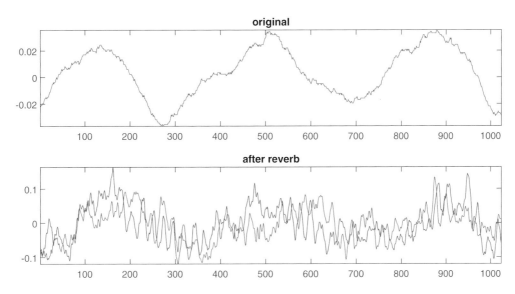

FIGURE 2.14
Output of Example 2.16.

```
    audior = reverb(audio);
    w(audior);
end
release(r)

figure,
subplot(211), plot(audio); axis tight; title('original');
subplot(212), plot(audior); axis tight; title('after reverb');
```

A **noise gate** is an electronic device used to attenuate a signal below a specified threshold. It is therefore frequently used to block unwanted noise below the threshold by *closing the gate* and allowing only the main sound above the level of the noise to pass through by *opening the gate*. Parameters used to control the opening and closing of the gate include the following: the *attack time* is the time it takes for the gate to change from closed to fully open condition, the *hold time* is the time the gate stays open even after the signal falls below the threshold, and the *release time* is the time it takes for the gate to change from open to fully closed condition. The AST function **noiseGate** is used to implement a dynamic gating object by suppressing signals below a given threshold, whose default value is −10 dB. In the following example, a audio signal is first corrupted by adding noise which is then reduced by using a noise gate (Figure 2.15).

Example 2.17: Write a program to remove noise from a signal using a noise gate.

```
clear; clc;
fn = 'Counting-16-44p1-mono-15secs.wav';
r = dsp.AudioFileReader(fn, 'SamplesPerFrame',1024);
fs = r.SampleRate;
w = audioDeviceWriter('SampleRate',fs);

while ~isDone(r)
    x = r();
    w(x);
end
release(r)
```

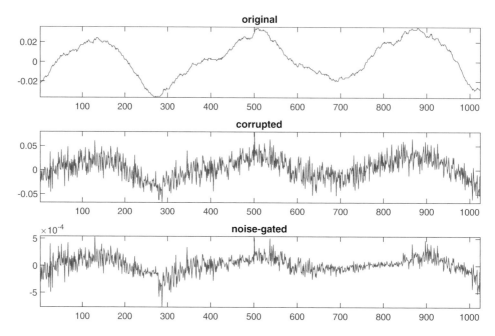

FIGURE 2.15
Output of Example 2.17.

```
while ~isDone(r)
    x = r();
    xc = x + (0.015)*randn(1024,1); %corrupted
    w(xc);
end
release(r)

g = noiseGate(-25, ...
    'AttackTime',0.01, ...
    'ReleaseTime',0.02, ...
    'HoldTime',0, ...
    'SampleRate',fs);

while ~isDone(r)
    x = r();
    xc = x + (0.015)*randn(1024,1); %corrupted
    y = g(xc);
    w(y);
end

release(r)
release(g)
release(w)

figure,
subplot(311), plot(x); axis tight; title('original');
subplot(312), plot(xc); axis tight; title('corrupted');
subplot(313), plot(y); axis tight; title('noise-gated');
```

The **dynamic range** of a signal is the ratio between the largest and smallest values. Signals with large dynamic range can be fitted in a narrower range of recording media using a process called dynamic range compression (DRC). A **dynamic range compressor** or simply compressor reduces the volume of large sounds and amplifies low sounds. Downward compression reduces loudness of sounds exceeding a certain threshold, while upward compression increases loudness of sounds below a certain threshold. Thresholds are specified in decibels, a lower threshold means more change, a higher threshold means less change. The amount of change is specified by a ratio e.g. a ratio of 3:1 means that if the input level is 3 dB above the threshold, then the output level is reduced to 1 dB over the threshold. The AST function **compressor** is used to implement a dynamic range compressor, which attenuates the volume of a loud sound that exceeds a specified threshold. The threshold value is specified in dB (–10 default), and the amount of compression is specified by a compression ratio (5 default). Given an input audio signal, the function returns the output compressed signal and the gain applied by the compressor in dB. The AST function **visualize** is used to graphically display the I/O characteristics of the compressor. In the following example, a compressor is defined with a specified threshold and ratio, its transfer function is visualized and then applied to an audio signal. Plots of the audio signal before and after the modification illustrate the change (Figure 2.16).

Example 2.18: Write a program to implement dynamic range compression of an audio signal using a specified threshold and ratio.

```
clear; clc;

frameLength = 1024; threshold = -30; cratio = 10;
fn = 'RockDrums-44p1-stereo-11secs.mp3';
r = dsp.AudioFileReader('Filename', fn, 'SamplesPerFrame', frameLength);
w = audioDeviceWriter('SampleRate',r.SampleRate);
c = compressor(threshold, cratio,'SampleRate',r.SampleRate);
visualize(c);
gg = zeros(1024,1); xx = zeros(1024,1); yy = zeros(1024,1);
while ~isDone(r)
    x = r();
    [y,g] = c(x);
    w(y);
    gg = [gg g(:,1)];
    xx = [xx x(:,1)];
    yy = [yy y(:,1)];
end
release(c)
release(w)

figure,
subplot(211), plot(xx(:)); hold on; plot(yy(:)); axis tight;
title('original & compressed signal');
subplot(212), plot(gg(:)); axis tight;
title('gain in dB');
```

A **dynamic range expander** is the opposite process of a compressor, i.e. it increases the level of an audio signal below a specified threshold. Apart from threshold and ratio, other parameters commonly used with compressors and expanders are attack time and release time. The *attack time* is the time period over which the compressor decreases the gain or the expander increases the gain to reach the final value specified by the ratio. The *release time* is the period over which the compressor increases the gain or the expander decreases the gain, once the input level falls below the threshold. The AST function **expander** is used to implement a dynamic range expander, which attenuates the volume of a quiet sound that is below a specified threshold. The threshold value is specified in dB (–10 default), and the amount of expansion is specified by a expansion ratio (5 default). Given an input audio signal, the function returns the output expanded signal and the gain applied by the expander in dB. In the following example, gain of the signal below a threshold of –40 dB increases by a factor of 10 i.e. as the input covers a range of 10 dB (–40 to –50), the output covers a range of 100 dB (–40 to –140) (Figure 2.17).

(a)

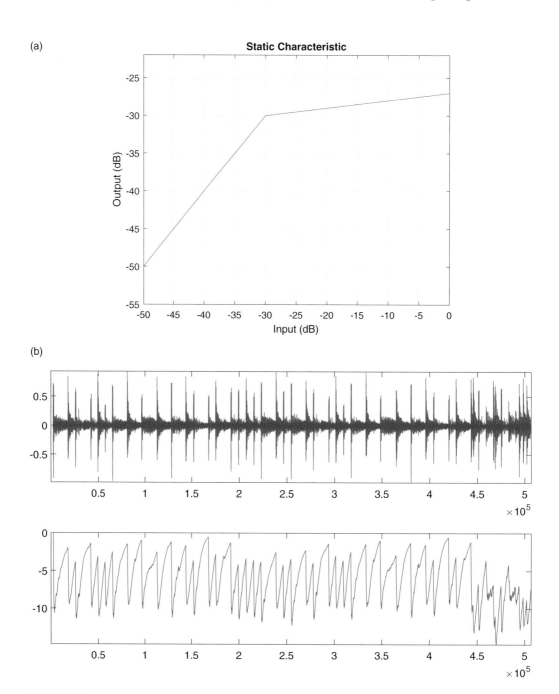

(b)

FIGURE 2.16
Output of Example 2.18 (a) filter (b) waveforms.

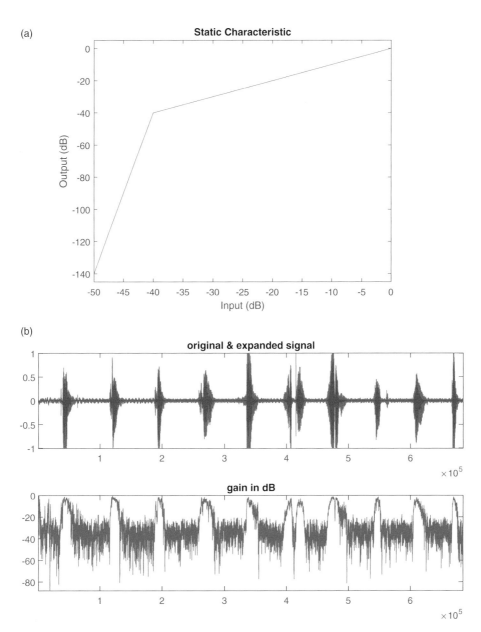

FIGURE 2.17
Output of Example 2.19 (a) filter (b) waveforms.

Example 2.19: Write a program to implement dynamic range expansion of an audio signal using a specified threshold and ratio.

```
clear; clc;

frameLength = 1024; threshold = -40; cratio = 10;
fn = 'Counting-16-44p1-mono-15secs.wav';
r = dsp.AudioFileReader('Filename', fn, 'SamplesPerFrame',frameLength);
w = audioDeviceWriter('SampleRate',r.SampleRate);
while ~isDone(r)
    x = r();
    xc = x + (1e-2/4)*randn(frameLength,1);
    w(xc);
end
release(r)

e = expander(-40,10, ...
    'AttackTime',0.01, ...
    'ReleaseTime',0.02, ...
    'HoldTime',0, ...
    'SampleRate',r.SampleRate);
visualize(e)
gg = zeros(1024,1); xx = zeros(1024,1); yy = zeros(1024,1);

while ~isDone(r)
    x = r();
    xc = x + (1e-2/4)*randn(frameLength,1);
    [y,g] = e(xc);
    w(y);
    gg = [gg g(:,1)];
    xx = [xx xc(:,1)];
    yy = [yy y(:,1)];
end

release(e)
release(w)

figure
subplot(311), plot(xx(.)); hold on; plot(yy(:)); axis tight;
title('original & expanded signal');
subplot(212), plot(gg(:)); axis tight;
title('gain in dB');
```

An audio **crossover filter** is an electronic filter used to split an audio signal into two or more frequency ranges. Crossover filters are usually two-way or three-way depending on number of frequency ranges. A common application of a three-way crossover filter is used in loudspeaker circuits where the audio frequency bands are routed to a woofer unit for handling low frequencies, a midrange unit for handling middle frequencies, and a tweeter unit for handling high frequencies since a single loudspeaker driver is not capable of covering the entire audio spectrum with acceptable quality. A two-way crossover filter consists of a low-pass and high-pass filter, while a three way crossover filter consists of a low-pass, band-pass, and high-pass filter. Parameters of a crossover filter include crossover frequencies, where the splits take place, and the slopes at which the filter gains increase or decrease. The AST function **crossoverFilter** implements an audio crossover filter which used an audio into two or more frequency bands. The following example creates a three band crossover filter with crossover frequencies at 100 Hz and 1000 Hz and a slope of 18 dB/octave and then applies it to an audio signal. The original signal along with the three bands is played back and plotted to display the result (Figure 2.18).

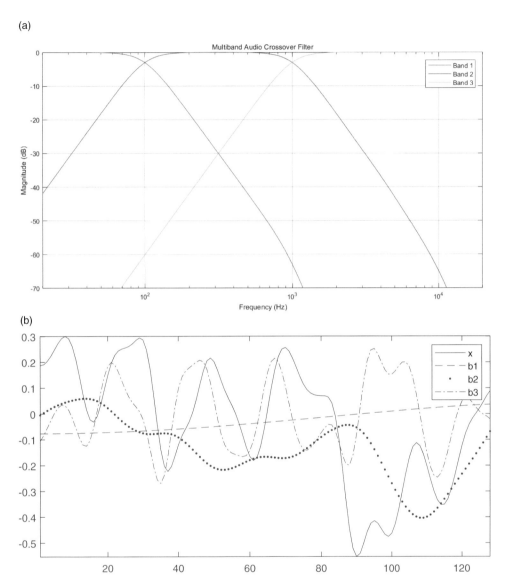

FIGURE 2.18
Output of Example 2.20 (a) filter (b) waveforms.

Example 2.20: Write a program to implement a three-way audio cross-over filter with specified cross-over frequencies and slopes.

```
clear; clc;

sf = 128; % Samples Per Frame
fn = 'RockGuitar-16-44p1-stereo-72secs.wav';
r = dsp.AudioFileReader(fn, 'SamplesPerFrame',sf);
fs = r.SampleRate;
w = audioDeviceWriter('SampleRate',fs);
fps = fs/sf; % frames per second
setup(r)
setup(w,ones(sf,2))

cf = crossoverFilter( ...
    'NumCrossovers',2, ...
    'CrossoverFrequencies',[100,1000], ...
    'CrossoverSlopes',18, ...
    'SampleRate',fs);

setup(cf,ones(sf,2))
visualize(cf)

fprintf('Original signal\n');
count = 0;
while count < fps*10
    x = r();
    w(x);
    count = count+1;
end
pause(5);

fprintf('Band 1\n');
count = 0;
while count < fps*10
    x = r();
    [band1,band2,band3] = cf(x);
    w(band1);
    count = count+1;
end
pause(5);

fprintf('Band 2\n');
count = 0;
while count < fps*10
    x = r();
    [band1,band2,band3] = cf(x);
    w(band2);
    count = count+1;
end
pause(5);

fprintf('Band 3\n');
count = 0;
while count < fps*10
    x = r();
    [band1,band2,band3] = cf(x);
    w(band3);
    count = count+1;
end

release(r)
release(cf)
release(w)
b1 = band1(:,1);
b2 = band2(:,1);
b3 = band3(:,1);

figure,
plot(x(:,1), 'b-'); hold on; plot(b1, 'r--');
plot(b2, 'r.'); plot(b3, 'r-.');
legend('x', 'b1', 'b2', 'b3'); axis tight;
```

2.6 Measurements and Feature Extraction

Pitch is a perceptual property of sound dependent on its frequency. Low-pitch sounds like drums have low frequencies, while high pitched sounds like whistles have high frequencies. Pitch is quantified in terms of fundamental frequency and the amount of overtones, harmonic or otherwise. The AST function **pitch** is used to estimate the fundamental frequency of an audio signal. The function returns the estimates along with their locations. The following example illustrates how the function is used to estimate the fundamental frequency $f0$ of an audio signal which is then plotted to display the result (Figure 2.19).

> **Example 2.21: Write a program to estimate the pitch in an audio signal.**
>
> ```
> clear; clc;
> [a,fs] = audioread('SpeechDFT-16-8-mono-5secs.wav');
> [f0, idx] = pitch(a, fs);
> subplot(211), plot(a); ylabel('Amplitude');
> subplot(212), plot(idx, f0); ylabel('Pitch (Hz)'); xlabel('Sample Number')
> ```

Voice activity detection (VAD) is a technique used in speech processing to detect the presence or absence of human speech in an audio signal. To analyze an audio signal whose characteristics can vary over time, the audio is partitioned into segments called audio frames, each of which is separately analyzed for the presence of speech sections. VAD systems typically extract features from each audio frame which are then subjected to classification rules to classify each segment as containing speech or not. Typical applications of VAD systems are in speech/speaker recognition and voice-activated systems. The AST function **voiceActivityDetector** detects the presence of speech in an audio segment. The function returns the probability that speech is present. The input audio signal is windowed into audio frames and then converted to the frequency domain. A scheme based on

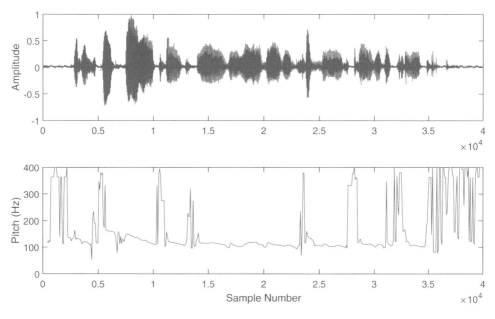

FIGURE 2.19
Output of Example 2.21.

log-likelihood ratio and Hidden Markov Model (HMM) is used to determine the probability that the current frame contains speech (Sohn et al., 1999). In the following example, the probability is calculated for each audio frame consisting of 441 samples, which is 10 ms in duration, and plotted (Figure 2.20).

Example 2.22: Write a program to detect the presence of speech segments in an audio signal.

```
clear; clc;
r = dsp.AudioFileReader('Counting-16-44p1-mono-15secs.wav');
fs = r.SampleRate;
r.SamplesPerFrame = ceil(10e-3*fs);
v = voiceActivityDetector;
w = audioDeviceWriter('SampleRate',fs);
aa = zeros(441, 1); pp = [];
nf = 0; % number of frames
while ~isDone(r)
    nf = nf + 1;
    x = r();
    probability = v(x);
    aa = [aa x];
    pp = [pp probability];
    w(x);
end

subplot(211), plot(aa(:)); axis tight;
xlabel('samples'); ylabel('amplitude');
subplot(212), plot(pp); axis([0 nf 0 1.2]);
xlabel('audio frames'); ylabel('probability');
```

Loudness is a perceptual property of sound dependent on its energy or pressure, measured in a unit called *decibels*. Sound pressure is the amount of pressure sound waves exert on a surface e.g. diaphragm and provides a way to compute its energy content. Sound pressure is measured as a deviation from atmospheric pressure and expressed in a unit called Pascals (1 Pa = 1 Nw/m^2). Sound pressure level (SPL) is the sound pressure expressed in

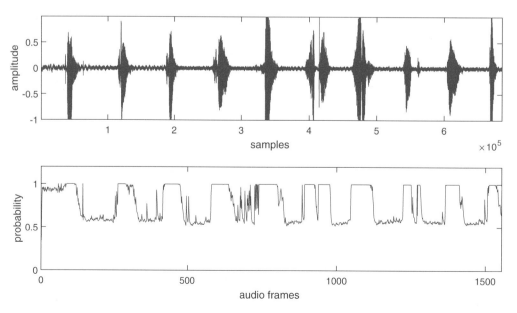

FIGURE 2.20
Output of Example 2.22.

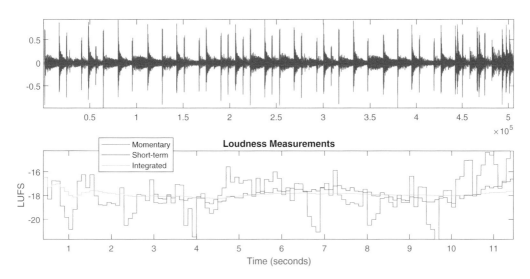

FIGURE 2.21
Output of Example 2.23.

decibels based on a reference sound pressure of 20 µPa, which is often considered the threshold of human hearing. To address the issues of discrepancy due to normalization based on the peak level of audio signals, the European Broadcasting Union in 2014 (EBU, 2014) and the International Telecommunication Union – Radiocommunication in 2015 (ITU-R, 2015) proposed the standards EBU R 128 and ITU-R BS 1770 for adoption of the Loudness Unit (LU), Loudness Unit Full Scale (LUFS), where 1 LU corresponds to a relative measurement of 1 dB on a digital scale and 0 LU = -23 LUFS, and the Loudness Range (LRA) which is a statistically determined value describing the loudness variation of a program. The R 128 standard recommends normalizing audio at the target level of -23 LUFS, and the measurement is calculated over the whole duration of a program transmitted on a broadcast channel, which is called "integrated loudness." To ensure that the loudness meters developed by different manufacturers provide consistent readings, EBU defines how to perform measurements via three distinct methods: (1) Momentary loudness, using a sliding time window of 400 ms to describe instantaneous loudness; (2) Short-term loudness, using a sliding time window of 3 seconds to describe a short time average loudness; and (3) Integrated loudness, to describe average loudness over the duration of the entire program. The AST function **loudnessMeter** calculates the momentary loudness, short-term loudness, integrated loudness, LRA, and true-peak value of an audio signal. The values are expressed in loudness units relative to full scale (LUFS). Peak value is measured in dB. The following example calculates the momentary, short-term, and integrated loudness of an audio signal and plots the results for visualization (Figure 2.21).

Example 2.23: Write a program to calculate the momentary, short-term and integrated loudness values of an audio signal.

```
clear; clc;
fn = 'RockDrums-44p1-stereo-11secs.mp3';
[a, sr] = audioread(fn);
r = dsp.AudioFileReader(fn);
lm = loudnessMeter('SampleRate',r.SampleRate);
momentary = [];
shortTerm = [];
integrated = [];
```

```
nf = 0;       % number of frames
sf = r.SamplesPerFrame;
fs = r.SampleRate;

while ~isDone(r)       % do for each frame
    nf = nf + 1;
    x = r();
    [m,s,i,n,p] = lm(x);
    momentary = [momentary; m];
    shortTerm = [shortTerm; s];
    integrated = [integrated; i];
end
release(r)

ns = nf * sf;          % number of samples
td = ns/fs; % time duration

t = linspace(0,td,length(momentary));
subplot(211), plot(a); axis tight;
subplot(212),
plot(t,[momentary,shortTerm,integrated])
title('Loudness Measurements')
legend('Momentary','Short-term','Integrated')
xlabel('Time (seconds)')
ylabel('LUFS')
axis tight;
```

The AST function **integratedLoudness** can also be used to compute the integrated loudness of an set of audio samples. The following example calculates integrated loudness in each audio frame in a file and plots the same (Figure 2.22).

Example 2.24: Write a program to calculate the integrated loudness of each audio frame consisting of 20000 samples.

```
clear; clc;
fn = 'RockDrums-44p1-stereo-11secs.mp3';
[xs, fs] = audioread(fn);
x = xs(:,1);                % isolate a single channel
```

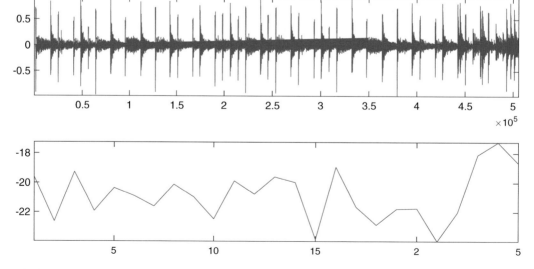

FIGURE 2.22
Output of Example 2.24.

```
sf = 20000;                  % no. of samples per frame
nf = floor(length(x)/sf);    % total no. of frames/file

for i = 0 : nf-1             % do for all frames
    frame{i+1} = x(i*sf+1:(i+1)*sf);
end

for i = 0 : nf-1             % do for all frames
    x1 = frame{i+1};         % samples in a frame
    ld(i+1) = integratedLoudness(x1,fs);
end

subplot(211), plot(x); axis tight;
subplot(212), plot(ld); axis tight;
```

The **MFCC** represents audio features derived from the power spectrum of sound expressed on a non-linear mel scale of frequency. These are widely used in applications like speech recognition, speaker recognition, and music recognition. First introduced during the 1980s (Davis and Mermelstein, 1980), these are used to approximate the shape of the human vocal tract and thereby estimate the sounds generated by the human voice box. A **cepstrum** (word formed by reversing the first four letters of "spectrum") is a result of taking the inverse of the logarithm of the spectrum of a signal i.e. signal $\xrightarrow{DFT}$ spectrum $\xrightarrow{\log}$ log-spectrum $\xrightarrow{IDFT}$ cepstrum. The **power cepstrum** is the squared magnitude of the cepstrum. A **mel scale** (Stevens et al., 1937) is a set of audio pitches uniformly separated from each other as perceived by human listeners. Studies have shown that equal frequency differences do not produce equal pitch differences to the average human listener. The mel scale is therefore non-linear, and mel values (m) are related to frequency values (f) using the following relations:

$$m = 2595 \cdot \log_{10}\left(1 + \frac{f}{700}\right) = 1127 \cdot \log_e\left(1 + \frac{f}{700}\right)$$

$$f = 700\left(10^{\frac{m}{2595}} - 1\right) = 700\left(e^{\frac{m}{1127}} - 1\right)$$

Although there are variations, the general process of computing MFCCs can be outlined by the following steps:

- Signal is divided into overlapping frames
- Each frame is windowed, usually using a Hamming window
- Each windowed frame is converted to the frequency domain using DFT
- Calculate power spectrum of each frame by squared magnitude of DFT coefficients
- Compute a set of 20–40 triangular mel-spaced filterbanks
- Multiply each filterbank with the power spectrum and add up the coefficients to get energy terms
- Take logarithm of each energy term to generate log filterbank energies
- Compute DCT of log filterbank energies to generate MFCCs
- Usually the first 12–13 coefficients are retained and the remaining discarded

The AST function **mfcc** returns the MFCCs for an audio input. The function partitions the speech into 1551 frames and computes the cepstral features for each frame. The first

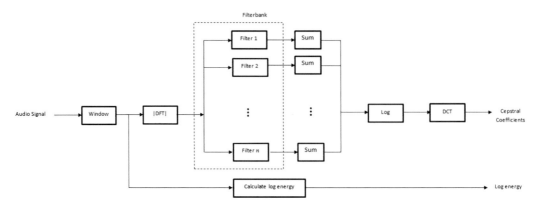

FIGURE 2.23
Block diagram of calculating cepstral coefficients.

element in the coefficients vector is the log-energy value and the remaining elements are the 13 cepstral coefficients computed by the function. The function also computes *frameloc* which is the location of the last sample in each frame. The log-energy E for each audio frame is calculated as follows, where $x(i)$ is the i-th audio sample in a frame

$$E = \log \Sigma \{x(i)\}^2$$

The cepstral coefficients are calculated by converting the input audio to the frequency domain by FFT and then using a filter bank to calculate energy in each band of frequencies. The log value of the filter outputs are calculated, and then a DCT operation is used to calculate the coefficient values. The Hamming window is used for the windowing operation (Figure 2.23). Window functions are discussed in Section 2.9.

In the following example, total number of audio samples in the audio file is: `size(A)` = 685056. This is divided into 1551 frames. So without overlap, the entire file can be divided into: `floor(size(A, 1)/1551)` = 441 sample sized frames. Amount of overlap by default: `round(fs*0.02)` = 882. So total samples in the first frame: 441 + 882 = 1323 which is equal to the first value in frameloc variable: `frameloc(1)` = 1323. The end point of the second frame is: 1323 + 441 = 1764 which is equal to the second value in frameloc variable: `frameloc(2)` = 1764. The end point of the third frame is: 1764 + 441 = 2205 which is equal to the third value in frameloc variable: `frameloc(3)` = 2205. The plots of the log-energy and the first three MFCC are displayed (Figure 2.24).

Example 2.25: Write a program to calculate the MFCC from each audio frame of a file.

```
[A,fs] = audioread('Counting-16-44p1-mono-15secs.wav');
[coeffs,d,dd,frameloc] = mfcc(A,fs);
nf = numel(frameloc);
c1 = coeffs(:,2);
c2 = coeffs(:,3);
c3 = coeffs(:,4);

subplot(121), plot(coeffs(:,1)); title('log-energy');
xlabel('frame index'); ylabel('magnitude');
subplot(122), plot(1:nf, c1, 1:nf, c2, 1:nf, c3); title('coeffs 1 2 3');
legend('c1', 'c2', 'c3');
xlabel('frame index'); ylabel('magnitude');
```

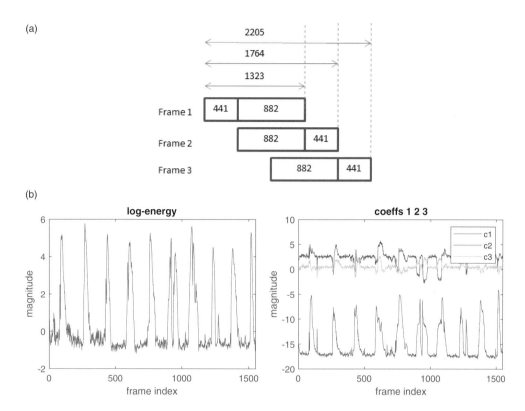

FIGURE 2.24
Output of Example 2.25. Plots for (a) log-energy (b) first three MFC coefficients.

2.7 Simulation, Tuning and Visualization

An oscilloscope is a device that graphically displays signal voltages as 2D plots usually as functions of time. The waveform can then be analyzed for measurement of various signal characteristics like amplitude, frequency, time period, peaks, and so on. At the heart of the oscilloscope is the oscillator which actually generates the waveforms, while the display scope produces graphical representations of the signals for visual perception. The DSPST function **dsp.TimeScope** is used for time-domain signal display and measurement, given the waveform type and frequency. Parameters are provided to specify the display limits in the X and Y directions of the plot. The following example illustrates time scope display of a square wave, a sine wave, and a sawtooth wave with specified parameters (Figure 2.25).

Example 2.26: Write a program to display sine, square and sawtooth waveforms using a time scope.

```
clear; clc;
osc1 = audioOscillator('square', 1000);
osc2 = audioOscillator('sine', 100);
osc3 = audioOscillator('sawtooth', 500);
scope = dsp.TimeScope(3, ...
    'SampleRate',osc1.SampleRate, ...
    'TimeSpan',0.01, ...
```

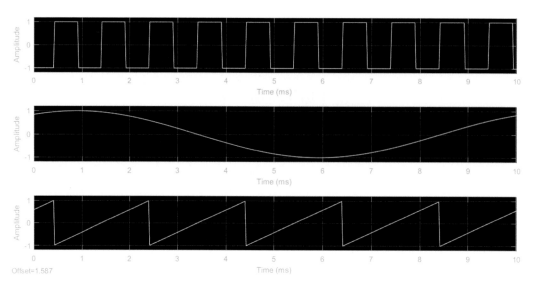

FIGURE 2.25
Output of Example 2.26.

```
          'YLimits',[-1.2, 1.2], ...
          'TimeSpanOverrunAction', 'Scroll', ...
          'ShowGrid',true);
    scope.LayoutDimensions = [3,1];
    scope(osc1(), osc2(), osc3());
```

The DSPST function **dsp.SineWave** can also be used to generate sine waves with varying amplitude, frequency, phase and can be used in conjunction with the time scope to display the curves. The following example illustrates that parameters of the time scope like the *TimeSpan* and *TimeDisplayOffset* can be used to specify the duration and the starting point of the wave (Figure 2.26).

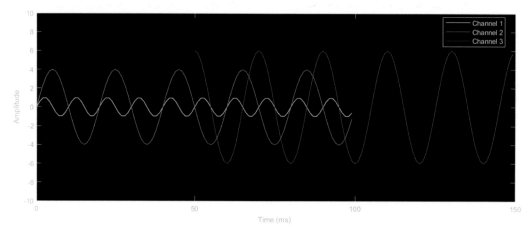

FIGURE 2.26
Output of Example 2.27.

Example 2.27: Write a program to display sinusoidal waveforms with specified amplitudes, frequencies, phase offsets and delay offsets using a timescope.

```
clear; clc;
sine1 = dsp.SineWave('Amplitude', 1, 'Frequency',100,'SampleRate',1000);
sine1.SamplesPerFrame = 10;

sine2 = dsp.SineWave('Amplitude', 4,'Frequency',50,'SampleRate',1000);
sine2.SamplesPerFrame = 10;

sine3 = dsp.SineWave('Amplitude', 6, 'Frequency',150,...
    'SampleRate',3000, 'PhaseOffset', pi/2);
sine3.SamplesPerFrame = 10;

scope = dsp.TimeScope(3, ...
    'SampleRate',1000,'TimeSpan',0.1, ...
    'TimeDisplayOffset',[0 0 0.05], ...
    'ShowLegend', true);

for ii = 1:10
    x1 = sine1();
    x2 = sine2();
    x3 = sine3();
    scope(x1, x2, x3);
end
```

`The frequency spectrum of a signal represents the frequency components of the signal. A spectrum analyzer is used to visually display the spectrum and examine the characteristics of the components like fundamental frequency, overtones, harmonics, as well as temporal parameters like attack, release, decay, and so on. The DSPST function **dsp. SpectrumAnalyzer** is used to display the frequency spectrum of time-domain signals. The following example shows the waveform and spectrum components of a sawtooth waveform (Figure 2.27).

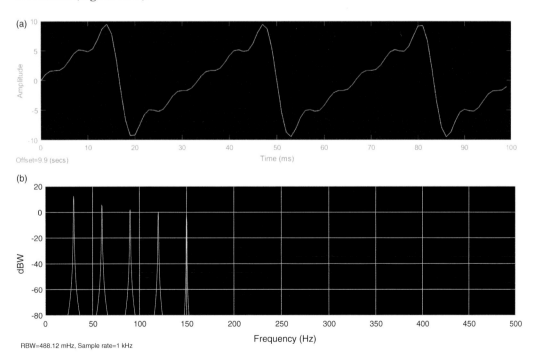

FIGURE 2.27
Output of Example 2.28 (a) time plot (b) frequency plot.

Example 2.28: Write a program to display a composite waveform and its frequency components using a spectrum analyzer.

```
clear; clc;

sine1 = dsp.SineWave('Amplitude', 6, 'Frequency', 30, 'SampleRate', 1000);
sine1.SamplesPerFrame = 1000;
sine2 = dsp.SineWave('Amplitude', 3, 'Frequency', 60, 'SampleRate', 1000);
sine2.SamplesPerFrame = 1000;
sine3 = dsp.SineWave('Amplitude', 2, 'Frequency', 90, 'SampleRate', 1000);
sine3.SamplesPerFrame = 1000;
sine4 = dsp.SineWave('Amplitude', 1.5, 'Frequency', 120, 'SampleRate', 1000);
sine4.SamplesPerFrame = 1000;
sine5 = dsp.SineWave('Amplitude', 1.2, 'Frequency', 150, 'SampleRate', 1000);
sine5.SamplesPerFrame = 1000;

scope = dsp.TimeScope(1, 'SampleRate',1000,'TimeSpan',0.1);
for ii = 1:10
    x1 = sine1() - sine2() + sine3() - sine4() + sine5();
    scope(x1);
end

Fs = 1000;      % Sampling freq
fs = 5000;      % Frame size

scope1 = dsp.SpectrumAnalyzer;
scope1.SampleRate = Fs;
scope1.ViewType = 'Spectrum';
scope1.SpectralAverages = 1;
scope1.PlotAsTwoSidedSpectrum = false;
scope1.RBWSource = 'Auto';
scope1.PowerUnits = 'dBW';
for idx = 1:1e2
    scope1(x1);
end
```

The DSPST function **dsp.ArrayPlot** is used to display vectors or arrays as stem plots. The following example illustrates using the function for visualization of a vector and an array. The vector is a plot of the function $[\sin(x) + \tan(x)]$, and the array is the 3 by 3 matrix as shown below, generated by the BM function **magic** such that summation along each row and column is the same:

$$\begin{bmatrix} 8 & 1 & 6 \\ 3 & 5 & 7 \\ 4 & 9 & 2 \end{bmatrix}$$

The color coding being used is 'yellow' for channel 1 (column 1) i.e. 8, 3, 4, 'blue' for channel 2 (column 2) i.e. 1, 5, 9, and 'red' for channel 3 (column 3) i.e. 6, 7, 2 (Figure 2.28).

Example 2.29: Write a program to display a vector and a matrix as stem plots.

```
clear; clc;
scope = dsp.ArrayPlot;
scope.YLimits = [-5 5];
scope.SampleIncrement = 0.1;
scope.Title = 'sin(x) + tan(x)';
scope.XLabel = 'X';
scope.YLabel = 'f(X)';
r = 0:.1:2*pi;
fx = sin(r) + tan(r);
scope(fx')
scope1 = dsp.ArrayPlot;
f = magic(3);
scope1(f)
scope1.ShowLegend = 1;
```

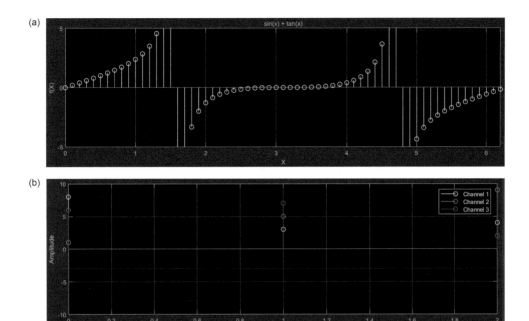

FIGURE 2.28
Output of Example 2.29 (a) vector (b) matrix.

2.8 Musical Instrument Digital Interface (MIDI)

Musical Instrument Digital Interface (MIDI) is the name of a protocol which provides a set of rules for interfacing digital musical instruments like synthesizers to digital computers. The MIDI protocol has three major portions: hardware, messages, and file format. The first part called hardware specifications outlines the cables, connectors, signal formats, and pin layouts required for connecting one MIDI device to another. A MIDI controller is a keyboard-like device which when played by pressing keys generates a set of digital instructions defining how sound is to be played. The sound module interprets the instruction and generates sound using either an FM synthesis method or a wavetable method. The synthesis method generates complex waveforms of musical notes by combining simple sinusoidal tones, while the wavetable method extracts pre-recorded digital sounds from a repository and uses filters to modify their loudness and pitch. These instructions travel from the controller through a MIDI-out port to a sound module through a MIDI-in port. The I/O ports use a 5-pin coaxial cable and connector which transmits the MIDI stream in an asynchronous mode at 31.25 Kbits/second. The second part, called the message format, defines how the musical information traveling from one device to another needs to be coded. The message format uses a 3-byte structure where the first byte defines the actual note to be played, while the remaining two bytes define additional related information like the note number and velocity (pressure). The messages activate multiple notes simultaneously using a 16-channel logical structure, where each channel can play a different note. The third part called file format defines how MIDI instructions are stored in a MIDI file. Sound cards of digital computers can interpret instructions from MIDI files and generate sounds from wavetable chips. Technical issues and updates are handled by the MIDI Manufacturers Association MMA (www.midi.org).

The AST function mididevinfo displays a table containing information about a list of MIDI devices attached to the system. The AST function mididevice is used to create an interface to one or more of the MIDI devices listed. The AST function midimsg is used to send and receive MIDI messages to one or more of the MIDI devices listed. The AST function midisend is used to send a MIDI message to one or more of the listed devices. The following example illustrates playing notes on a MIDI device by sending messages.

Example 2.30: Write a program to identify the MIDI devices available and play a series of musical notes on one of the devices.

```
clear; clc;
mididevinfo
device = mididevice('Microsoft MIDI Mapper');

channel = 1; velocity = 64; duration = 1; timestamp = 1;
note = 60;
msg = midimsg('Note', channel, note, velocity, duration, timestamp)
midisend(device, msg);
note = 61;
msg = midimsg('Note', channel, note, velocity, duration, timestamp)
midisend(device, msg);
note = 62;
msg = midimsg('Note', channel, note, velocity, duration, timestamp)
midisend(device, msg);
note = 63;
msg = midimsg('Note', channel, note, velocity, duration, timestamp)
midisend(device, msg)
note = 64;
msg = midimsg('Note', channel, note, velocity, duration, timestamp)
midisend(device, msg);
note = 65;
msg = midimsg('Note', channel, note, velocity, duration, timestamp)
midisend(device, msg);
note = 66;
msg = midimsg('Note', channel, note, velocity, duration, timestamp)
midisend(device, msg);
note = 67;
msg = midimsg('Note', channel, note, velocity, duration, timestamp)
midisend(device, msg);
note = 68;
msg = midimsg('Note', channel, note, velocity, duration, timestamp)
midisend(device, msg);
```

Apart from playing each note individually, notes specified using note numbers can be played using a loop using an array structure. The *ProgramChange* option of the midimsg function can be used to change the sound of the instruments. The General MIDI (GM-1) standard defines a set of 128 sounds each with a unique number to identify it. The names of the instruments indicate what sort of sound will be heard for a specific program change number (PC#). The instruments are grouped by families, as indicated below (*https://www. midi.org/specifications-old/item/gm-level-1-sound-set*):

PC#	Family Name	PC#	Family Name
1–8	Piano	65–72	Reed
9–16	Percussion	73–80	Pipe
17–24	Organ	81–88	Synth Lead
25–32	Guitar	89–96	Synth Pad
33–40	Bass	97–104	Synth Effects
41–48	Strings	105–112	Ethnic
49–56	Ensemble	113–120	Percussive
57–64	Brass	121–128	Sound Effects

The following example plays a set of eight notes forward and reverse in a loop using different instruments. Although the GM1 does not define the actual characteristics of each sound, the names in parenthesis are intended only as guides: 6 (harp), 8 (celesta), 14 (bells), 18 (organ), 48 (string), 55 (orchestra), 61 (brass), 73 (flute), 76 (blown bottle), and so on. The sound effects include 122 (sea shore), 123 (bird tweet), 124 (telephone ring), 126 (applause), 127 (gunshot), and so on.

Example 2.31: Write a program to illustrate how musical notes can be played using a variety of instrument sounds.

```
clear; clc;
device = mididevice('Microsoft GS Wavetable Synth');
melody = [61,62,63,64,65,66,67,68];
channel = 5;
velocity = 120;
duration = 1;
timestamp = 1;
n = numel(melody);
pc = [6,8,14,18,48,55,61,73,76,80,91,98,100,102,112,117,118,115,122,123,124,126,127];
counter=1;
for count=1:numel(pc)
    programChangeMessage = midimsg('ProgramChange',1, pc(counter));
    midisend(device,programChangeMessage);

    for i = 1:n
        idx = (2*i-1):(2*i);
        msgArray1(idx) = midimsg('Note',1,melody(i),velocity,duration,i);
        msgArray2(idx) = midimsg('Note',1,melody(9-i),velocity,duration,i);
    end
    midisend(device,msgArray1)
    midisend(device,msgArray2)
    counter=counter+1;
end
```

2.9 Temporal Filters

A **temporal filter** takes in a time-domain signal, modifies the frequency content, and returns a new time-domain signal, usually for the purpose of removing high-frequency noise. Functionally filters are of four types: high pass, low pass, band pass, and band stop. High-pass filters are used to remove low frequencies below a specified threshold, low-pass filters are used to remove high frequencies above a specified threshold, band-pass filters are used to allow a range of frequencies within a lower and upper threshold, and band-stop filters block a range of frequencies within a lower and upper threshold. Structurally the filters can be of two types: finite impulse response (FIR) and infinite impulse response (IIR).

A FIR filter is a filter whose response to an impulse input is of finite duration after which it becomes zero. It is represented by the relation $y = H(b,x)$ and takes an n-element input signal x i.e. $x = [x(1),..., x(n)]$ and uses an m-element filter coefficient b i.e. $b = [b(1),..., b(m)]$. The n-element output produced $y = [y(1), ..., y(n)]$ is a linear combination of the input and filter coefficients such that the i-th output element is given by:

$$y(i) = b(1)\cdot x(i) + b(2)\cdot x(i-1) + ... + b(m)\cdot x(i-m+1) = \sum_{k=1}^{m} b(k)\cdot x(i-k+1)$$

An IIR filter is a filter whose response to an impulse input does not become exactly zero in a finite time but continues indefinitely with decaying amplitudes. This is possible due to an internal feedback mechanism whereby a part of the output is fed back to the input and causes it to respond indefinitely. It is represented as $y = H(b,a,x)$, takes an n-element input signal x i.e. $x = [x(1),..., x(n)]$, and uses m-element filter coefficients a and b i.e. $a = [a(1), ..., a(m)]$ and $b = [b(1),..., b(m)]$. The n-element output produced $y = [y(1),..., y(n)]$ is a non-linear combination of the input and filter coefficients given by:

$$a(1) \cdot y(i) + a(2) \cdot y(i-1)... + a(m) \cdot y(i-(m-1))$$

$$= b(1) \cdot x(i) + b(2) \cdot x(i-1) + ... + b(m) \cdot x(i-(m-1))$$

Written in the compact form, where by convention $a(1) = 1$:

$$y(i) = \sum_{k=1}^{m} b(k) \cdot x(i-k+1) - \sum_{k=2}^{m} a(k) \cdot y(i-k+1)$$

The BM function **filter** is used to implement a 1-D filter on a 1-D signal. A moving-average filter is a common method used for smoothing noisy data. The following example computes averages along a vector of data by sliding a window of width ω and computing averages of the data contained within each window, according to the relation: $y(i) = (1/\omega)\{x(i) + x(i-1) + ... + x(i-(\omega-1))\}$. The data are sinusoid corrupted by random noise (Figure 2.29).

Example 2.32: Write a program to remove noise from a signal using a FIR and an IIR filter.

```
clear; clc;
t = linspace(-pi,pi,100);
rng default   %initialize random number generator
x = sin(t) + 0.25*rand(size(t));   % add noise to signal
```

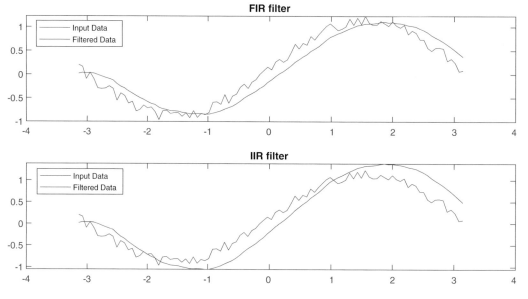

FIGURE 2.29
Output of Example 2.32.

```
% FIR filter
a1 = 1;
w = 10; b = (1/w)*ones(1,w);
y1 = filter(b,a1,x);
figure,
subplot(211),
plot(t,x); hold on; plot(t,y1);
legend('Input Data','Filtered Data')
title('FIR filter'); hold off;

% IIR filter
a2 = [1, -0.2];
y2 = filter(b,a2,x);
subplot(212),
plot(t,x); hold on; plot(t,y2);
legend('Input Data','Filtered Data')
title('IIR filter'); hold off;
```

A **window function** is a function to isolate and analyze a small part of a signal. Usually the values of the signal outside the window are zero or rapidly tend to be zero. Some of the most frequently used windows of width N are discussed below:

The simplest window is the **rectangular window** which has the value 1 over the width and 0 elsewhere

$$\omega(x) = \begin{cases} 1, & 0 \le x \le N \\ 0, & x\langle 0, x\rangle N \end{cases}$$

The **Bartlett window** is a triangular window defined as follows:

$$\omega(x) = \begin{cases} \dfrac{2x}{N}, & 0 \le x \le \dfrac{N}{2} \\ 2 - \dfrac{2x}{N}, & \dfrac{N}{2} \le x \le N \end{cases}$$

The **Gaussian window** is defined as follows, where σ is the standard deviation of the Gaussian curve:

$$\omega(x) = \exp\left(-\frac{x^2}{2\sigma^2}\right), \frac{-(N-1)}{2} \le x \le \frac{(N-1)}{2}$$

The **Blackman window** is defined as follows, where $M = N/2$ for N even and $M = (N+1)/2$ for N odd:

$$\omega(x) = 0.42 - 0.5\cos\left(\frac{2\pi x}{N-1}\right) + 0.08\cos\left(\frac{4\pi x}{N-1}\right), 0 \le x \le M-1$$

The **flat-top window** is defined as follows:

$$\omega(x) = 0.215 - 0.417\cos\left(\frac{2\pi x}{N-1}\right) + 0.277\cos\left(\frac{4\pi x}{N-1}\right)$$

$$- 0.084\cos\left(\frac{6\pi x}{N-1}\right) + 0.007, 0 \le x \le N-1$$

The **Hamming window** is defined as follows:

$$\omega(x) = 0.54 - 0.46 \cos\left(\frac{2\pi x}{N}\right), 0 \le x \le N$$

The SPT function **window** is used to generate a window of the specified type and width. The following example shows plots of a number of window functions (Figure 2.30).

Example 2.33: Write a program to plot different window functions.

```
clear; clc;
t = 0:500;
y = sin(2*pi*t/100);
h1 = window(@rectwin,300);
h1(1:100) = 0; h1(451:501)= 0; y1 = y.*h1';
h2 = window(@hamming,501); y2 = y.*h2';
h3 = window(@bartlett,501); y3 = y.*h3';
h4 = window(@flattopwin,501); y4 = y.*h4';
h5 = window(@blackman,501); y5 = y.*h5';

figure,
subplot(341), plot(t,y);
axis([0 500 -1.2 1.2]);
title('Original Waveform');
subplot(343), plot(h1); title('Rectangular Window'); axis tight;
subplot(344), plot(y1); title('Windowed Waveform'); axis tight;
subplot(345), plot(h2); title('Hamming Window'); axis tight;
subplot(346), plot(y2); title('Windowed Waveform'); axis tight;
subplot(347), plot(h3); title('Bartlett Window'); axis tight;
subplot(348), plot(y3); title('Windowed Waveform'); axis tight;
subplot(349), plot(h4); title('Flat-top Window'); axis tight;
subplot(3,4,10), plot(y4); title('Windowed Waveform'); axis tight;
subplot(3,4,11), plot(h5); title('Blackman Window'); axis tight;
subplot(3,4,12), plot(y5); title('Windowed Waveform'); axis tight;
```

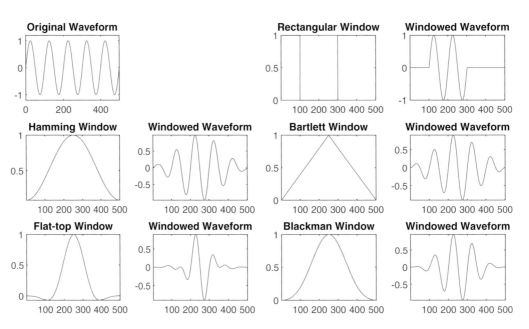

FIGURE 2.30
Output of Example 2.33.

2.10 Spectral Filters

It has already been discussed in Chapter 1 that the forward DFT expresses frequency-domain value $X(k)$ in terms of time-domain value $x(i)$:

$$X(k) = \sum_{i=0}^{N-1} x(i) \cdot e^{-j \cdot \frac{2\pi ki}{N}}$$

Here $j = \sqrt{-1}$, and e is an exponential operator given by: $e = \lim_{n \to \infty} \left(1 + \frac{1}{n}\right)^n \approx 2.718$.

The inverse DFT expresses time-domain value $x(i)$ in terms of frequency-domain value $X(k)$:

$$x(i) = \frac{1}{N} \sum_{k=0}^{N-1} X(k) \cdot e^{j \cdot \frac{2\pi ki}{N}}$$

The BM functions **fft** and **ifft** are used to implement the forward DFT and inverse DFT. The following example shows a compsite note being generated by adding three sinusoidal tones with frequencies of 50 Hz, 120 Hz, and 150 Hz and amplitudes of 0.6, 0.4, and 0.2, respectively. The composite note is transformed to the frequency domain using forward DFT and its frequency components are extracted, after which it is converted back to time domain using inverse DFT (Figure 2.31).

Example 2.34: Write a program to display frequency components of a composite note using Discrete Fourier Transform.

```
clear; clc;
k = 1000;
Fs = k;             % Sampling frequency
T = 1/Fs;           % Sampling period
L = 1500;           % Length of signal
```

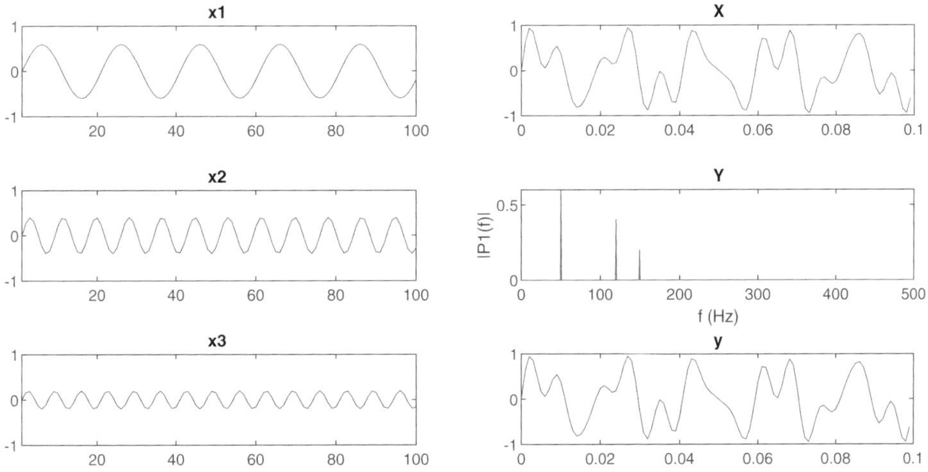

FIGURE 2.31
Output of Example 2.34.

```
t = (0:L-1)*T;          % Time vector

x1 = 0.6*sin(2*pi*50*t);          % First component
x2 = 0.4*sin(2*pi*120*t);         % Second component
x3 = 0.2*sin(2*pi*150*t);         % Third component

figure,
subplot(321),
plot(x1(1:k/10)); axis([1, k/10, -1, 1]); grid; title('x1');

subplot(323),
plot(x2(1:k/10)); axis([1, k/10, -1, 1]); grid; title('x2');

subplot(325),
plot(x3(1:k/10)); axis([1, k/10, -1, 1]); grid; title('x3');

subplot(322),
X = [x1+x2+x3];
plot(t(1:k/10),X(1:k/10));  grid; title('X');

subplot(324),
Y = fft(X);
P2 = abs(Y/L);
P1 = P2(1:L/2+1);
P1(2:end-1) = 2*P1(2:end-1);
f = Fs*(0:(L/2))/L;
plot(f,P1);  grid; title('Y');
xlabel('f (Hz)')
ylabel('|P1(f)|')

subplot(326),
y = ifft(Y);
plot(t(1:k/10),y(1:k/10));  grid; title('y');
```

A **spectrogram** is a visual representation of the frequency of a signal. Specifically it depicts how the frequency varies with time i.e. what frequencies occurs at what instants of time. Usually the spectrogram is represented as a 3-D plot, the x-axis depicts time, the y-axis depicts frequency, and the z-axis depicts the amplitude of the frequency i.e. what is the amplitude of a specific frequency occurring at a specific instant of time. When displayed as a 2-D plot, the amplitude is represented by varying the brightness or color values of pixels, brighter values depicting larger values of amplitude. In digital signal analysis, spectrograms are generated using DFT. Audio signals are usually segmented or windowed into audio frames before calculating DFT for each frame. The windows generated can be either non-overlapping rectangular type or overlapping type using Hamming or Gaussian functions. The DFT calculated for each frame is called short-time Fourier transforms (STFT). The size (duration) of the window can be varied suited to specific applications. A smaller (shorter) window can generate results with high temporal precision (i.e. at specific instants of time) but low-frequency precision, as the small number of samples within the window cannot be used to accurately compute frequency of the signal. On the other hand, a larger (longer) window can generate results with high-frequency precision but low timing precision. This follows from the uncertainty principle which implies that one cannot achieve high-temporal resolution and high-frequency resolution at the same time. The SPT function **spectrogram** is used to plot the frequency against time of the input signal by splitting it into overlapping segments and using STFT for each segment. The general syntax is the following, where x is the input signal, window is the window size, noverlap is the amount of overlap between audio frames, and nfft is the number of sampling points to calculate the DFT: spectrogram(x, window, noverlap, nfft).

In addition to instantaneous frequency values, spectrograms are also extensively used for signals with variable frequencies. These signals called **chirp** have frequencies which vary in a linear or non-linear way over time. Given an instantaneous frequency f_0 at time t_0

and instantaneous frequency f_1 at time t_1, variations can be linear, quadratic, logarithmic, and so on. For a linear variant, frequency at any specified time is given by:

$$f_i = f_0 + \left(\frac{f_1 - f_0}{t_1 - t_0} \right) t$$

For a quadratic variant, frequency at any specified time is given by:

$$f_i = f_0 + \left\{ \frac{f_1 - f_0}{\left(t_1 - t_0 \right)^2} \right\} t^2$$

If $f_1 > f_0$, the frequency values increase over time and the shape of the spectrogram is concave, if $f_1 < f_0$, the frequency values decrease over time and the shape of the spectrogram is convex. The SPT function **chirp** generates a waveform with variable frequency. The general syntax is the following, where f_0 is the instantaneous frequency at time t_0, f_1 the instantaneous frequency at time t_1, and method specifies the interpolation type: `chirp(t,f_0,t_1,f_1,'method')`. In the following example, X is the composite waveform of the previous example, while c1 and c2 are signals of variable frequencies whose spectrograms are quadratic (Figure 2.32).

Example 2.35: Write a program to generate spectrograms of frequency domain signals.

```
clear; clc;
figure,
t = 0:1/1e3:2;
```

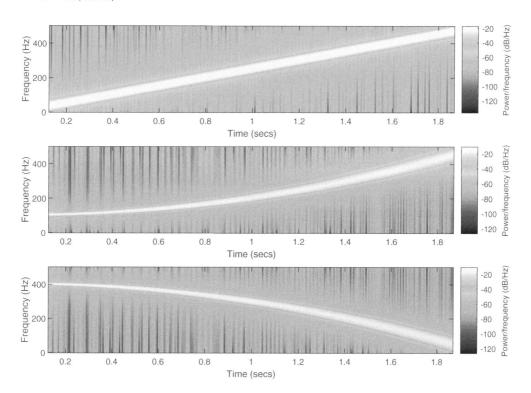

FIGURE 2.32
Output of Example 2.35.

```
y1 = chirp(t,0,1,250);
subplot(311),spectrogram(y1,256,250,256,1e3,'yaxis')
y2 = chirp(t,100,1,200,'quadratic');
subplot(312),spectrogram(y2,256,250,256,1e3,'yaxis')
y3 = chirp(t,400,1,300,'quadratic');
subplot(313),spectrogram(y3,256,250,256,1e3,'yaxis')
```

Filter design implies finding the order and coefficients of the filter that meet specific requirements. Recalling the FIR and IIR filter characteristics defined earlier and using slightly different notations, if $x(n)$ be the input signal, $y(n)$ be the output signal, and M is the filter order, then for a FIR filter, we have the following where b denotes the coefficient:

$$y(n) = \sum_{k=0}^{M} b(k) \cdot x(n-k)$$

To find the transfer function of the filter, we take the Z-transform of each side and re-arrange:

$$H(z) = \sum_{k=0}^{M} b(k) \cdot z^{-k}$$

For an IIR filter, the output is given by the following where a and b are the coefficients:

$$a(0) \cdot y(n) = \sum_{k=0}^{M} b(k) \cdot x(n-k) - \sum_{k=1}^{N} a(k) \cdot y(n-k)$$

Taking the Z-transform of both sides and assuming $a(0) = 1$, we have:

$$H(z) = \frac{\sum_{k=0}^{M} b(k) \cdot z^{-k}}{\sum_{k=1}^{N} a(k) \cdot z^{-k}}$$

If $H_d(z)$ is the desired frequency response of the filter, then we minimize the mean square error between the current filter and the desired filter through an iterative procedure i.e. $|H(z) - H_d(z)|^2$. The SPT function **designfilt** is used to design digital filters of various types by specifying magnitude and frequency constraints. A **low-pass FIR filter** can be designed by specifying the option *lowpassfir* and including the passband frequencies and stopband frequencies. The **passband** is the region of the frequency band where the filter should pass its input through to the output, and the **stopband** is the region of the frequency band over which the filter is intended not to transmit its input. The DSPST function **fvtool** can be used to visualize the frequency response of the designed filter. The following example illustrates design of two low-pass FIR filters, in the first case, the difference between the pass and stop frequencies being small (100 Hz) and in the second case, the difference being larger (600 Hz), and their filtering action on a noisy sinusoid (Figure 2.33).

Example 2.36: Write a program to design a low-pass FIR filter.

```
clear; clc;

% part 1
lpfir1 = designfilt('lowpassfir', ...
    'PassbandFrequency',300, ...
```

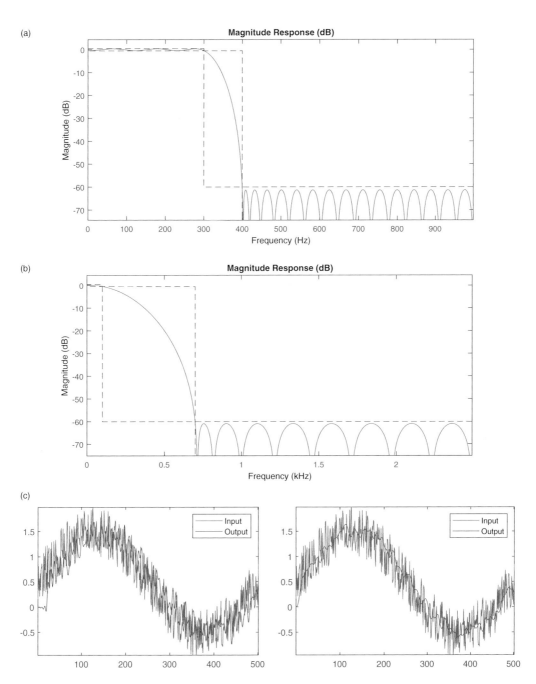

FIGURE 2.33
Output of Example 2.36 (a) part 1 (b) part 2 (c) part 3.

```
    'StopbandFrequency',400, ...
    'SampleRate',2000)
fvtool(lpfir1)

% part 2
lpfir2 = designfilt('lowpassfir', ...
    'PassbandFrequency',100, ...
```

```
    'StopbandFrequency',700, ...
    'SampleRate',5000)
fvtool(lpfir2)

% part 3
t = 0:500;
dataIn = sin(2*pi*t/500) + rand(1,501);
dataOut1 = filter(lpfir1,dataIn);
dataOut2 = filter(lpfir2,dataIn);

figure,
subplot(121), plot(dataIn); hold; plot(dataOut1, 'r');
axis tight; legend('Input', 'Output');
subplot(122), plot(dataIn); hold; plot(dataOut2, 'r');
axis tight; legend('Input', 'Output');
```

A **low-pass IIR filter** can be designed by specifying the option *lowpassiir* and including the passband frequencies and stopband frequencies. The following example illustrates design of two low-pass IIR filters, in the first case, the difference between the pass and stop frequencies being small (100 Hz) and in the second case, the difference being larger (600 Hz), and their filtering action on a noisy sinusoid (Figure 2.34).

Example 2.37: Write a program to design a low-pass IIR filter.

```
clear; clc;

% part 1
lpiir1 = designfilt('lowpassiir', ...
    'PassbandFrequency',300, ...
    'StopbandFrequency',400, ...
    'SampleRate',2000)
fvtool(lpiir1)

% part 2
lpiir2 = designfilt('lowpassiir', ...
    'PassbandFrequency',100, ...
    'StopbandFrequency',700, ...
    'SampleRate',5000)
fvtool(lpiir2)

% part 3
t = 0:500;
dataIn = sin(2*pi*t/500) + rand(1,501);
dataOut1 = filter(lpiir1,dataIn);
dataOut2 = filter(lpiir2,dataIn);

figure,
subplot(121), plot(dataIn); hold; plot(dataOut1, 'r');
axis tight; legend('Input', 'Output');
subplot(122), plot(dataIn); hold; plot(dataOut2, 'r');
axis tight; legend('Input', 'Output');
```

A **high-pass FIR filter** can be designed by specifying the option *highpassfir* and including the passband frequencies and stopband frequencies. The following example illustrates design of two high-pass FIR filters, in the first case, the difference between the pass and stop frequencies being small (100 Hz) and in the second case, the difference being larger (600 Hz), and their filtering action on a noisy sinusoid (Figure 2.35).

Example 2.38: Write a program to design a high-pass FIR filter.

```
clear; clc;

% part 1
hpfir1 = designfilt('highpassfir', ...
```

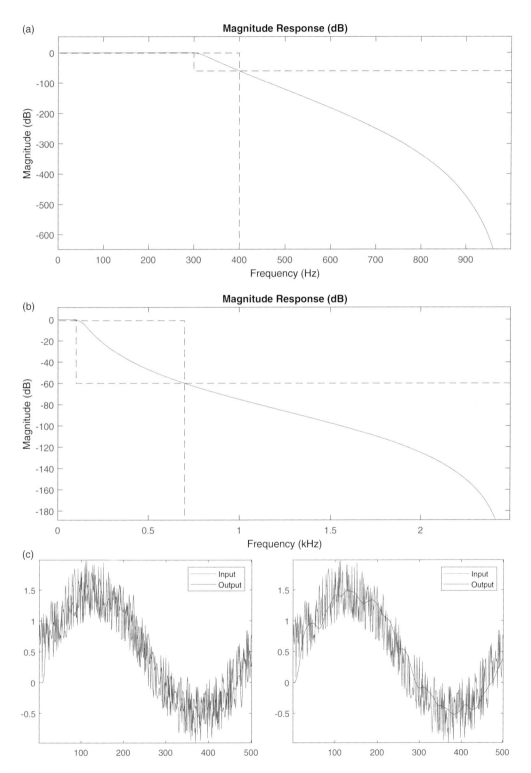

FIGURE 2.34
Output of Example 2.37 (a) part 1 (b) part 2 (c) part 3.

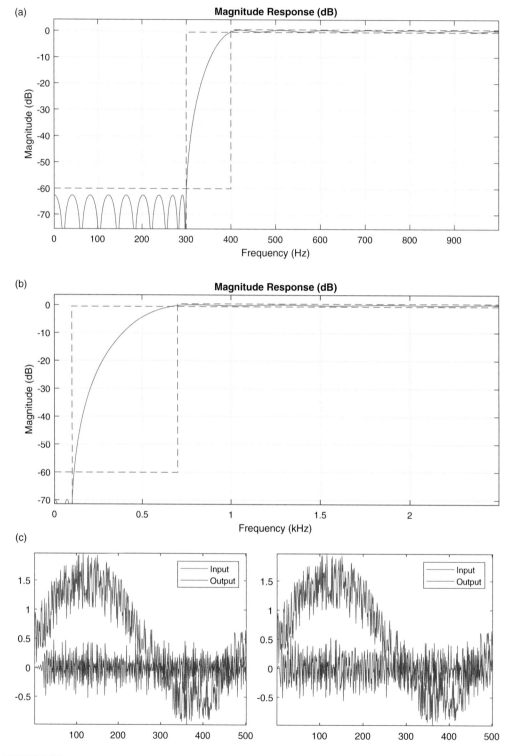

FIGURE 2.35
Output of Example 2.38 (a) part 1 (b) part 2 (c) part 3.

```
        'PassbandFrequency',400, ...
        'StopbandFrequency',300, ...
        'SampleRate',2000)
    fvtool(hpfir1)

    % part 2
    hpfir2 = designfilt('highpassfir', ...
        'PassbandFrequency',700, ...
        'StopbandFrequency',100, ...
        'SampleRate',5000)
    fvtool(hpfir2)

    % part 3
    t = 0:500;
    dataIn = sin(2*pi*t/500) + rand(1,501);
    dataOut1 = filter(hpfir1,dataIn);
    dataOut2 = filter(hpfir2,dataIn);

    figure,
    subplot(121), plot(dataIn); hold; plot(dataOut1, 'r');
    axis tight; legend('Input', 'Output');
    subplot(122), plot(dataIn); hold; plot(dataOut2, 'r');
    axis tight; legend('Input', 'Output');
```

A **high-pass IIR filter** can be designed by specifying the option *highpassiir* and including the passband frequencies and stopband frequencies. The following example illustrates design of two high-pass IIR filters, in the first case, the difference between the pass and stop frequencies being small (100 Hz) and in the second case, the difference being larger (600 Hz), and their filtering action on a noisy sinusoid (Figure 2.36).

Example 2.39: Write a program to design a high-pass IIR filter.

```
clear; clc;

% part 1
hpiir1 = designfilt('highpassiir', ...
    'PassbandFrequency',400, ...
    'StopbandFrequency',300, ...
    'SampleRate',2000)
fvtool(hpiir1)

% part 2
hpiir2 = designfilt('highpassiir', ...
    'PassbandFrequency',700, ...
    'StopbandFrequency',100, ...
    'SampleRate',5000)
fvtool(hpiir2)

% part 3
t = 0:500;
dataIn = sin(2*pi*t/500) + rand(1,501);
dataOut1 = filter(hpiir1,dataIn);
dataOut2 = filter(hpiir2,dataIn);

figure,
subplot(121), plot(dataIn); hold; plot(dataOut1, 'r');
axis tight; legend('Input', 'Output');
subplot(122), plot(dataIn); hold; plot(dataOut2, 'r');
axis tight; legend('Input', 'Output');
```

A **band-pass FIR filter** can be designed by specifying the option *bandpassfir* and including the lower and upper cutoff frequencies. A **band-pass IIR filter** can be designed by specifying the option *bandpassiir* and including the lower and upper cutoff frequencies. The following example illustrates design of a band-pass FIR and band-pass IIR filter and their filtering action on a noisy sinusoid (Figure 2.37).

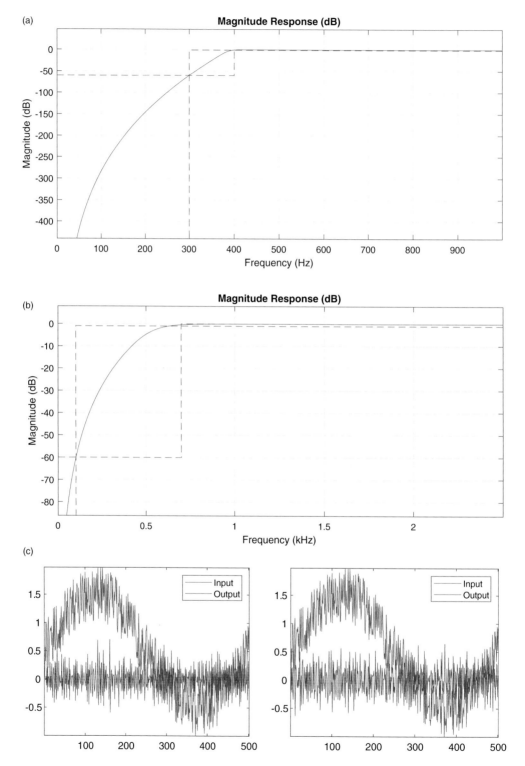

FIGURE 2.36
Output of Example 2.39 (a) part 1 (b) part 2 (c) part 3.

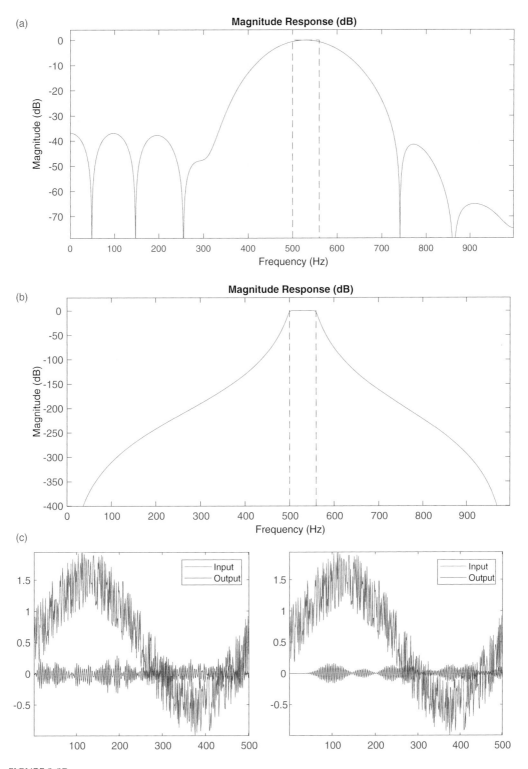

FIGURE 2.37
Output of Example 2.40 (a) part 1 (b) part 2 (c) part 3.

Example 2.40: Write a program to design a band-pass FIR filter and band-pass IIR filter.

```
clear; clc;

% part 1
bpfir = designfilt('bandpassfir', 'FilterOrder',20, ...
    'CutoffFrequency1',500,'CutoffFrequency2',560, ...
    'SampleRate',2000)
fvtool(bpfir)

% part 2
bpiir = designfilt('bandpassiir', 'FilterOrder',20, ...
    'HalfPowerFrequency1',500,'HalfPowerFrequency2',560, ...
    'SampleRate',2000)
fvtool(bpiir)

% part 3
t = 0:500;
dataIn = sin(2*pi*t/500) + rand(1,501);
dataOut1 = filter(bpfir,dataIn);
dataOut2 = filter(bpiir,dataIn);

figure,
subplot(121), plot(dataIn); hold; plot(dataOut1, 'r');
axis tight; legend('Input', 'Output');
subplot(122), plot(dataIn); hold; plot(dataOut2, 'r');
axis tight; legend('Input', 'Output');
```

A **band-stop FIR filter** can be designed by specifying the option *bandstopfir* and including the lower and upper cutoff frequencies. A **band-stop IIR filter** can be designed by specifying the option *bandstopiir* and including the lower and upper cutoff frequencies. The following example illustrates design of a band-stop FIR and band-stop IIR filter and their filtering action on a noisy sinusoid (Figure 2.38).

Example 2.41: Write a program to design a band-stop FIR filter and band-stop IIR filter.

```
clear; clc;

% part 1
bsfir = designfilt('bandstopfir', 'FilterOrder',20, ...
    'CutoffFrequency1',500,'CutoffFrequency2',600, ...
    'SampleRate',2000)
fvtool(bsfir)

% part 2
bsiir = designfilt('bandstopiir', 'FilterOrder',20, ...
    'HalfPowerFrequency1',500,'HalfPowerFrequency2',600, ...
    'SampleRate',2000)
fvtool(bsiir)

% part 3
t = 0:500;
dataIn = sin(2*pi*t/500) + rand(1,501);
dataOut1 = filter(bsfir,dataIn);
dataOut2 = filter(bsiir,dataIn);

figure,
subplot(121), plot(dataIn); hold; plot(dataOut1, 'r');
axis tight; legend('Input', 'Output');
subplot(122), plot(dataIn); hold; plot(dataOut2, 'r');
axis tight; legend('Input', 'Output');
```

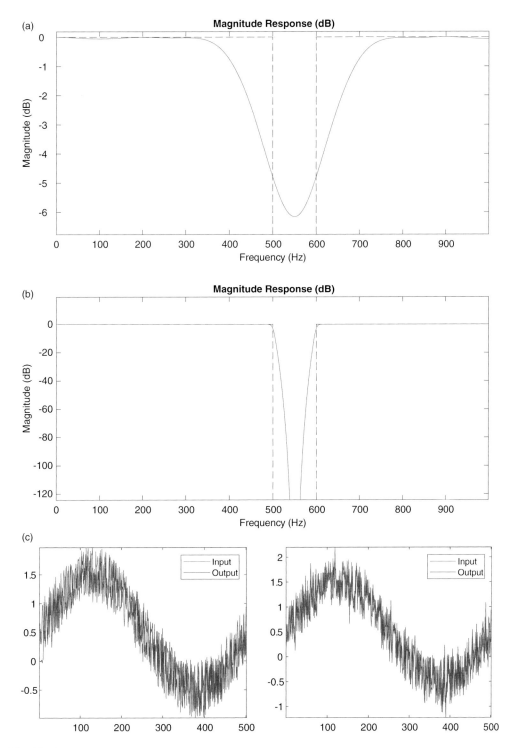

FIGURE 2.38
Output of Example 2.41 (a) part 1 (b) part 2 (c) part 3.

2.11 Audio Processing Using Simulink

To create a simulink model, type *simulink* at the command prompt or choose Simulink > Blank Model > Library Browser > AST. The audio source files reside in the folder *(matlab-root)/toolbox/audio/samples*. Within the AST toolbox, the following libraries and blocks are included for audio processing:

1. Dynamic Range Control: Compressor, Expander, Limiter, Noise Gate
2. Effects: Reverberator
3. Filters: Crossover filter, Graphic EQ, Octave filter, Parametric EQ filter, Weighting filter
4. Measurements: Cepstral feature, Loudness meter, Voice activity detector
5. Sinks: Audio device writer, Spectrum analyzer, To multimedia file
6. Sources: Audio device reader, From multimedia file, MIDI controls

Example 2.42: Create a Simulink model to generate an audio reverb effect.

- AST > Sources > From multimedia file
- AST > Sinks > Audio Device writer
- AST > Effects > Reverberator (Figure 2.39).

Example 2.43: Create a Simulink model to implement a crossover filter.

- AST > Sources > From multimedia file (Filename : RockGuitar-16–44p1-stereo-72secs.wav)
- AST > Sinks > Audio Device writer
- AST > Filters > Crossover filter (Number of crossover : 2, crossover frequency : [100, 1000], crossover order : [3, 3] (18 dB/octave)) > Visualize Response
- Simulink > Signal Routing > Manual Switch (Figure 2.40).

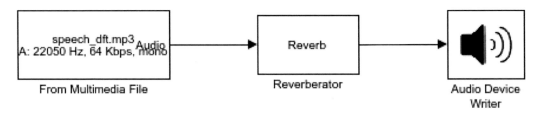

FIGURE 2.39
Output of Example 2.42.

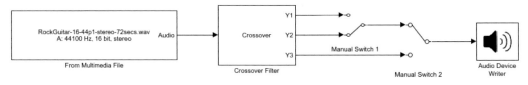

FIGURE 2.40
Output of Example 2.43.

Example 2.44: Create a Simulink model to implement a voice activity detector.

- AST > Sources > From multimedia file (Filename : Counting-16–44p1-mono-15secs.wav)
- AST > Sinks > Audio Device writer
- DSPST > Sinks > Time Scope (File > Number of Input Ports > 2)
- AST > Measurements > VAD (Figure 2.41).

Example 2.45: Create a Simulink model to implement a loudness meter.

- AST > Sources > From multimedia file (Filename : RockDrums-44p1-stereo-11secs.mp3)
- AST > Sinks > Audio Device writer
- AST > Measurements > Loudness meter
- DSPST > Sinks > Time Scope (File > Number of Input Ports > 2)
- DSPST > Math Functions > Matrices & Linear Algebra > Matrix Operations > Matrix Concatenate (Figure 2.42).

Example 2.46: Create a Simulink model to sum up multiple sine waves and display the resultant waveform.

- DSPST > Sources > Sine wave
- DSPST > Sinks > Time Scope (File > Number of Input Ports > 2)

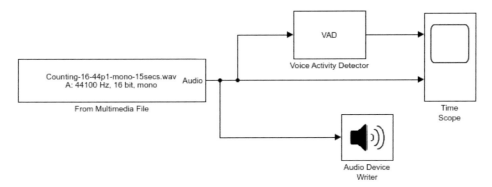

FIGURE 2.41
Output of Example 2.44.

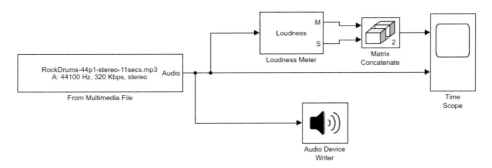

FIGURE 2.42
Output of Example 2.45.

- DSPST > Sinks > Time Scope (File > Number of Input Ports > 1)
- Simulink > Commonly Used Blocks > Sum (Figure 2.43).

Example 2.47: Create a Simulink model to implement a noise gate.

- AST > Sources > From multimedia file (Filename : Counting-16–44p1-mono-15secs.wav)
- AST > Sinks > Audio Device writer
- DSPST > Sinks > Time Scope (File > Number of Input Ports > 1)
- DSPST > Math Functions > Matrices & Linear Algebra > Matrix Operations > Matrix Concatenate
- Simulink > Signal Routing > Manual Switch
- Simulink > Commonly Used Blocks > Sum
- DSPST > Sources > Random Source (Source Type : Gaussian, Mean : 0, Variance : 1/1000) (Figure 2.44).

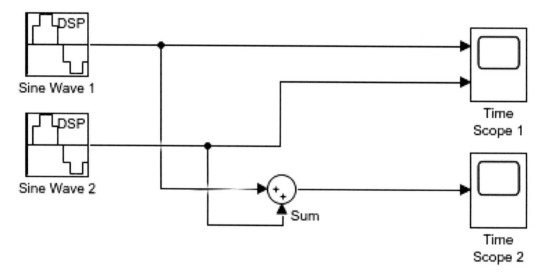

FIGURE 2.43
Output of Example 2.46.

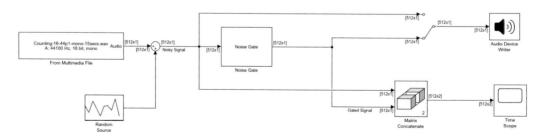

FIGURE 2.44
Output of Example 2.47.

Review Questions

1. Differentiate between BM functions and AST functions for audio processing.

2. Mention how the postulates of the Nyquist sampling theorem is relevant for digital audio processing.

3. Specify how the accuracy of digital audio waveforms can be controlled using Fourier series.

4. Explain how a time-domain signal can be reconstructed from DFT coefficients and basis functions.

5. How can a digital audio file be read into the system and its waveform can be plotted against time?

6. Discuss the main functions of the AST and DSPST toolboxes for audio processing tasks.

7. How can composite waveforms be displayed on a time scope display using AST functions?

8. How can an audio signal be split into non-overlapping audio frames of a specified duration?

9. How can a noise gate filter and a crossover filter be implemented on an audio signal?

10. How can a loudness meter be used to measure the momentary, short-term, and integrated loudness?

11. What is the utility of MFCC and how can it be computed from an audio signal?

12. How can MIDI messages be used to generate sounds of different instruments?

13. How can noise be removed from an audio signal using FIR and IIR filters?

14. How can spectral filters be implemented in the frequency domain?

15. How can audio processing tasks be done using the Simulink environment?

3

Video Processing

3.1 Introduction

Video is a combination of image and audio, so as a general rule whatever theory and applications were valid for those, are usually valid for video as well. Video consists of a set of still images called **frames** which are displayed to the user one after another at a specific speed called the **frame rate** measured in frames per second, shortened to fps. If displayed at a fast enough rate, the human eye cannot distinguish the individual images as separate entities but merges them together creating an illusion of moving images. This phenomenon is called **persistence of vision** (PoV). It has been observed that the frame rate should be around 25 to 30 fps for perceiving a smooth motion without gaps or jerks. Audio is added and synchronized with the apparent movement of images to create a complete video sequence. A video file therefore consists of multiple images and one or more audio tracks. One disadvantage of handling so much information together is an increase in file size and consequently large amount of processing resources to handle them. For example, a one minute video file consisting of 30 frames each 640 by 480 pixels in size and using 24-bits color information takes up more than 1582 MB of space. Audio sampled at 44,100 Hz adds another 10 MB to the file every minute. Moreover, playback of the video file requires a bandwidth of around 30 MB/s. Schemes of compression therefore are very important for video to handle such large overheads.

For creating digital video, we first of all need the visual and audio information to be recorded in the form of electrical signals onto magnetic tapes or disks. The term for specifying this form of representation is **motion video** to distinguish it from another kind of representation called **motion picture** used in cinema theaters where video frames are recorded onto celluloid film using a photo-chemical process. Motion video in the form of electrical signals is generated from an **analog video camera** and stored in magnetic tapes like video cassettes and later played back using video cassette player (VCP). TV transmission is also a popular example of motion video display. Earlier generation analog video cameras used vacuum tubes called **cathode ray tubes** (CRT) to generate these signals, which can then be fed to a monitor to display the video, while audio was recorded separately using a microphone and fed to loud-speakers for generation of sound. Monochrome or grayscale video required a single intensity signal from the camera for the visual information and one or two audio signals, depending on whether the sound being played is mono or stereo. For displaying an image on a CRT monitor screen, the electron beam from the cathode is activated and focused on the phosphor-coated screen for emitting light. **Phosphor** is a chemical substance which emits a glow of light when it comes in contact with charged particles like electrons. To generate an image on the screen, the electron beam starts at the upper-left corner of the screen and sequentially traces over the first row of phosphor dots from left to right. At the end of each horizontal line, the beam moves diagonally to the beginning of the next row and starts the tracing operation. At the lower-right corner, the beam moves diagonally to the starting point at the upper-left corner and repeats the operation once again. This process is called **raster scanning** and usually completed about 60 times each second for a steady picture on screen, which is denoted as the **refresh rate** of the monitor, and each image produced on the screen is called a **frame**. A monitor which supports 60 frames per second produces a non-flickering image and is called a **progressive scan** monitor. An alternative technique, used especially for monitors of lower refresh rates, is called **interlacing** and the corresponding monitor is referred to as **interlace scan** monitor. In this case, one frame is split into two halves, each called a **field**. The first field, made of odd-numbered rows, is called the odd-field and the second row, made up of even-numbered rows, is called the even-field. Each field contains only half the number of rows and is scanned 60 times per second, reducing the effective refresh rate to 30 frames per second. Due to PoV, this kind of arrangement leads to a smooth blending of the rows of each field and helps to produce a non-flickering image even at low refresh rates. Notations which include the letters "*p*" and "*i*" are used to distinguish between progressive and interlaced monitors e.g. 720p and 1080i, the number denoting the total number of horizontal rows in the monitor.

New generation video cameras replace CRTs with electronic photo-sensors called **charge coupled devices** (CCD) which generate electrical signals roughly proportional to the intensity of light falling on them. Signals from a CCD array are collected sequentially and sent to a monitor for display. Modern day monitors use **liquid crystal display** (LCD) elements instead of CRT and electron beam. LCD elements are small transparent blocks filled with a liquid organic chemical substance consisting of long rod-like molecules which has properties of manipulating the direction of light rays flowing through the substance. LCD elements along with pair of polarizing filters allow light to flow from a backlight source like an LED to the observer in front creating the perception of a lighted pixel. Current flowing through the LCD elements change the orientation of molecules and prevent the light from reaching the observer. Switching on and off specific dots helps to create an image on screen. In case of color video cameras, three separate **RGB signals** corresponding to the primary colors red, green, and blue, are used to create composite colors on screen, and these signals are fed to a color monitor using three separate cables, a scheme which came to be known as **component video**. Inside the monitor, these signals are used to activate a color generation system like CRT electron guns or LCD elements for color reproduction.

RGB signals for color reproduction worked well when transmitted over short distances, typically over a few meters. However, when these signals were needed to be transmitted over large distances spanning several kilometers like in **TV transmission**, engineers ran into a separate set of problems. First, three separate copper cables running over several kilometers made the system costly. Second, even if the cost was ignored, three separate signals transmitted along the three cables did not arrive exactly at the same instant of time at the receiving end due to differences in attenuation factors, with the result that the images were frequently out of sync and distorted. Third, when color TV transmission started in several countries, the earlier monochrome system of black & white (B/W) TV also continued side by side, and so the engineers had to come up with a system such that the same transmitted signals could cater to both the B/W TV sets and color TV sets at the same time, which was not possible using the existing RGB signal format. To deal with all these problems, a new signal format was developed which was called **composite video** and instead of the RGB signals, it used a different form of signal called **YC signals**.

Here, Y indicated the **luminance** or intensity signal and C indicated the **chrominance** or color signal. The advantages of this format over the RGB format included the fact that both Y and C could be transmitted over a single cable or channel, and therefore arrived exactly at the same instant at the receiving end. This was done by splitting the typical 6 MHz bandwidth of a TV channel into two parts, 0 to 4 MHz was allotted to Y and 1.5 MHz for C, the remaining 0.5 MHz being used for audio. The reason behind this unequal distribution was because of the fact that the human eye was more sensitive toward the Y information and less toward the C information. Another advantage of the YC signal format was that only the Y part of the signal could be used to cater to the B/W TV sets while both Y and C could be used to cater to the color TV sets. This meant that only a single transmission system was required and only a filter at the receiving end could be used to remove the color signal for monochrome viewing. Due to the usage of a single transmission cable, cost of the system could also be reduced.

The Y signal representing the grayscale intensity can be computed using a linear combination of the RGB signals. After experimenting with a number of combinations and keeping in mind that the human eye was more sensitive toward the green part of the color spectrum, it was finally decided that Y would be composed of 60% of G, 30% of R, and 10% of B resulting in the relation:

$$Y = 0.299R + 0.587G + 0.114B$$

The color information was represented using a circular scale instead of a linear scale and needs two components for identifying a specific color value as a point on a plane, which was the color wheel. The color sub-components are called Cb and Cr and defined as follows:

$$Cb = B - Y$$

$$Cr = R - Y$$

Accordingly, the composite video format, requiring a single cable or channel for transmission, is specifically referred to as using the **YCbCr** signal format. Most of the analog video equipment used today for transmitting video signals use the composite video cable and connector for interfacing e.g. between VCP and TV.

Conversion of RGB to the YC signal format has another important advantage: the possibility of using less bandwidth by reducing color information. Studies have shown that the human eye is more sensitive to luminance (brightness) information than to chrominance (color) information. This finding is exploited to reduce color information during video transmission, a process referred to as **chroma sub-sampling**. The reduction in color information is denoted by a set of three numbers expressed as a ratio *A:B:C*. The numbers denote the amount of luminance and chrominance information within a window on the screen usually 4 pixels wide and 2 pixels high. Common values include 4:2:2 which imply that within a sliding 4×2 window, there are 4 pixels containing Y information along the first row, 2 pixels of C information along the first row, and 2 pixels of C information along the second row. Essentially, this means while all the pixels contain brightness information, color information is reduced to half along the horizontal direction. Other values frequently used are 4:1:1 and 4:2:0. The first set indicates one-fourth color information horizontally, while the second set indicates half-color information both along the horizontal and vertical directions. Obviously, a value of 4:4:4 indicates no reduction in color information.

Like images and audio, compression schemes called CODEC (coder/decoder) can be used to reduce the size of digital video files. Because of the large size of video files, lossless algorithms are not used much. Lossy compression algorithms delete information from the image and audio components to reduce size of video files. **File formats** in which the video is saved depend on the compression scheme used. The Windows native audio file format is AVI which is typically uncompressed. Lossy compression algorithms are associated with file formats like MPEG (MPEG-1), Window Media Video (WMV), MPEG-4 (MP4), Apple Quicktime Movie (MOV), and 3rd Generation Partnership Project (3GPP) for mobile platforms).

3.2 Toolboxes and Functions

Video processing functions in MATLAB® can be divided into two broad categories: basic MATLAB functions and Computer Vision System Toolbox (CVST) functions. The BM functions are a set of basic tools used for performing preliminary mathematical matrix operations and graphical plotting operations. The CVST, which provides algorithms, functions, and apps for designing and simulating computer vision and video processing systems, includes a set of more advanced tools for specialized processing tasks like blob analysis, object detection, and motion tracking. Some functions like import/export and basic playback, however are common to both BM and CVST collections. For solution of a specific task at hand, functions from both the BM set and the CVST set might be necessary. The source of these functions is accordingly mentioned as and when it is used to illustrate examples throughout this book. The MATLAB features for video processing tasks are illustrated as solutions to specific examples throughout this chapter. This book has been written using MATLAB version 2018b; however, most of the functions can be used with versions 2015 and later with little or no change. The video file formats supported by MATLAB includes MPEG-1 (.mpg), Windows Media® Video (.wmv), MPEG-4 including H.264 encoded video (.mp4, .m4v), Apple QuickTime Movie (.mov), 3GPP, and Ogg Theora (.ogg). Most of the video files used in the examples are included in the MATLAB package and do not need to be supplied by the user. The inbuilt media files are included in the folders: *(matlab-root)\toolbox\vision\visiondata*.

3.2.1 Basic MATLAB® (BM) Functions

The BM functions used in this chapter fall into five different categories: Language Fundamentals, Mathematics, Graphics, Data Import and Analysis, Programming Scripts, and Functions. A list of these is provided below along with their hierarchical structure and a one line description of each. The BM set consists of thousands of functions out of which a subset has been used in this chapter keeping in view the scope and length of this book.

1. *Language Fundamentals:*
    1. clc: clear command window
2. *Graphics*
    1. figure: Create a figure window
    2. frame2im: Return image data associated with movie frame

3. getFrame: Capture axes or figure as movie frame

4. im2frame: Convert image to movie frame

5. plot: two-dimensional (2-D) line plot

6. subplot: Multiple plots in a single figure

7. title: Plot title

8. xlabel, ylabel: Label x-axis, y-axis

3. *Data Import and Analysis*

1. clear: Remove items from workspace memory

2. hasFrame: Determine if frame is available to read

3. mmfileinfo: Information about multimedia file

4. readFrame: Read video frame from video file

5. VideoReader: Read video files

6. VideoWriter: Write video files

4. *Programming Scripts and Functions*

1. for ... end: Repeat statements specified number of times

2. load: Load variables from file into workspace

3. pause: Stop MATLAB execution temporarily

4. while ... end: Execute statements while condition is true

5. *Advanced Software Development*

1. release: release resources

2. step: Run System object algorithm

3.2.2 Computer Vision System Toolbox (CVST) Functions

CVST provides algorithms, functions, and apps for designing and simulating computer vision and video processing systems. CVST includes libraries of video processing algorithms for modifying digital video signals. These are divided into a number of categories: (1) Input, Output, and Graphics; (2) Object Detection and Recognition; and (3) Object Tracking and Motion Estimation. Some of the most commonly used functions from CVST discussed in this chapter are listed below:

1. *Input, Output, and Graphics:*

1. insertObjectAnnotation: Annotate image or video

2. insertText: Insert text in image or video

3. vision.VideoFileReader: Read video frames from video file

4. vision.VideoPlayer: Play video or display image

5. vision.DeployableVideoPlayer: Display video

6. vision.VideoFileWriter: Write video frames to video file

2. *Object Detection and Recognition:*

1. vision.BlobAnalysis: Properties of connected regions

2. vision.ForegroundDetector: Foreground detection using Gaussian mixture models (GMM)

3. vision.PeopleDetector: Detect upright people using Histogram of Oriented Gradient (HOG) features

4. vision.CascadeObjectDetector: Detect objects using the Viola–Jones algorithm

5. ocr: Recognize text using optical character recognition (OCR)

3. *Object Tracking and Motion Estimation:*

1. configureKalmanFilter: Create Kalman filter for object tracking

2. opticalFlowFarneback: Estimate optical flow using Farneback method

3. opticalFlowHS: Estimate optical flow using Horn–Schunck method

4. opticalFlowLK: Estimate optical flow using Lucas–Kanade method

5. opticalFlowLKDoG: Estimate optical flow using Lucas–Kanade derivative of Gaussian method

6. vision.BlockMatcher: Estimate motion between images or video frames

7. vision.HistogramBasedTracker: Histogram-based object tracking

8. vision.PointTracker: Track points in video using Kanade–Lucas–Tomasi (KLT) algorithm

9. vision.KalmanFilter: Kalman filter for object tracking

3.3 Video Input Output and Playback

The most basic way to play a video file is to use the IPT function implay with a specified filename. The BM function mmfileinfo can be used to extract information about the file e.g. the audio and video structures. In the following example, this is indicated as method-1. The BM function VideoReader can be used to read a video file, which creates a *VideoReader* object. Properties of the object can be used to retrieve video information like duration, frame height and width, file path, frame rate, number of frames, file format, and so on. The property *getFileFormats* returns the video file formats supported. The *read* method can be used to read the video data which is stored in a 4-D array containing the frame height, frame width, color channel index, and frame index. The IPT function *implay* can then be used to play the video data. For playing a subset of the frames, the start-frame and end-frame can be specified. A value of *Inf* for the end-frame plays the video until the end. In the following example, this is indicated as method-2 (Figure 3.1).

Example 3.1: Write a program to read and play video files

```
clear; clc;
fv = 'rhinos.avi';
fa = 'RockDrums-44p1-stereo-11secs.mp3';

% method-1
implay(fv);

ia = mmfileinfo(fa); ia.Audio
iv = mmfileinfo(fv); iv.Video

% method-2
vr = VideoReader(fv); % video reader object
d = vr.Duration
h = vr.Height
```

FIGURE 3.1
Output of Example 3.1.

```
w = vr.Width
p = vr.Path

bp = vr.BitsPerPixel
fr = vr.FrameRate
nf = vr.NumberOfFrames
fo = vr.VideoFormat
ff = vr.getFileFormats

v = read(vr);          % 4-D array containing video frames
implay(v)
sf = 11;     % start frame
ef = 50;     % end frame
cv1 = read(vr,[sf, ef]);
cv2 = read(vr,[50, Inf]);

implay(cv1) % Total frame numbers indicated in status bar
implay(cv2) % Total frame numbers indicated in status bar
```

Alternatively, the *readFrame* method can be used to read each frame individually and play them back using a loop. Each frame returned (vf) is a 3-D array for a color image, and the operation is to be repeated depending on the total number of frames to be displayed. The total number of frames can be extracted from the property *NumberOfFrames* of the video object. The video frames are shown as a series of images in a figure window, the pause between each image being the reciprocal of the frame rate, which is the time gap between each consecutive frame. A counter can be used to display the corresponding frame number in the title field of the figure window. This syntax is to be used when the video needs to be played from a specified offset time instead of the beginning. The starting time for playback can be specified using the option *CurrentTime* while creating the video object. In that case since the total number of frames to be played is not known without computation, the loop for displaying the frames needs to use the *while* structure instead of the *for* structure, with the *hasFrame* method to determine whether a frame is available for playing. The following example displays two instances of a video, one for playing all the frames and the other for playing the video with an offset of 2.5 seconds from the beginning. The number of frames actually played is displayed on top of each image. In the first case all 114 frames are played while in the second case the number of frames played can be calculated

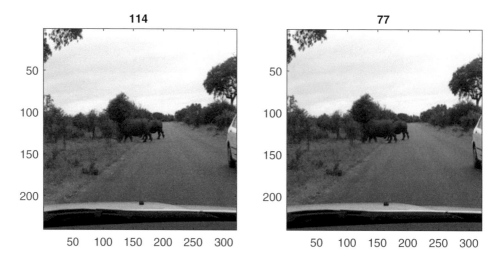

FIGURE 3.2
Output of Example 3.2.

by subtracting the offset time from the total duration and multiplying the result by the frame rate, with a rounding off to the next integer, which in this case becomes $(7.6 - 2.5) \times 15 \approx 77$ (Figure 3.2).

**Example 3.2: Write a program to read and play a subset
of video frames using a starting time offset.**

```
clear; clc;
fn = 'rhinos.avi';
vr = VideoReader(fn);   % video reader object
nt = vr.NumberOfFrames; % number of frames
fr = vr.FrameRate;      % frame rate
du = vr.Duration;       % video duration
of = 2.5;               % time offset in seconds from beginning

subplot(121)
vr = VideoReader(fn);
fi = 0;
for i = 1:nf
    vf = readFrame(vr);     % video frame
    fi = fi + 1;            % frame index number
    image(vf); title(fi); axis square;
    pause(1/fr);
end

subplot(122)
vr = VideoReader(fn, 'CurrentTime', of);
fi = 0;
while hasFrame(vr)
    vf = readFrame(vr);
    fi = fi + 1;
    image(vf); title(fi); axis square;
    pause(1/fr);
end

nof = (du - of)*fr; % actual number of frames to be played
round(nof)
```

The *read* method stores the video in a 4-D array containing the frame height, frame width, color index, and frame index. It is possible to manipulate the frame index and select a

FIGURE 3.3
Output of Example 3.3.

subset of the total frames for display and playback. The following example displays every 10-th frame of a video and plays back the subset of the frames. To display the frames a 3×3 grid is created where frames 10, 20, 30, 40, 50, 60, 70, 80, 90 are displayed, and these frames are collected into a new 4-D array and played back (Figure 3.3).

Example 3.3: Write a program to display every 10-th frame of a video and play back these frames

```
clear; clc;
vr = VideoReader('rhinos.avi');
v = read(vr);    % original video array

figure

f = 0;               % frame index number
for r = 1:3          % row index of display grid
    for c = 1:3      % column index of display grid
        f = f + 10;
        subplot(3, 3, c + (r-1)*3, imshow(v(:,:,:,f)); title(f);
        w(:,:,:, c + (r-1)*3) = v(:,:,:,f);
    end
end

implay(w)    % new video array
```

The IPT function **implay** can also be used to playback a sequence of individual images that do not form part of a video, while the IPT function **immovie** can be used to create a movie from a sequence of frames. The following example loads a sequence of images in memory and plays them back. For indexed images, the corresponding color map is to be specified (Figure 3.4).

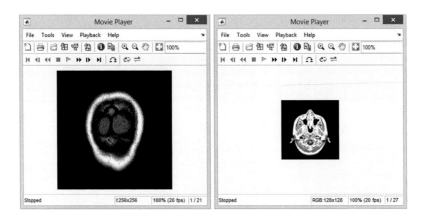

FIGURE 3.4
Output of Example 3.4.

> **Example 3.4: Write a program to playback a sequence of individual images, and to create a movie from a sequence of images.**
>
> ```
> clear; clc;
>
> load mristack;
> implay(mristack);
>
> load mri;
> mov = immovie(D,map);
> implay(mov)
> ```

The BM function **getframe** is used to capture the current figure as a movie frame. The BM function **movie** is used to play the collection of frames as a movie. The following example creates a 3-D surface from the BM functions **peaks** and **surf** and then changes the view of the surface using the BM function **view** by changing the azimuth and elevation from 1° to 360°. Each of these views is captured as a video frame which are then played back in sequence (Figure 3.5).

> **Example 3.5: Write a program to play back a sequence of graphic figures as a video.**
>
> ```
> clear; clc;
> for i = 1:360
> ```

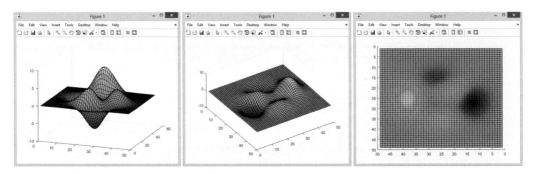

FIGURE 3.5
Output of Example 3.5.

```
    j = 1;
    surf(peaks); view(i,i); f(j) = getframe;
    j = j+1;
  end
movie(f)
```

While the *read* method automatically creates a 4-D structure to store the entire video, the *readFrame* method only reads one frame at a time. To store the entire video, a structure can be explicitly created by using the BM function **struct** by specifying the frame dimensions, datatype, and colormap, if required. The field *cdata* contains the actual video data returned by *readFrame*. The following example illustrates the creation of the structure and shows that the BM function **movie** can be used to playback video stored in the structure by specifying the frame rate and optionally a repeat value. The BM function **set** positions the playback window with respect to the screen coordinates (Figure 3.6).

Example 3.6: Write a program to create a movie structure and repeat playback of a video two times.

```
clear; clc;
vr = VideoReader('xylophone.mp4');
w = vr.Width;          % frame width
h = vr.Height;         % frame height
r = 2;                 % repeat playback
m = struct('cdata',zeros(h,w,3,'uint8'),'colormap',[]);

% read each frame and store in structure
k = 1;
while hasFrame(vr)
    m(k).cdata = readFrame(vr);
    k = k+1;
end

hf = figure;           % figure handle
x = 150; y = 200;      % position of playback window
set(hf,'position', [x y w h]);

movie(hf, m, r, vr.FrameRate);      % playback r times
```

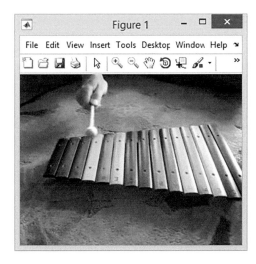

FIGURE 3.6
Output of Example 3.6.

The BM function **VideoWriter** can be used to write video files and creates a *VideoWriter* object . Properties of the object can be used to specify filename, file format, compression ratio, duration, and so on. The methods *open* and *close* can be used to open a file for writing video data and close the file after writing video data. The method *writeVideo* is to write the actual video data to the file. The following example is used to create video frames from plots of random numbers and write these frames to a video file. The BM function **getframe** is used to capture the current figure as a movie frame. The frame numbers are displayed on top of the images. After being created, the IPT function **implay** is used to play the video file (Figure 3.7).

Example 3.7: Write a program to create video frames from graphical plots.

```
clear; clc;
v = VideoWriter('newv.avi');
open(v);

for i = 1:15
    plot(rand(10,5), 'o'); title(i);
    f(i) = getframe;
    writeVideo(v, f);
end;
close(v);
fps = 12;
implay('newv.avi', fps);
```

Apart from the BM functions discussed above, the CVST function **vision.VideoFileReader** is used to read video frames from a file, the CVST function **vision.VideoPlayer** is used to playback the frames sequentially, and the CVST function **vision.VideoFileWriter** is used to write video to a file. The following example writes the first 50 frames of a video file to a new file whose name is specified. The reader creates a *VideoReader* object (*vr*) from which the frame rate is extracted (*fr*). The writer creates a *VideoWriter* object (*vw*) by specifying the filename and frame rate. The first 50 frames are read from *vr*, one frame at a time by using the *step* method and written to *vw* one frame at a time. After the operation is completed, the reader and writer are closed using the *release* method. Finally, the player creates a *VideoPlayer* object (*vp*) and each frame read by the reader is displayed by the player, with a pause in between equal to the reciprocal of the frame rate, which is the time gap between each consecutive frame. After the playback completes, both the reader and player are closed using the *release* method. Note that the new AVI file created contains only the first 50 frames read from the original video which contained 337 frames. This can be verified

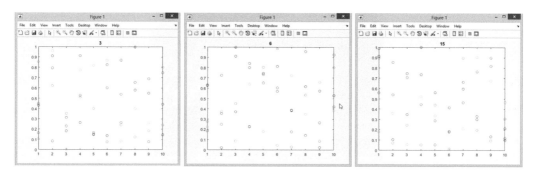

FIGURE 3.7
Output of Example 3.7.

FIGURE 3.8
Output of Example 3.8.

by invoking the Movie Player using the IPT function **implay** in which the total number of frames are displayed at the bottom-right corner (Figure 3.8).

Example 3.8: Write a program to use a subset of frames in a video to create a new video.

```
clear; clc;
fn = 'viplanedeparture.mp4';
vr = vision.VideoFileReader(fn);
fr = vr.info.VideoFrameRate;
vw = vision.VideoFileWriter('myFile.avi','FrameRate',fr);
for i = 1:50
    frame = step(vr);
    step(vw, frame);
end
release(vr);
release(vw);

clear;
fn = 'myFile.avi';
vr = vision.VideoFileReader(fn);
vp = vision.VideoPlayer;
fr = vr.info.VideoFrameRate;
while ~isDone(vr)
    frame = step(vr);
    step(vp, frame);
    pause(1/fr);
end
release(vr);
release(vp);
```

Before we end this section, it is worthwhile to mention that an additional means to play a video file is by using the CVST function **vision.DeployableVideoPlayer** which unlike the other functions mentioned earlier can generate C code and is capable of displaying high definition video at high frame rates. The following example illustrates this function. This function creates a *DeployableVideoPlayer* object which reads one frame at a time from a VideoReader object and displays one frame at a time using the Deployable Video Player, with a pause between each frame equal to the reciprocal of the frame rate, which is the time gap between each consecutive frame (Figure 3.9).

FIGURE 3.9
Output of Example 3.9.

Example 3.9: Write a program to play a video using the Deployable Video Player.

```
clear; clc;

vr = vision.VideoFileReader('atrium.mp4');
vp = vision.DeployableVideoPlayer;
fr = vr.info.VideoFrameRate;
cont = ~isDone(vr);
  while cont
    frame = step(vr);
    step(vp, frame);
    pause(1/fr);
    cont = ~isDone(vr) && isOpen(vp);
  end

release(vr);
release(vp);
```

3.4 Processing Video Frames

Since video frames are individual images, image processing techniques can be applied to each frame and compiled back to the video. The following example shows specific frames of a video being modified (inverted) and then written back to another video sequence. Using a video reader object, the frames are read and stored in a 4-D video array, from which every 5-th frame is read and image adjustment algorithms are applied to invert them. The modified frames are written back to another 4-D array, displayed in an image grid and then played back (Figure 3.10).

Example 3.10: Write a program to invert every 5-th frame of a video.

```
clear; clc;
vr = VideoReader('rhinos.avi');
nf = vr.NumberOfFrames;

v = read(vr);

g = v;
for k = 1:5:nf
    g(:,:,:,k) = imadjust(v(:,:,:,k), [0, 1], [1, 0]);
end

k = 0;
```

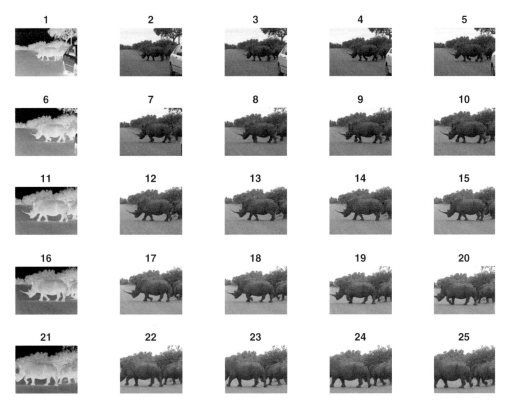

FIGURE 3.10
Output of Example 3.10.

```
for r = 1:5
    for c = 1:5
        k = k + 1;
        subplot(5, 5, c+(r-1)*5), imshow(g(:,:,:,k)); title(k)
    end
end

implay(g);
```

The BM function **im2frame** converts an image to a video frame, and the BM function **frame2im** converts the frames back to images. The following example converts five images to frames of a video and plays back the video in a Movie Player. Since the frames of a video usually all have the same dimensions, the images are resized to have the same width and height; however, this is not an essential requirement. An alternate way to playback the images like a video is to display them sequentially in a figure window with a pause in between (Figure 3.11).

Example 3.11: Write a program to convert a set of images to frames of a video.

```
clear; clc;

h = 300; w = 400;
a = imread('peppers.png'); a = imresize(a, [h w]);
b = imread('coloredChips.png'); b = imresize(b, [h w]);
c = imread('pears.png'); c = imresize(c, [h w]);
d = imread('football.jpg'); d = imresize(d, [h w]);
```

FIGURE 3.11
Output of Example 3.11.

```
e = imread('saturn.png'); e = imresize(e, [h w]);

f(1) = im2frame(a);
f(2) = im2frame(b);
f(3) = im2frame(c);
f(4) = im2frame(d);
f(5) = im2frame(e);

implay(f);

for i = 1:5
    imshow(frame2im(f(i)));
    pause(0.5)
end
```

The CVST function **insertText** is used to insert text in images and video frames. The following example displays frame numbers at the bottom right of each frame of a video. The position, color, opacity, and font size are specified as arguments for each frame, and the process is repeated for all frames in a loop structure (Figure 3.12).

Example 3.12: Write a program to insert text at specific locations of video frames.

```
clear; clc;

vr = VideoReader('xylophone.mp4');
nf = vr.NumberOfFrames;
fw = vr.Width;
fh = vr.Height;

v = read(vr);
position = [0.75*fw, 0.75*fh];

for k = 1:nf
    g(:,:,:,k) = insertText(v(:,:,:,k), position, num2str(k), ...
        'FontSize',20,'BoxColor','red',...
        'BoxOpacity',0.5,'TextColor','white');
end
implay(g);
```

FIGURE 3.12
Output of Example 3.12.

FIGURE 3.13
Output of Example 3.13.

The frames of a video can be selectively played at a specified frame rate. The following example shows only the odd frames, and even frames of a video are selected and played back at a frame rate 10% of the original fps. The frame numbers reflect the odd and even sets (Figure 3.13).

> **Example 3.13: Write a program to play only the odd frames and even frames of a video at a specified frame rate.**
>
> ```
> clear; clc;
> vr = VideoReader('xylophone.mp4');
> nf = vr.NumberOfFrames;
> fw = vr.Width;
> fh = vr.Height;
> ```

```
fr = vr.FrameRate;

v = read(vr);
position = [0.75*fw, 0.75*fh];

for k = 1:nf
    g(:,:,:,k) = insertText(v(:,:,:,k), position, num2str(k), ...
        'FontSize',20,'BoxColor','red',...
        'BoxOpacity',0.5,'TextColor','white');
end

j=1;
for k = 1:2:nf     % select odd frames
    m(:,:,:,j) = g(:,:,:,k);
    j = j+1;
end
implay(m, fr/10);

j=1;
for k = 2:2:nf     % select even frames
    m(:,:,:,j) = g(:,:,:,k);
    j = j+1;
end
implay(m, fr/10);
```

In a similar way, any kind of image processing task can be applied to each frame of a video and the processed frames can be written back to a new video. The following example shows a color video being converted to grayscale, then to binary and finally an edge detection filter being applied to the binary frames. The processed frames are then compiled back to a video structure (Figure 3.14). The figure shows frame 20 of each version.

Example 3.14: Write a program to convert a color video to grayscale and binary versions and apply an edge detection filter to each frame.

```
clear; clc;
fn = 'viplanedeparture.mp4';
vr = VideoReader(fn);
nf = vr.NumberOfFrames;

v = read(vr);

for k = 1:nf
    w(:,:,:,k) = rgb2gray(v(:,:,:,k));
end
```

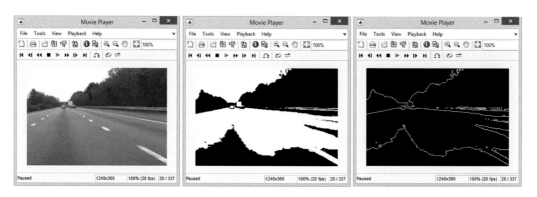

FIGURE 3.14
Output of Example 3.14.

```
implay(w);

for k = 1:nf
    u(:,:,:,k) = im2bw(w(:,:,:,k));
end

implay(u);

for k = 1:nf
    x(:,:,:,k) = edge(u(:,:,:,k), 'sobel');
end

implay(x);
```

The final example of this section shows noise being applied to all frames of a video and then removed using a noise filter (Figure 3.15).

Example 3.15: Write a program to apply a median filter to each frame of a video for noise reduction.

```
clear; clc;
fn = 'vipbarcode.mp4';
r = VideoReader(fn);
nf = r.NumberOfFrames;
fps = r.FrameRate;

v = read(r);

for k = 1:nf
    x(:,:,:,k) = rgb2gray(v(:,:,:,k));
end

implay(x);

for k = 1:nf
    y(:,:,:,k) = imnoise(x(:,:,:,k), 'salt & pepper', 0.1);
end

implay(y);

for k = 1:nf
    z(:,:,:,k) = medfilt2(y(:,:,:,k));
end

implay(z);
```

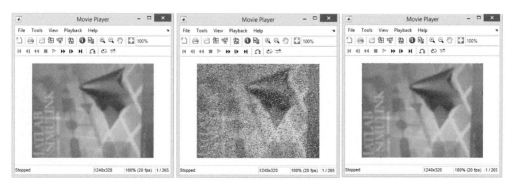

FIGURE 3.15
Output of Example 3.15.

3.5 Video Color Spaces

RGB signals can be converted to YCbCr color space using specified color conversion matrices (Poynton, 1996), as discussed in Section 3.1. Y, Cb, and Cr are components of a family of color spaces used to represent digital video frames. Y is the luminance (brightness) component of the signal, and Cb and Cr are the chrominance (color) components of the signal. Taking into account the fact that the human vision more sensitive to the green part of the spectrum than red or blue, YCbCr is related to RGB, according to the following relations:

$$Y = 0.299R + 0.587G + 0.114B$$

$$Cb = B - Y$$

$$Cr = R - Y$$

Substituting the value of Y in the second and third equations, we derive the following:

$$Y = 0.299R + 0.587G + 0.114B$$

$$Cb = -0.299R - 0.587G + 0.886B$$

$$Cr = 0.701R - 0.587G - 0.114B$$

The inverse relations are derived as follows;

$$R = Y + Cr$$

$$B = Y + Cb$$

$$G = \frac{Y - 0.299R - 0.114B}{0.587} = \frac{Y - 0.299(Y + Cr) - 0.114(Y + Cb)}{0.587} = Y - 0.1942Cb - 0.5094Cr$$

The IPT functions **rgb2ycbcr** and **ycbcr2rgb** convert RGB image to YCbCr color space and vice versa. The following example converts an RGB color image to YCbCr color space and displays it along with the individual color channels (Figure 3.16). To convert an entire video, all frames need to be converted in a likewise manner.

Example 3.16: Write a program to convert an RGB image and video to YCbCr color space and display the channels individually.

```
clear; clc;
RGB = imread('peppers.png');
R = RGB(:,:,1); G = RGB(:,:,2); B = RGB(:,:,3);
YCBCR = rgb2ycbcr(RGB);
Y = YCBCR(:,:,1); CB = YCBCR(:,:,2); CR = YCBCR(:,:,3);

figure,
subplot(241), imshow(RGB); title('RGB')
subplot(242), imshow(R, []); title('R')
subplot(243), imshow(G, []); title('G')
subplot(244), imshow(B, []); title('B')
subplot(245), imshow(YCBCR); title('YCbCr')
```

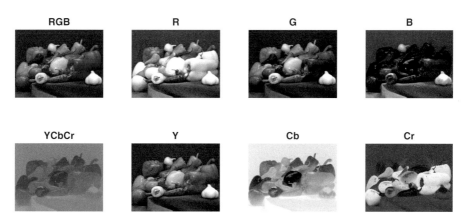

FIGURE 3.16
Output of Example 3.16.

```
subplot(246), imshow(Y, []); title('Y')
subplot(247), imshow(CB, []); title('Cb')
subplot(248), imshow(CR, []); title('Cr')

vr = VideoReader('xylophone.mp4');
nf = vr.NumberOfFrames;
v = read(vr);
for k = 1:nf
    YCbCr(:,:,:,k) = rgb2ycbcr(v(:,:,:,k));
end
implay(YCbCr);
```

The **National Television Systems Committee** (NTSC) established in 1940, is a world standard body for analog television broadcasting in a number of countries including USA, Canada, Korea, and Japan. It has specified a standard using 525 horizontal lines, 30 frames per second (fps), two interlaced fields per frame, an aspect ratio of 4:3, and frequency modulation for the audio signal. In 1950, the committee was reconstituted to establish a standard for color TV transmission compatible with the existing format for B/W TV. As per the recommendations of the committee a chroma sub-sampling scheme of 4:2:2 was chosen with a slight reduction in the frame rate from 30 fps to 29.97 fps due to technical issues related to the addition of a color sub-carrier signal. The luminance information is calculated from the RGB signals using the Y equation given above, while the chrominance part is represented using two sub-components named I (in-phase) and Q (quadrature-phase). YIQ is related to RGB, according to the following relations:

$$\begin{bmatrix} Y \\ I \\ Q \end{bmatrix} = \begin{bmatrix} 0.299 & 0.587 & 0.114 \\ 0.597 & -0.274 & -0.321 \\ 0.211 & -0.523 & 0.311 \end{bmatrix} \begin{bmatrix} R \\ G \\ B \end{bmatrix}$$

$$\begin{bmatrix} R \\ G \\ B \end{bmatrix} = \begin{bmatrix} 1.000 & 0.956 & 0.621 \\ 1.000 & -0.272 & -0.647 \\ 1.000 & -1.106 & 1.706 \end{bmatrix} \begin{bmatrix} Y \\ I \\ Q \end{bmatrix}$$

Currently these conversion relations are part of the ITU-R BT.1700 standard *(https://www.itu.int/rec/R-REC-BT.1700-0-200502-I/en)*.

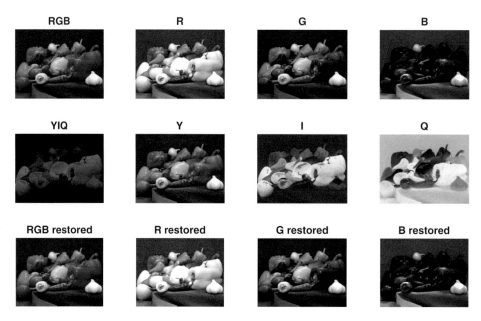

FIGURE 3.17
Output of Example 3.17.

The IPT functions **rgb2ntsc** and **ntsc2rgb** convert RGB image to NTSC color space and vice versa. The following example illustrates the conversion using still images. Each channel is separated out and displayed individually (Figure 3.17).

> **Example 3.17: Write a program to convert an RGB image to NTSC color space and display the channels individually. Also convert the image back to RGB color space.**
>
> ```
> clear; clc;
> RGB = imread('peppers.png');
> R = RGB(:,:,1); G = RGB(:,:,2); B = RGB(:,:,3);
> YIQ = rgb2ntsc(RGB);
> Y = YIQ(:,:,1); I = YIQ(:,:,2); Q = YIQ(:,:,3);
>
> figure,
> subplot(341), imshow(RGB); title('RGB')
> subplot(342), imshow(R, []); title('R')
> subplot(343), imshow(G, []); title('G')
> subplot(344), imshow(B, []); title('B')
> subplot(345), imshow(YIQ); title('YIQ')
> subplot(346), imshow(Y, []); title('Y')
> subplot(347), imshow(I, []); title('I')
> subplot(348), imshow(Q, []); title('Q')
>
> rgb = ntsc2rgb(YIQ);
> subplot(349), imshow(rgb); title('RGB restored')
> subplot(3,4,10), imshow(rgb(:,:,1)); title('R restored')
> subplot(3,4,11), imshow(rgb(:,:,2)); title('G restored')
> subplot(3,4,12), imshow(rgb(:,:,3)); title('B restored')
> ```

Phase Alternation Lines (PAL) is the other major TV transmission standard used in Europe, Asia, Africa, and Australia. It recommends a system of 625 horizontal lines at 25 frames per second, two interlaced fields per frame, an aspect ratio of 4:3, and a chroma sub-sampling scheme of 4:2:2. PAL uses the same luminance definition as NTSC but its chrominance sub-components named U and V are different. YUV is related to RGB, according to the following relations:

$$
\begin{bmatrix} Y \\ U \\ V \end{bmatrix} = \begin{bmatrix} 0.299 & 0.587 & 0.114 \\ -0.147 & -0.289 & 0.436 \\ 0.615 & -0.515 & -0.100 \end{bmatrix} \begin{bmatrix} R \\ G \\ B \end{bmatrix}
$$

$$
\begin{bmatrix} R \\ G \\ B \end{bmatrix} = \begin{bmatrix} 1.000 & 0 & 1.140 \\ 1.000 & -0.395 & -0.581 \\ 1.000 & 2.032 & 0 \end{bmatrix} \begin{bmatrix} Y \\ U \\ V \end{bmatrix}
$$

Currently these conversion relations are part ITU-R BT.601 standard *(https://www.itu.int/rec/R-REC-BT.601-7-201103-I/en)*. Since there is no inbuilt function to convert RGB to PAL color space, the following example uses the above transformation matrix for the conversion. It also implements the RGB to the NTSC transformation matrix to compare the results with the previous example which used the inbuilt function *rgb2ntsc*. The casting to the *double* datatype is necessary since the numerical calculations can change the values of the R, G, B components to more than 255 or less than 0, which would be truncated if the default datatype of *uint8* was used. The BM function **cat** concatenates the three matrices along the third dimension i.e. along the color channels (Figure 3.18).

Example 3.18: Write a program to convert RGB image to NTSC and PAL color spaces using transformation matrices.

```
clear; clc;
RGB = imread('peppers.png');
RGB = double(RGB);
R = RGB(:,:,1); G = RGB(:,:,2); B = RGB(:,:,3);

Y = 0.299 * R + 0.587 * G + 0.114 * B ;
I = 0.596 * R - 0.274 * G - 0.322 * B ;
Q = 0.211 * R - 0.523 * G + 0.312 * B ;
YIQ = cat(3,Y,I,Q);
```

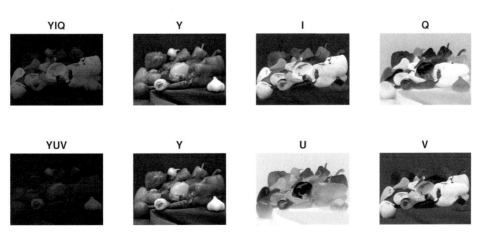

FIGURE 3.18
Output of Example 3.18.

```
Y = 0.29900  * R + 0.58700 * G + 0.11400 * B;
U = -0.14713 * R - 0.28886 * G + 0.43600 * B;
V = 0.61500  * R - 0.51499 * G - 0.10001 * B;
YUV = cat(3,Y,U,V);


figure
subplot(241), imshow(uint8(YIQ)); title('YIQ')
subplot(242), imshow(Y, []); title('Y')
subplot(243), imshow(I, []); title('I')
subplot(244), imshow(Q, []); title('Q')

subplot(245), imshow(uint8(YUV)); title('YUV')
subplot(246), imshow(Y, []); title('Y')
subplot(247), imshow(U, []); title('U')
subplot(248), imshow(V, []); title('V')
```

Note: It needs to be noted here that the NTSC and PAL standards developed for TV transmission were in a way responsible for creating the standard sampling rate of 44,100 Hz for CD quality digital audio in the later part of 1970. This value of 44.1 kHz was thought to be compatible with both the standards due to the following reasoning. Although the NTSC proposes 525 horizontal lines per frame i.e. 262 lines per field, taking into account the landing zones for the electron beam during raster scanning, there are actually 245 active lines per field which together with 30 frames per second i.e. 60 fields per second, and three audio samples being encoded per video line, leads to a total of $245 \times 60 \times 3 = 44,100$ audio samples per second. Again for PAL which proposes 625 lines per frame i.e. 312 lines per field, out of which 294 are active lines, and 25 frames per second i.e. 50 fields per second, and three audio samples being encoded per video line, leads to a total of $294 \times 50 \times 3 = 44,100$ audio samples per second. Because of these computations, 44.1 kHz was considered to the best usable value compatible with the existing standards during the 1970s.

3.6 Object Detection

3.6.1 Blob Detector

In computer vision, the term object detection relates to detecting the presence of well-known semantic entities like cars, chairs, tables, people, and so on in images and videos (Dasiopoulou et al., 2005), usually for the purpose of annotation, classification, and retrieval. Each class of object is characterized by a set of unique features and machine learning algorithms like neural networks are trained to look for those features in images and video frames (Zhang, 2018). The first step toward object detection is called **blob detection**, where a blob is a region in a digital image that differs in intensity or color from its surrounding regions. Once a blob is detected, it can be segmented from the rest of the image and then further processed to identify the semantic nature of the object represented by the region. One of the most common blob detectors is the **Laplacian of Gaussian** (LoG) operator. The input image $f(x,y)$ is first convolved with a Gaussian kernel $g(x,y)$

$$g(x,y) = \left(\frac{1}{2\pi\sigma}\right)\exp\left(-\frac{x^2+y^2}{2\sigma^2}\right)$$

$$L(x,y) = g(x,y) \otimes f(x,y)$$

The Laplacian operator ∇^2 is the double derivative of the argument, in this case the operator is applied to the Gaussian convolved image:

$$\nabla^2 L(x,y) = \frac{\partial^2 L(x,y)}{\partial x^2} + \frac{\partial^2 L(x,y)}{\partial y^2} = Lxx + Lyy$$

This usually results in strong positive responses for dark regions and strong negative responses for bright regions. However, the main problem in applying the blob detector is that the operator response is strongly dependent on the relative sizes of the blob regions and the size of the Gaussian kernel. Since the size of a prospective blob region is unknown for an arbitrary image, a multi-scale approach is adopted i.e. a number of different sizes of the LoG kernel is used to observe which scale produces the strongest response. A collection of such different sizes or scales is called the **scale space**. The CVST function **vision. BlobAnalysis** detects objects in a binary image and analyzes a number of properties for each object like area, centroid, bounding box, major axis length, minor axis length (of the ellipses that have the same normalized second central moments as the object itself), orientation (angle between major axis and X-axis), eccentricity, perimeter, and so on. The following example detects bright regions over a dark background and computes some shape properties of these regions. The computed values are displayed beside each region after enclosing them within bounding boxes (Figure 3.19).

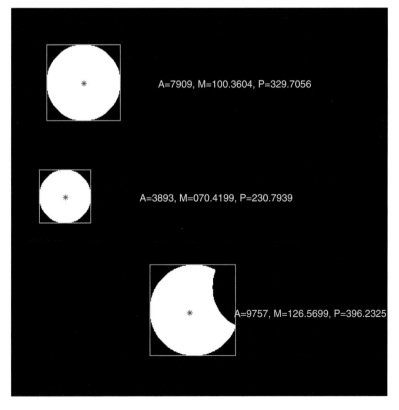

FIGURE 3.19
Output of Example 3.19.

Example 3.19: Write a program to detect presence of blobs in an image and compute their shape properties.

```
clear; clc;

h = vision.BlobAnalysis;
I = imread('circlesBrightDark.png'); b = im2bw(I, 0.7);

h.MajorAxisLengthOutputPort = true;
h.MinorAxisLengthOutputPort = true;
h.OrientationOutputPort = true;
h.EccentricityOutputPort = true;
h.EquivalentDiameterSquaredOutputPort = false;
h.ExtentOutputPort = false;
h.PerimeterOutputPort = true;


[area, ctrd, bbox, maja, mina, orie, ecce, peri] = step(h, b)

c = insertShape(double(b), 'Rectangle', bbox, 'Color', 'green');
figure, imshow(c);
hold on;

for i = 1:3
    plot(ctrd(i,1), ctrd(i,2), 'r*');
end

x = [ctrd(1,1)+100; ctrd(2,1)+100; ctrd(3,1)+60];
y = [ctrd(1,2); ctrd(2,2); ctrd(3,2)];
A1 = num2str(area(1)); A2 = num2str(area(2)); A3 = num2str(area(3));
M1 = num2str(maja(1)); M2 = num2str(maja(2)); M3 = num2str(maja(3));
P1 = num2str(peri(1)); P2 = num2str(peri(2)); P3 = num2str(peri(3));
S
s1 = strcat('A= ', A1, ', M=0', M1, ', P=', P1);
s2 = strcat('A= ', A2, ', M= ', M2, ', P=', P2);
s3 = strcat('A= ', A3, ', M= ', M3, ', P=', P3);
txt = [ s1; s2; s3];
text(x, y, txt, 'Color', 'yellow');
```

3.6.2 Foreground Detector

Real-time segmentation of moving regions in a video involves background subtraction. A common method of modeling pixels of the region is to assume that they have been derived from a mixture of adaptive Gaussians (Stauffer and Grimson, 1999). Based on the persistence and variance of each of the Gaussians of the mixture, it can be determined which Gaussians correspond to the background, and pixels which do not fit the background distributions are considered as foreground. The values of a particular pixel over time are considered as a time series i.e. at any time t, the states of a particular pixel (x_0, y_0) for a video V is given by

$$\{X_1, ..., X_t\} = \{V(x_0, y_0, i) : 1 \le i \le t\}$$

Each pixel of each frame is then used to update the adaptive distributions to differentiate the background from the foreground. Every new pixel value X_t is checked against the K Gaussian distributions using a k-means approximation until a match is found. A match is defined as a pixel value within 2.5 standard deviations of a distribution. The probability of observing the current pixel value is given by the following, where $\omega_{i,t}$ is an estimate of the weight, which determines what portion of the data is accounted by this Gaussian, of the i-th Gaussian in the mixture at time t, $\mu_{i,t}$ is the mean value of the i-th Gaussian at time t, and $c_{i,t}$ is the covariance matrix of the i-th Gaussian at time t, and η is a Gaussian probability density function.

$$P(X_t) = \sum_{i=1}^{K} \omega_{i,t} \eta \left(X_t, \mu_{i,t}, c_{i,t} \right)$$

The above method enables identifying the foreground pixels in each new frame, which can subsequently be segmented into regions by a two-pass connected component algorithm. To identify blobs from grayscale or color images, the CVST function **vision. ForegroundDetector** computes a foreground mask by determining whether individual pixels are part of the background or foreground. It uses a specified number of initial video frames for training the background model and GMM to create a foreground mask. The mask is used to create bounding box parameters based on which the CVST function **insertShape** draws designated shapes with specified colors on to the video frames. The following example detects a foreground object from each frame of a video and inserts a bounding box to demarcate it (Figure 3.20).

> **Example 3.20: Write a program to detect presence of foreground objects in a video and draw a bounding box around any if found.**
>
> ```
> clear; clc;
>
> fn = 'visiontraffic.avi';
> vr = vision.VideoFileReader(fn, 'VideoOutputDataType','uint8');
> fd = vision.ForegroundDetector;
> ba = vision.BlobAnalysis('CentroidOutputPort', false, ...
>     'AreaOutputPort', false, ...
>     'BoundingBoxOutputPort', true, ...
>     'MinimumBlobArea', 100);
>
> vp = vision.VideoPlayer();
> while ~isDone(vr)
>     frame = vr();
>     mask = fd(frame);
>     bbox = ba(mask);
>     out = insertShape(frame,'FilledRectangle', bbox, 'Color', 'white', 'Opacity',0.3);
>     vp(out);
> end
> release(vp);
> release(vr);
> ```

3.6.3 People Detector

To identify unoccluded people in upright positions, the human detector uses the HOG features and a trained Support Vector Machine (SVM) classifier (Dalal and Triggs, 2005).

FIGURE 3.20
Output of Example 3.20.

The method is based on evaluating normalized local histograms of image gradient orientations. For a pixel at location (x,y) with intensity $I(x,y)$, the gradient magnitude G and orientation θ is given as follows:

$$Gx = \frac{\partial I(x,y)}{\partial x} = \{I(x+1,y) - I(x-1,y)\}$$

$$Gy = \frac{\partial I(x,y)}{\partial y} = \{I(x,y+1) - I(x,y-1)\}$$

$$G = \sqrt{Gx^2 + Gy^2}$$

$$\theta = \arctan\left(\frac{Gy}{Gx}\right)$$

An image of an upright person, standing or walking, is typically assumed to be 64×128 pixels in dimensions, which is divided into rectangular cells, each 8×8 pixels in size. Hence, there would be eight cells along the width and 16 cells along the height. A collection of 2×2 cells constitute one block i.e. each block is 16×16 pixels in size. Blocks are considered with 50% overlap and within each block, the pixel orientations are quantized into nine bins ranging over angles 0°–180° i.e. with 20° angular separation. The vote for the histogram bins is the gradient magnitude. The final feature vector is a concatenation of all the histograms in each block. The CVST function **vision.PeopleDetector** can be used to detect people in an image in the following example and draws bounding boxes around them along with detection scores. The function **insertObjectAnnotation** is used to draw rectangles around the detected people by using the coordinates of the bounding boxes returned by the detector and display the matching scores returned by the detector (Figure 3.21).

> **Example 3.21: Write a program to detect presence of people in a video frame and draw a bounding box around any if found.**

```
clear; clc;
I = imread('visionteam1.jpg');
[row, col, cch] = size(I);
I = imresize(I, [0.6*row, 0.9*col]);
pd = vision.PeopleDetector;
[bb, scores] = pd(I);
I = insertObjectAnnotation(I,'rectangle', bb, scores);
figure, imshow(I);
title('Detected people and detection scores');
```

3.6.4 Face Detector

The **cascade object detector** uses the Viola–Jones algorithm (Viola and Jones, 2001) to detect people's faces, noses, eyes, mouth, or upper body. It is most suitable to detect object categories whose aspect ratio does not vary too much. It uses a sliding window to decide whether the window contains the object of interest. The size of the window may vary to accommodate objects at different scales but its aspect ratio remains fixed. The cascade classifier consists of a collection of learners, and each stage is trained using a technique called boosting. Each learner labels the region within the current window as either

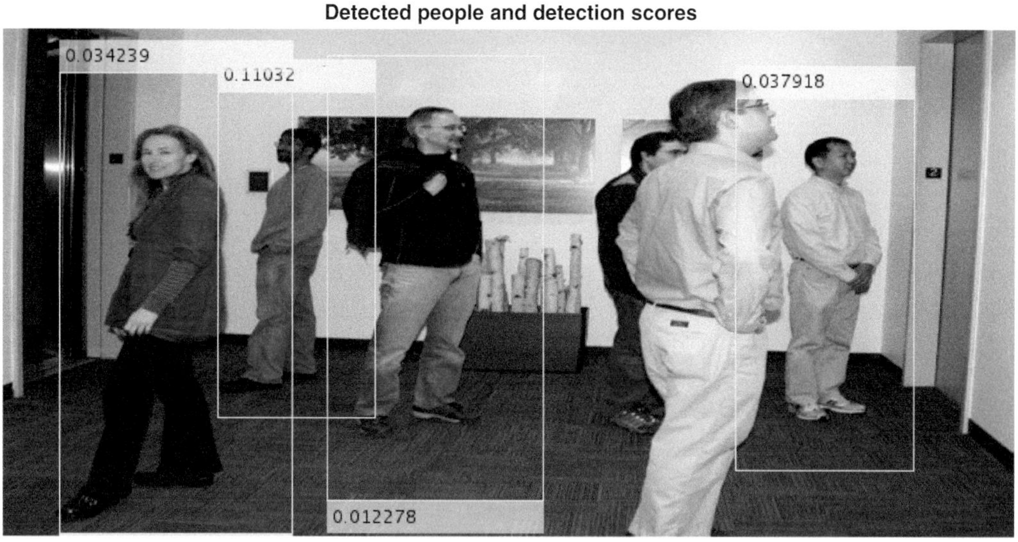

FIGURE 3.21
Output of Example 3.21.

positive or negative. If the label is positive, the classifier passes the region to the next stage, and if the label is negative, the classification of the region completes and the window is shifted to the next position. The classifier reports an object found at the current window location when all the stages of the classifier reports the region as positive. A cascade classifier needs to be trained using a set of positive samples and a set of negative samples. The Viola–Jones algorithm can detect a variety of objects; however, it was primarily motivated by the problem of human face detection. A constraint for this is that the faces should be frontal, upright, and unoccluded. The algorithm uses the Haar-like features developed from the concept of Haar wavelets (Haar, 1910), take the form of rectangular regions at specific locations in a detection window. Features consist of sums of pixel intensities in different regions and differences between these sums, which are then used to categorize subsections of the image. For example, for a human face, regions like the eyes are darker than neighboring regions like the cheeks, so sum of pixel intensities within a rectangle placed above the eyes would be different than those placed above the cheeks by a specified threshold, which if true, signals to the system what part of the face this might be. Because Haar-like features are weak learners with low classification rates, a large number of such features are organized to form a "cascade" for good classification. However, because of their simplicity, the calculation speed is quite high. Moreover, due to the use of summed-up tables or integral images (Crow, 1984), a Haar-like feature of any size can be calculated in constant time, depending on the number of rectangles present. An **integral image** is a 2-D lookup table in the form of a matrix with the same size as the original image, such that the value at any point is the sum of all pixels located on the upper-left region of the original image. This allows computation of rectangular areas at any position or scale using only four lookups. For example, to calculate the area enclosed within the rectangle bounded by the upper-left vertex A, upper-right vertex B, lower-right vertex C, and lower-left vertex D of an integral image I we can perform $sum = I(C) + I(A) - I(B) - I(D)$. Viola–Jones uses three types of features in their computations: 2-rectangle, 3-rectangle, and 4-rectangle which respectively needs 6, 8, and 9 lookups to calculate their sums. In the following example,

Detected faces

FIGURE 3.22
Output of Example 3.22.

a cascade object detector *fd* is used to generate bounding box parameters *bb* of detected faces in an image which are demarcated using rectangles with annotations (Figure 3.22).

> **Example 3.22: Write a program to use a cascade classifier to detect human faces in an image.**
>
> ```
> clear; clc;
> fd = vision.CascadeObjectDetector;
> I = imread('visionteam.jpg');
> bb = fd(I);
> df = insertObjectAnnotation(I,'rectangle', bb,'Face');
> figure,
> imshow(df)
> title('Detected faces');
> ```

3.6.5 Optical Character Recognition (OCR)

OCR refers to techniques for converting images of text to the actual editable text. The image is usually acquired from scanned documents or digital photographs and has wide applications in data entry from printed documents like books, receipts, bank statements, invoices, business cards, car number plates, and so on. After the OCR process, the text become computer recognizable and can be searched and edited using word processing software. Major challenges in OCR include handling various fonts, styles, sizes of typed text as well as almost unlimited variations in the handwritten text. The general techniques for text recognition involve identification of each character or word from a text stream and matching the isolated glyph with stored patterns within a database. The CVST function **ocr** is used to identify text from images, and the results returned by the function include a list of words recognized from the input document and the location of the characters in the form of coordinates of bounding boxes. In the following example, the function returns an *ocrText* object ot containing the actual words, number of words, and

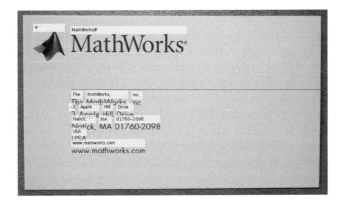

FIGURE 3.23
Output of Example 3.23.

word bounding boxes detected. Based on this the CVST function **insertObjectAnnotation** returns an annotated image ai where the words are demarcated with bounding box rectangles and the actual words on top of each (Figure 3.23).

Example 3.23: Write a program to recognize text from a document using OCR.

```
clear; clc;
I = imread('businessCard.png');
ot = ocr(I);
nw = size(ot.Words, 1);

for i=1:nw
    bb(i,:) = ot.WordBoundingBoxes(i,:);
    wd{i,:} = ot.Words{i};
end

ai = insertObjectAnnotation(I, 'rectangle', bb, wd);
imshow(ai)
```

3.7 Motion Tracking

3.7.1 Histogram Based Tracker

The **histogram based tracker** uses the histogram of pixel values of an object in an initial video frame to track it over the subsequent frames using the mean-shift algorithm. The **mean-shift** algorithm (Comaniciu and Meer, 2002) is a non-parametric technique (i.e. cannot be calculated from parameters of an equation) proposed for finding an area of a feature space where the density of the feature points is the highest. The process starts with an initial window in the feature space with a known center and a centroid which is calculated based on the distribution of the data points within the window area. Typically these two will not match, and the window is shifted so that its center coincides with the calculated centroid. Based on the new data distribution within the window, the centroid is re-calculated and the window re-shifted. This process continues in an iterative manner until the window center coincides with the centroid when the window is deemed to be over the region of maximum density. However, it requires knowing the size of the object to be tracked, which is defined by the user at the initialization by specifying a window size and location. The data distribution map is obtained after a back-projection of a color model which is typically the histogram of the color that needs to be identified. An example would be to use the skin color of a person's face for tracking it over the video frames. The color is usually represented in the HSV color space, and the skin histogram based on the hue-saturation (h,s) values becomes the model histogram in the initial frame. The model histogram depicts that for each value of (h,s) pair in the initial window, what is the probability of the pixel occurring i.e. how many pixels have that value of (h,s) in the initial window. This histogram is then applied over other frames to find the areas containing the same skin hue and saturation. Thus for a test frame, for each pixel find the value of (h,s) and then using a lookup of the model histogram find the probability value for that color and substitute this value for the pixel value. This process is called **histogram back-projection**. After the process completes, the brightest regions in the test frame correspond to the highest probability of finding the object of the initial frame in the test frame. The mean-shift algorithm is then used to move the window to the region containing the brightest areas in the test frame. This process is continued for all frames. Now, one problem with the mean-shift algorithm is that the window size and orientation remain the same over all the frames, whereas the actual object being tracked may change in both size (e.g. moving closer or farther from the camera) and orientation (e.g. upright or rotated). To handle this, a modified version of the mean-shift known as the **continuously adaptive mean-shift** (CAMshift) has been proposed (Bradski, 1998) which updates the size and orientation of the window in each frame by computing the major axis and its angle of inclination of the best fitting ellipse to the window. The CVST function **vision.HistogramBasedTracker** incorporates the CAMShift algorithm for object tracking. The following example shows a human face being tracked in each frame of an input video. Based on the HSV equivalent of an initial frame and an initial search region, the *initializeObject* method generates a 16 bin histogram, which is used by the tracker to return the bounding box (*bbox*) coordinates of the target object in each frame of the video. These coordinates are used to generate a rectangular shape with specified border color and inserted into the frame which is then displayed by the player object (*vp*) within a Video Player window. Each frame is displayed likewise with a pause in between (Figure. 3.24).

Example 3.24: Write a program to track movement of an object in a video and draw a bounding box around any if found.

```
clear; clc;

vr = vision.VideoFileReader('vipcolorsegmentation.avi');
```

FIGURE 3.24
Output of Example 3.24.

```
vp = vision.VideoPlayer();
si = vision.ShapeInserter('BorderColor','Custom', ...
    'CustomBorderColor',[1 0 0]);
objectFrame = vr();
objectHSV = rgb2hsv(objectFrame);
objectRegion = [40, 45, 25, 25];
tr = vision.HistogramBasedTracker;
initializeObject(tr, objectHSV(:,:,1) , objectRegion);
while ~isDone(vr)
  frame = vr();
  hsv = rgb2hsv(frame);
  bbox = tr(hsv(:,:,1));
  out = si(frame,bbox);
  vp(out);
  pause(0.1)
end
```

3.7.2 Optical Flow

Optical flow is a technique used for determining motion of objects in a video by measuring the change in pixel intensities over a set of frames. The underlying assumption is that a pixel at location (x,y) and time t in one video frame will have moved by amount Δx and Δy in some other frame at Δt if there is a relative motion between an object and the observer, but its intensity or brightness will remain the same. If $I(x,y,t)$ be the intensity of a pixel in one frame, then the brightness constraint can be written as:

$$I\left(x,y,t\right) = I\left(x + \Delta x, y + \Delta y, t + \Delta t\right)$$

Assuming the movement to be small, using the Taylor series expansion and ignoring the higher order terms we have:

$$I\left(x + \Delta x, y + \Delta y, t + \Delta t\right) = I\left(x,y,t\right) + \left(\frac{\partial I}{\partial x}\right)\Delta x + \left(\frac{\partial I}{\partial y}\right)\Delta y + \left(\frac{\partial I}{\partial t}\right)\Delta t$$

$$\frac{\partial I}{\partial x}\Delta x + \frac{\partial I}{\partial y}\Delta y + \frac{\partial I}{\partial t}\Delta t = 0$$

$$\left(\frac{\partial I}{\partial x}\right)\left(\frac{\Delta x}{\Delta t}\right) + \left(\frac{\partial I}{\partial y}\right)\left(\frac{\Delta y}{\Delta t}\right) + \left(\frac{\partial I}{\partial t}\right)\left(\frac{\Delta t}{\Delta t}\right) = 0$$

$$Ix \cdot Vx + Iy \cdot Vy + It = 0$$

Here, Ix, Iy, and It are the partial derivatives of the image and Vx, Vy are the components of the optical flow velocity. The two unknowns Vx and Vy cannot be solved from one equation and hence there are a number of methods which introduce additional constraints for estimating the optical flow. The Horn–Schunck method (Horn and Schunck,1981) assumes that the optical flow is smooth over the entire image and estimates the velocities Vx and Vy by minimizing a global energy functional using an iterative process. The Farneback method (Farneback, 2003) approximates each neighborhood of both frames by quadratic polynomials, and then a method to estimate displacement fields from the polynomial expansion coefficients is derived. The Lucas–Kanade method (Lucas and Kanade, 1981) assumes that the flow is constant in a local neighborhood of each pixel under consideration and solves the flow equation using the least squares criterion. The plain least square solution provides the same importance to all pixels within a neighborhood window. In some cases, better results have been obtained by using a weighted version of the least square solution by using derivative of a Gaussian function (Barron et al., 1992). The CVST functions **opticalFlowFarneback**, **opticalFlowHS**, **opticalFlowLK**, and **opticalFlowLKDoG** are used to estimate the direction and speed of a moving object from one video frame to another using the Farneback, Horn–Schunck, Lucas–Kanade, and Lucas–Kanade derivative of Gaussian methods. The *estimateFlow* method estimates the optical flow of the video frame with respect to the previous frame. The following example illustrates each of the above methods creating optical flow objects and using five frames for estimation in each case. The CVST function **plot** plots the optical flow as a quiver plot. The parameter *DecimationFactor* controls the number of arrows, and the parameter *ScaleFactor* controls the length of arrows (Figure 3.25).

Example 3.25: Write a program to estimate the direction and speed of a moving object in a video using optical flow analysis.

```
clear; clc;
fn = 'visiontraffic.avi';
vr = VideoReader(fn);
v = read(vr);

for k = 155:160
    w(:,:,:,k-154) = rgb2gray(v(:,:,:,k));
end;

% Farneback method

figure,
off = opticalFlowFarneback;
for i = 1:5
    vf = w(:,:,:,i);
    of1 = estimateFlow(off,vf);
    subplot(221),
```

Farneback method

Horn-Schunck method

Lucas-Kanade method

Lucas-Kanade DoG method

FIGURE 3.25
Output of Example 3.25.

```
        imshow(vf); hold on; title('Farneback method');
        plot(of1,'DecimationFactor',[10 10],'ScaleFactor',2);
        hold off;
end;

% Horn-Schunck method

ofh = opticalFlowHS;
for i = 1:5
        vf = w(:,:,:,i);
        of2 = estimateFlow(ofh,vf);
        subplot(222),
        imshow(vf); hold on; title('Horn-Schunck method');
        plot(of2,'DecimationFactor',[2 2],'ScaleFactor',30);
        hold off;
end;

% Lucas-Kanade method

ofl = opticalFlowLK;
for i = 1:5
        vf = w(:,:,:,i);
        of3 = estimateFlow(ofl,vf);
        subplot(223),
        imshow(vf); hold on; title('Lucas-Kanade method');
        plot(of3,'DecimationFactor',[5 5],'ScaleFactor',10);
        hold off;
end;

% Lucas-Kanade DoG method

ofd = opticalFlowLKDoG;
for i = 1:5
        vf = w(:,:,:,i);
        of4 = estimateFlow(ofd,vf);
        subplot(224),
        imshow(vf);hold on; title('Lucas-Kanade DoG method');
        plot(of4,'DecimationFactor',[3 3],'ScaleFactor',25);
        hold off;
end;
```

3.7.3 Point Tracker

The point tracker uses the KLT algorithm to track a set of points over video frames (Lucas and Kanade, 1981; Tomasi and Kanade, 1991). Given a set of feature points in the current frame, the KLT algorithm attempts to find corresponding points in the subsequent frames such that the difference between them can be minimized. The feature points initially can be generated by algorithms like the Harris Corner Detector. For each point, the corresponding geometric transformation is computed between consecutive frames. Finally, motion vectors are generated using optical flow to get a track for each point over frames. From the optical flow equation, we have the following, where A^T represents the transpose of:

$$Ix \cdot Vx + Iy \cdot Vy = -It$$

$$\begin{bmatrix} Ix, Iy \end{bmatrix} \cdot \begin{bmatrix} Vx \\ Vy \end{bmatrix} = -It$$

$$A \cdot V = B$$

$$A^T \cdot A \cdot V = A^T \cdot B$$

$$V = \mathrm{inv}\left(A^T \cdot A\right) \cdot A^T \cdot B$$

The CVST function **vision.PointTracker** is used to track points in a video using the KLT algorithm. The CVST function **detectMinEigenFeatures** is used to detect corners using

FIGURE 3.26
Output of Example 3.26.

the minimum eigenvalue algorithm (see Section 4.5 for details on the minimum eigenvalue algorithm). The initial frame *vf* and the specified region of interest *rg* are used to return feature points which is used by the point tracker *pt* to track the corresponding points in the subsequent frames. The points are designated by markers in each of the frames and sent to the player for display (Figure 3.26).

> **Example 3.26: Write a program to track corner points of an object in a video.**
>
> ```
> clear; clc;
>
> vr = vision.VideoFileReader('visionface.avi');
> vp = vision.VideoPlayer('Position',[100,100,680,520]);
> vf = vr();
> rg = [264,122,93,93];
> points = detectMinEigenFeatures(rgb2gray(vf),'ROI',rg);
> pt = vision.PointTracker('MaxBidirectionalError',1);
> initialize(pt,points.Location,vf);
> while ~isDone(vr)
>       frame = vr();
>       [points,validity] = pt(frame);
>       out = insertMarker(frame,points(validity, :),'+');
>       vp(out);
> end
> release(vp);
> release(vr);
> ```

3.7.4 Kalman Filter

The Kalman filter algorithm is used for tracking an object by using a series of measurements observed over time and produces estimates of unknown variables by calculating a joint probability distribution over the variables for each timeframe (Kalman, 1960). The constraint that needs to be satisfied is that the object must be moving at constant velocity or constant acceleration. The algorithm involves two steps: prediction and correction. The first step uses previous steps to predict the current state of the object, while the second step uses actual measurements of the object location to correct the predicted state. The CVST function **vision.KalmanFilter** is used to track objects in a video using the Kalman filter. The method *predict* uses previous states to predict the current state, and the method *correct* uses the current measurement, such as object location, to correct the state. The CVST function **configureKalmanFilter** is used to configure a Kalman filter, by setting up the filter for tracking a physical object in a Cartesian coordinate system, moving with constant velocity or constant acceleration. The CVST function **insertObjectAnnotation** is used to annotate

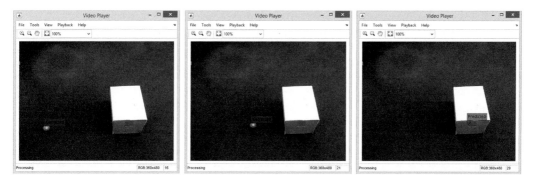

FIGURE 3.27
Output of Example 3.27.

images with specified text at specified locations. In the following example, a *KalmanFilter* object *kf* is initialized based on a detected object location *dloc* in an initial frame *vf* of a video generated using a *ForegroundDetector* object *fd* and a *BlobAnalysis* object *ba*. Using a motion model of constant acceleration, and the initial location the Kalman filter is used to predict and track the object location *tloc* in subsequent frames of the video. If the object is detected in a frame, then its predicted location is corrected, otherwise the predicted location is assumed to be correct. A circle and a label are used to annotate the object over the frames (Figure 3.27).

Example 3.27: Write a program to use a Kalman filter to track moving objects in a video.

```
clear; clc;

vr = vision.VideoFileReader('singleball.mp4');
vp = vision.VideoPlayer('Position',[100,100,500,400]);
fd = vision.ForegroundDetector('NumTrainingFrames',10,'InitialVariance',0.05);
ba = vision.BlobAnalysis('AreaOutputPort',false,'MinimumBlobArea',70);
kf = []; isTrackInitialized = false;

while ~isDone(vr)
    vf  = step(vr);
    mask = step(fd, rgb2gray(vf));
    dloc = step(ba,mask); % detected location
    isObjectDetected = size(dloc, 1) > 0;

    if ~isTrackInitialized
        if isObjectDetected
            kf = configureKalmanFilter('ConstantAcceleration',...
                dloc(1,:), [1 1 1]*1e5, [25,10,10], 25);
            isTrackInitialized = true;
        end;
        label = ''; circle = zeros(0,3);
    else
        if isObjectDetected
            predict(kf);
            tloc = correct(kf, dloc(1,:)); % tracked location
            label = 'Corrected';
        else
            m tloc = predict(kf);
            label = 'Predicted';
        end;
        circle = [tloc, 5];
    end;
```

```
        vf = insertObjectAnnotation(vf,'circle', circle,label,'Color','red');
        step(vp,vf);
        pause(0.2);
end;
release(vp);
release(vr);
```

3.7.5 Block Matcher

The block matching algorithm provides a way to locate matching blocks or patterns across different frames of a video and thereby estimate motion of objects and background in a video sequence. It involves dividing the frame into macroblocks and comparing them across consecutive frames. A motion vector is generated to depict the motion of the macroblock over frames. Typical size of a macroblock is 16 × 16 pixels, and the search for a measure of motion involves an area of about 7–8 pixels on all four sides of a previous macroblock. A metric for matching two macroblocks can be the mean absolute difference (MAD) or the mean square error (MSE), as shown below, where N is each side of the macroblock and $A(i,j)$ and $B(i,j)$ are pixels belonging to the current and previous macroblocks:

$$\mathrm{MAD} = \frac{1}{N^2}\sum_{i=0}^{N-1}\sum_{j=0}^{N-1}\left|A(i,j)-B(i,j)\right|$$

$$\mathrm{MSE} = \frac{1}{N^2}\sum_{i=0}^{N-1}\sum_{j=0}^{N-1}\left\{A(i,j)-B(i,j)\right\}^2$$

There can be a number of possible search methods to locate matching blocks between two consecutive frames. The best results are obtained by an exhaustive search mode in which the block is moved over the search region one pixel at a time; however, this is computationally expensive and can potentially take a lot of time. An alternative faster matching method frequently used is called the three-step method and proceeds as follows: (1) start with the search location at center, (2) set step size as four and search area as seven, (3) search all eight locations around the center, (4) pick the location with the minimum difference, (5) set the new search origin to the above location with the minimum difference, (6) set the new step size half the previous value and repeat the procedure, and (7) continue until the step size reduces to 1. The resulting location for step size 1 is the one with the best match. The CVST function **vision.BlockMatcher** compares the position of objects in two different frames of a video and draws a motion vector to represent motion of the objects, if any. In the following example, a block size of 145 and an exhaustive search method (which is the default option) are used (Figure 3.28).

FIGURE 3.28
Output of Example 3.28.

Example 3.28: Write a program to estimate the motion of an object in a video.

```
clear; clc;

vr = VideoReader('singleball.mp4');
v = read(vr);
frame1 = v(:,:,:,17);
frame2 = v(:,:,:,24);
figure,
subplot(131), imshow(frame1);
subplot(132), imshow(frame2);
img1 = rgb2gray(frame1);
img2 = rgb2gray(frame2);
k = 145; %blocksize

bm = vision.BlockMatcher('ReferenceFrameSource','Input port','BlockSize',[k k]);
bm.OutputValue = 'Horizontal and vertical components in complex form';
ab = vision.AlphaBlender;
motion = bm(img1,img2);
img12 = ab(img2,img1);
[X,Y] = meshgrid(1:k:size(img1,2),1:k:size(img1,1));
subplot(133),
imshow(img12); hold on;
quiver(X(:),Y(:),real(motion(:)),imag(motion(:)));
hold off;
```

3.8 Video Processing Using Simulink

To create a simulink model, type **simulink** at the command prompt or choose Simulink > Blank Model > Library Browser > Computer Vision System Toolbox (CVST). The video source files reside in the folder *(matlab-root)/toolbox/vision/visiondata/*. Within the CVST toolbox, the following libraries and blocks are included for image and video processing:

- Analysis and Enhancement: Block Matching, Contrast adjustment, Corner detection, Edge detection, Histogram equalization, Median filter, Optical flow, Template matching, Trace boundary
- Conversions: Autothreshold, Chroma resampling, Color space conversion, Image complement, Image data-type conversion
- Filtering: 2-D convolution, 2-D FIR filter, Median filter
- Geometric Transformations: Resize, Rotate, Shear, Translate, Warp, Affine
- Morphological Operations: Close, Dilate, Erode, Open, Top-hat, Bottom-hat
- Sinks: To multimedia file, To video display, Video Viewer, Frame rate display, Video to workspace
- Sources: From multimedia file, Image from file, Image from workspace, Read binary file, Video from workspace
- Statistics: 2-D autocorrelation, 2-D correlation, 2-D histogram, 2-D mean, 2-D median, 2-D STD, Blob analysis, PSNR
- Text & Graphics: Insert text, Draw shapes, Draw markers, Compositing
- Transforms: 2-D DCT, 2-D FFT, 2-D IDCT, 2-D IFFT, Hough transform, Hough lines, Gaussian pyramid
- Utilities: Block processing, Image pad

Example 3.29: Create a Simulink model for edge detection in a video file

- CVST > Sources > From multimedia file (Filename : viplanedeparture.mp4)
- CVST > Sinks > Video viewer
- CVST > Conversions > Color Space Conversion (Conversion : RGB to intensity)
- CVST > Analysis & Enhancement > Edge Detection (Figure 3.29)

Example 3.30: Create a Simulink model to convert color space of a video

- CVST > Sources > From multimedia file (Filename : visiontraffic.avi)
- CVST > Conversion > Color space conversion (RGB to YCbCr)
- CVST > Sinks > Video viewer (Figure 3.30)

Example 3.31: Create a Simulink model to geometrically transform a video file

- CVST > Sources > From multimedia file (Filename : visiontraffic.avi)
- CVST > Geometric Transformation > Resize (Resize Factor % : [150 120])
- CVST > Geometric Transformation > Rotate (Angles radians : pi/6)
- CVST > Sinks > Video viewer (Figure 3.31)

Example 3.32: Create a Simulink model to demonstrate optical flow

- CVST > Sources > From multimedia file (Filename : visiontraffic.avi)
- CVST > Sinks > Video viewer
- CVST > Analysis & Enhancements > Optical Flow (Method : *L-K, N* = 5) (Figure 3.32)

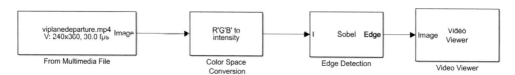

FIGURE 3.29
Output of Example 3.29.

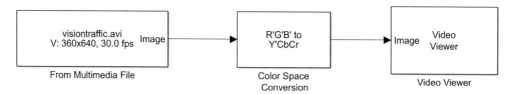

FIGURE 3.30
Output of Example 3.30.

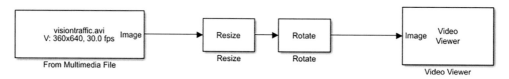

FIGURE 3.31
Output of Example 3.31.

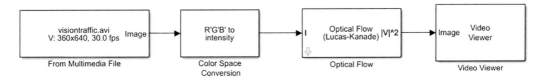

FIGURE 3.32
Output of Example 3.32.

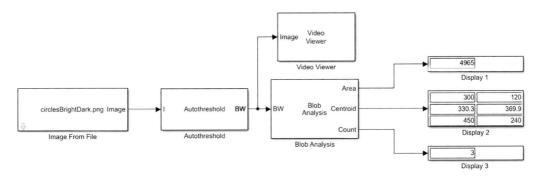

FIGURE 3.33
Output of Example 3.33.

Example 3.33: Create a Simulink model to demonstrate blob analysis.

- CVST > Sources > Image from file (Filename : circlesBrightDark.png)
- CVST > Sinks > Video viewer
- CVST > Conversions > Autothreshold (Thresholding operator : <=, Threshold scaling factor : 0.7)
- CVST > Statistics > Blob Analysis (Statistics : Area Centroid, Blob Properties : Output number of blobs found)
- Simulink > Sinks > Display (Figure 3.33)

Example 3.34: Create a Simulink model to count video frame numbers using data flow from Workspace to Simulink.

Variables from workspace need to be sent to the Simulink model for passing the following parameter values: number of video frames (*nf*), frame rate (*fps*), and video time duration (*vd*). The following script extracts parameters from a specified video file, copies them to Workspace memory, and initializes a simulink model named *ex030806.slx* which utilizes those variables to execute the model. Also, prior to running the model, the *Simulation Stop time* field in the toolbar of the Simulink Project window should be set to *vd*-1/*fps* in order to stop the execution once all the frames of the video have been counted.

```
clear; clc;
fn = 'rhinos.avi';
r = VideoReader(fn);
nf = r.NumberOfFrames;
w = r.Width;
h = r.Height;
fps = r.FrameRate;
vd = nf/fps;
sim('ex030806')
```

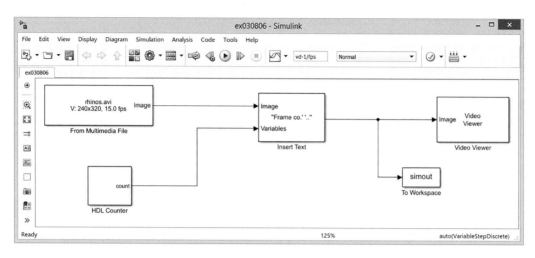

FIGURE 3.34
Output of Example 3.34.

- CVST > Sources > From Multimedia File (Filename : rhinos.avi, Number of times to play file : 1)
- HDL Coder > Sources HDL Counter (Initial value : 1, Count to value : nf, Sample time : 1/fps)
- CVST > Text & Graphics > Insert Text (Text : ['Frame count : %d'], Color value : [1,1,1], Location : [10,220])
- Sinks > Video viewer (Figure 3.34)

To capture output from Simulink to workspace for the previous problem, a To Workspace block called *simout* is added to the model whose input is taken from the output generated from the model. For example, to display the last frame of the video with the frame count annotation shown, the following script can be executed from workspace:

```
imshow(simout.Data(:, :, nf))
```

Review Questions

1. How to read and play a video file using a specific subset of frames?
2. How to create a 4-D structure for storing the frames of a video file?
3. How to capture a sequence of images and write them down as frames of a video?
4. How to insert textual captions at specific locations of video frames?
5. How can image processing filters be applied to selected frames of a video file?
6. Differentiate between component video and composite video. How is RGB video converted to YCbCr?
7. How can an RGB video be converted to NTSC and PAL color spaces?
8. What is a blob detector and how can it be used to detect presence of blobs in video frames?

9. How to detect presence of foreground objects in a video and demarcate them with bounding boxes?

10. What is HOG and how can it be used to detect humans in a video file?

11. How can be histogram based tracker be used to track movement of an object over video frames?

12. What is optical flow and how can it be used to estimate direction and speed of a moving object?

13. How can a Kalman filter be used to track occluded objects in a video frame?

14. How can a block matcher be used to generate a motion vector for a moving object?

15. How can a Simulink model be created to geometrically transform video frames?

16. How to use a Simulink model to convert video frames to binary and perform edge detection?

17. How can feature points in an initial frame be tracked over subsequent frames using a point tracker?

18. How can text be recognized in a video frame using OCR?

19. How can human faces be detected in frames of a video?

20. How to remove noise from video frames?

4

Pattern Recognition

4.1 Introduction

The term **pattern recognition** is related to the automated discoveries of patterns and regularities in data using computer algorithms and thereafter utilizing these patterns for classification of the data into different categories. This chapter will discuss how the media processing techniques described in the earlier chapters can be applied for pattern recognition and classification of image audio and video into categories, where each category is characterized by similarity in the underlying data patterns. Pattern recognition algorithms are nowadays extensively applied to machine learning, computer vision, and data mining applications among others like medical imaging, voice-activated systems, video surveillance, and biometrics. Pattern recognition is intimately connected with the concept of machine learning because an automated system is made to *learn* about the patterns in the data which enable it to group the data into classes. Based on the learning procedure, pattern recognition can be divided into two broad types: supervised and unsupervised. **Unsupervised learning** is the more generalized form where the data are grouped based on the value of the data only without any additional or prior information about the data. In this case, the grouping is more commonly termed as **clustering** which implies categorization of similar data into one cluster, while differentiation of dissimilar data into other clusters. **Supervised learning** is a more specialized type of grouping where the data are accompanied with additional information about the characteristics of each class. In this case, the grouping is termed as **classification** which implies using the prior information to classify similar data into one class and dissimilar data into other classes. In reality, the supervised system learning goes through two phases: the **training phase** where the system is fed with information about the characteristic of each class (i.e. what patterns of data characterizes each class) and subsequently the **testing phase** where the trained system is asked to classify unknown data into pre-defined classes based on what it has learned earlier during the training phase. The patterns about the data are represented as a collection of values referred to as a **feature vector**. The term *feature* indicates a mathematical representation, typically numerical, about some visual or aural characteristics perceptible by the human audio-visual system. An example is an image histogram which is a numerical representation about the distribution of color or intensity levels in an image. The term *vector* indicates the representation of the values in the form of a set or a one-dimensional matrix. The number of values (or elements) in the vector is called its **dimension**. A typical gray-level image histogram has 256 elements since it uses an 8-bit encoding, while a three-dimensional (3D) vector can be used to represent a single color, specifying the R, G, and B values. The underlying basis of pattern recognition is to determine which data are similar to each other by comparing their feature vectors since the feature vector are representations of the properties of the data. To compare two *n*-dimensional feature vectors, it is assumed that each of the vectors has been plotted as a point in an *n*-dimensional space, called the **feature space**. The distance between these two *n*-dimensional points is used to represent the difference between the vectors themselves. If the values of two vectors are similar, the difference between them will be small and they will be nearer to each other in the feature space. On the other hand, if the vectors are dissimilar, then they would be farther apart in the feature space. The distance is therefore a measure of the similarity and called a **similarity metric**. There are several established ways of computing distance metrics or similarity metrics, which is discussed in details in Section 4.4. Note however that feature vector representation of data is not unique, meaning that the same data can be represented in multiple ways using different feature vectors, some might be more effective than others depending on the specific applications, and it is usually up to the researchers to find the best feature representations, usually by a trial and error process. The amount of effectiveness is calculated using some performance evaluation methods, which can be used to compare similar research efforts. The entire process of pattern recognition can be divided into a number of steps: (1) data acquisition, which involves reading the digital data into the system; (2) pre-processing, which involves making the data more suitable for feature computation and comparisons; (3) feature extraction, which involves generating the feature vectors and their dimensionalities; (4) comparison of the feature vectors for finding similarities between data, resulting in clustering or classification; and (5) performance evaluation, which shows how effective the comparison process has been (Bishop, 2006).

4.2 Toolboxes and Functions

Computer Vision System Toolbox™ (CVST) provides algorithms, functions, and apps for designing and simulating feature extraction and representation, while the **Statistics and Machine Learning Toolbox**™ (SMLT) provides algorithms, functions, and apps to describe, analyze, and compare the data. In this chapter, we will discuss some commonly used functions of both of these toolboxes as solutions to specific examples related to pattern recognition problems. In addition, we explore some functions from the **Neural Network Toolbox** (NNT) which provides algorithms, pre-trained models, and apps to create, train, visualize, and simulate neural networks. This book has been written using MATLAB® version 2018b; however, most of the functions can be used with versions 2015 and later with little or no change. Although each of these toolboxes contains a large number of functions, a subset of these have been described here keeping in mind the scope and coverage of this book and these have been listed below along with a one line description for each.

4.2.1 Computer Vision System Toolbox (CVST)

The functions in the CVST described here belong to the category Feature Detection and Extraction and involves detection, extraction, and representation of various kind of features used for clustering and classification.

1. detectBRISKFeatures: Detect features using Binary Robust Invariant Scalable Keypoints (BRISK) algorithm

2. detectMinEigenFeatures: Detect corners using minimum eigenvalue algorithm
3. detectHarrisFeatures: Detect corners using Harris–Stephens algorithm
4. detectFASTFeatures: Detect corners using Features from Accelerated Segment Test (FAST) algorithm
5. detectMSERFeatures: Detect features using Maximally Stable Extremal Regions (MSER) algorithm
6. detectSURFFeatures: Detect features using Speeded-Up Robust Features (SURF) algorithm
7. detectKAZEFeatures: Detect features using the KAZE algorithm
8. extractFeatures: Extract interest point descriptors
9. extractLBPFeatures: Extract local binary pattern (LBP) features
10. extractHOGFeatures: Extract histogram of oriented gradients (HOG) features

4.2.2 Statistics and Machine Learning Toolbox (SMLT)

The functions in the SMLT described here belong to a number of categories viz. Descriptive Statistics, Probability Distributions, Probability Distributions, Cluster Analysis, Regression, Dimensionality Reduction, and Classification. A list of these have been provided below along with a one line description of each.

1. Descriptive Statistics and Visualization:
    1. gscatter: Scatter plot by group
    2. tabulate: Frequency table
2. Probability Distributions:
    1. mvnrnd: Multivariate normal random numbers
    2. mahal: Mahalanobis distance
    3. fitgmdist: Fit Gaussian mixture model to data
    4. pdf: Probability density function for Gaussian mixture distribution
3. Cluster Analysis:
    1. kmeans: k-means clustering
    2. kmedoids: k-medoids clustering
    3. knnsearch: Find k-nearest neighbors (k-NN)
    4. linkage: Hierarchical cluster tree
    5. cluster: Construct agglomerative clusters from linkages
    6. dendrogram: Dendrogram plot
    7. pdist: Pairwise distance between pairs of observations
    8. silhouette: Silhouette plot
    9. evalclusters: Evaluate clustering solutions
4. Regression:
    1. statset: Create statistics options structure
5. Dimensionality Reduction:
    1. pca: Principal component analysis (PCA)

6. Classification:
   1. fitctree: Fit binary classification decision tree for multiclass classification
   2. fitcdiscr: Fit discriminant analysis classifier
   3. predict: Predict labels using classification model
   4. cvpartition: Data partitions for cross-validation
   5. training: Training indices for cross-validation
   6. test: Test indices for cross-validation
   7. fitcnb: Train multiclass naive Bayes model
   8. fitcsvm: Train binary support vector machine (SVM) classifier
   9. confusionmat: Confusion matrix

4.2.3 Neural Network Toolbox (NNT)

1. Feedforwardnet: Create a feedforward neural network
2. Patternnet: Create a pattern recognition network
3. Perceptron: Create a perceptron
4. train: Train neural network
5. view: View neural network

4.3 Data Acquisition

Data acquisition involves reading the data into the system. There can be two broad categories of data acquisition methods: one involves directly reading the alpha numeric features and the other involves reading the raw media like image audio or video. Directly reading the features is a more compact process requiring less time as it involves only alpha-numeric data, while reading media files can involve additional steps described later. Features thus acquired can be directly used in the computations of clusters or classes, although in some cases post-processing in the form of dimension reduction can be done to improve performance or visualization. MATLAB includes more than 30 data sets built within the software package which can be invoked using the BM function **load** to read the variables from a .MAT file into the workspace. A MAT file is a binary file containing the data stored in variables and can be created using the BM function **save**, while new data can be appended to an existing MAT file. Both the above functions can be used either in the command syntax or the function syntax, as shown in the example below. The BM function **ones** creates a matrix of the specified size filled with 1s. The BM function **whos** can be used to view the contents in a MAT file.

Example 4.1: Write a program to save and load variables to and from a MAT file.

```
clear; clc;

a = rand(1,10);
b = ones(10);
save sample1.mat a b            % command syntax
```

```
save('sample2.mat','a','b')          % function syntax

load sample1                         % command syntax
load('sample2', 'b');                % function syntax

c = magic(10);
save('sample1.mat','c','-append')    % append

whos('-file','sample1')              % view
```

A list of the sample data sets included within MATLAB can be obtained using the following navigations:

- Documentation Home>Statistics and Machine Learning Toolbox>Descriptive Statistics and Visualization>Managing Data>Data Import and Export>Sample Data Sets
- Documentation Home>Neural Network Toolbox>Getting Started with Neural Network Toolbox>Neural Network Toolbox Sample Data Sets for Shallow Networks

The first method of data acquisition involves directly reading alpha-numeric data from data sets. We will discuss this using an in-built data set, which is used extensively in the coming sections, called the **Fisher-Iris data set**. This data set introduced by Ronald Fisher in 1936 (Fisher, 1936) consists of 50 samples each from three classes or species of Iris flowers (Iris setosa, Iris virginica, and Iris versicolor). Four features were measured from each sample in centimeters: sepal length, sepal width, petal length, and petal width. In the following example, the two variables load consist of (a) 150×4 matrix called *meas* which lists the four measurements for 150 samples and (b) a 150×1 list called *species* specifying the class of the flower species. The following example shows a visual representation of the data set using a separate color for each class and the centroids of the classes being depicted by crosses (Figure 4.1). Since the measurements are four-dimensional i.e. having four parameters, two measurements are used at a time to generate 2-D plots. The first figure shows a plot of the classes using the first and second measurements i.e. sepal length

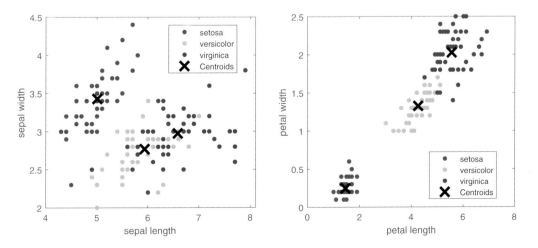

FIGURE 4.1
Output of Example 4.2.

and sepal width, and the second figure shows a plot of the classes using the third and fourth measurements i.e. petal length and petal width. The SMLT function **gscatter** generates a scatter plot using a grouping variable, in this case *species* i.e. it displays each species of data points using a different color. It can be seen that the second figure provides a better discrimination of the classes than the first figure, which indicates that the petal-based features are better features for classification than the sepal-based features. In Chapter 4, we would see how all the four parameters can be used to generate a 2-D plot using PCA.

Example 4.2: Write a program to display the Fisher-iris dataset using a separate color for each class. Also find and plot the centroids of the three classes.

```
load fisheriris;

x = meas(:,1:2);
c1x = mean(x(1:50,1));      c1y = mean(x(1:50,2));
c2x = mean(x(51:100,1));    c2y = mean(x(51:100,2));
c3x = mean(x(101:150,1));   c3y = mean(x(101:150,2));
c = [c1x, c1y ; c2x, c2y ; c3x, c3y];

figure;
subplot(121),
gscatter (x(:,1), x(:,2), species); hold on;
plot(c(:,1),c(:,2),'kx','MarkerSize',15,'LineWidth',3);
legend('setosa','versicolor','virginica','Centroids');
xlabel('sepal length'); ylabel('sepal width');

x = meas(:,3:4);
c1x = mean(x(1:50,1));      c1y = mean(x(1:50,2));
c2x = mean(x(51:100,1));    c2y = mean(x(51:100,2));
c3x = mean(x(101:150,1));   c3y = mean(x(101:150,2));
c = [c1x, c1y ; c2x, c2y ; c3x, c3y];

subplot(122),
gscatter (x(:,1), x(:,2), species); hold on;
plot(c(:,1),c(:,2),'kx','MarkerSize',15,'LineWidth',3);
legend('setosa','versicolor','virginica','Centroids');
xlabel('petal length'); ylabel('petal width');
```

The second method of data acquisition involves reading raw media files and using feature extraction algorithms to compute and store the features. This scheme involves reading multiple media files from a folder and storing them in a matrix for subsequent processing. The simplest way of doing this is to specify the pathname of each image file separately for reading them. However, for large number of files, a number of more efficient techniques are available, as shown in the example below.

Method-1: If the image files are located in the current folder and are of a specified format like JPG, then method-1 illustrates how to read each file using a loop and store them in a structure. The image files can then be accessed individually by specifying a subscript in the structure. The example shows each image being displayed separately in a figure window. The BM function **dir** lists all the files in a specified folder using additional constraints like file extensions. The BM function **cell** creates an array of cells where each cell can contain any type of data.

Method-2: If the images are located in an arbitrary folder, then the full pathnames of the images need to be specified while reading them. The images are then displayed in a figure window by using a subplot grid with approximately equal number of rows and columns, each of which is equal to the square root of the total number of images. The figure window is finally enlarged to cover the full screen. The BM function **ceil** rounds toward positive infinity.

In the following example, multiple image files are read and stored in a cell structure. Each image can be accessed by specifying a subscript to the cell variable *img* using braces { }. In the second method, files are read from a specified folder using the full pathname. The images are then displayed using a grid whose rows and columns are derived from the square root of the total number of files read. The grid is viewed in a figure window which is enlarged to cover the entire screen. The BM function **matlabroot** returns the full path to the folder where MATLAB is installed, while the BM function **fullfile** builds the full pathname from parts (Figure 4.2).

Example 4.3: Write a program to read multiple images from a folder without specifying their filenames separately

```
clear; clc;

% method-1

files = dir('*.jpg');
num = length(files);
img = cell(1,num);
for k = 1:num
    img{k} = imread(files(k).name);
end
% display images
for k = 1:num
    figure;
    imshow(img{k});
end

% method-2

clear;
folder = fullfile(matlabroot, 'toolbox', 'images', 'imdata', '*.png');
files = dir(folder);
num = length(files);
rows = ceil(sqrt(num));
```

FIGURE 4.2
Output of Example 4.3.

```
% display images
for k = 1: num
  pathname = files(k).name;
  subplot(rows, rows, k), imshow(pathname);
  title(files(k).name);
end

% enlarge figure window to full screen.
pause(5)
set(gcf, 'Units', 'Normalized', 'OuterPosition', [0, 0, 1, 1]);
```

Method-3: For a classification scheme, if the data items are named according to the syntax *class(sample)* i.e. 1(1), 1(2), 1(3), ... be the samples of the first class, 2(1), 2(2), 2(3), ... be the samples of the second class and so on, then the images can be stored in a 2-D matrix where the rows denote the classes and the columns denote the samples of each class. Here, *NC* denotes number of classes and *NT* denotes number of training samples per class. The BM function **strcat** concatenates strings horizontally, and the BM function **num2str** converts numbers to strings.

Method-4: For very large collection of images, the BM function **imageDatastore** provides a way to manage a collection of images at a specified location by creating a datastore. The images to be included as part of the datastore can also be specified using additional filters like file extensions. In this case, the images are read from the image collection of MATLAB located at *toolbox/images/imdata* folder under MATLAB root folder (Figure 4.3).

Example 4.4: Write a program to read multiple images for creating a training dataset and an image datastore.

```
clear; clc;

% method-3

path = 'data/train/';   ext = '.jpg';
NC = 3;       % number of classes
NT = 5;       % number of training samples per class
```

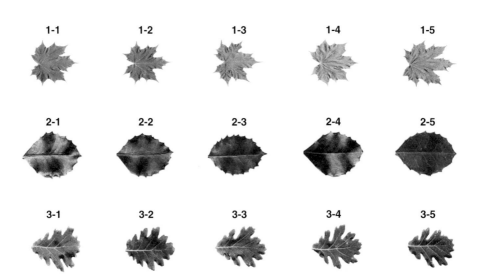

FIGURE 4.3
Output of Example 4.4.

```
for CL = 1: NC
    for SA = 1: NT
        class = num2str(CL);
        sample = num2str(SA);
        I{CL,SA} = imread(strcat(path,class,' (',sample,')',ext));
    end;
    fprintf(' %d\n', CL)
end;
for i = 1:NC          % do for all classes
    for j = 1:NT     % do for all samples per class
        k = j+(i-1)*NT;
        subplot(NC, NT, k),imshow(I{i,j}), title(strcat(num2str(i),'-',num2str(j)));
    end;
end;

% method-4

clear;
loc = fullfile(matlabroot, 'toolbox', 'images', 'imdata');
imds = imageDatastore(loc,'FileExtensions',{'.jpg','.tif'});
montage(imds);
figure,
num = length(imds.Files);
rows = ceil(sqrt(num));
for k = 1: num
  subplot(rows, rows, k), imshow(imds.Files{k});
end;
set(gcf, 'Units', 'Normalized', 'OuterPosition', [0, 0, 1, 1]);
```

4.4 Pre-processing

Pre-processing is not a single operation but a collection of different activities with a common objective of making the media more suitable for pattern recognition. The term **pre-processing** implies that these activities are carried out before the feature extraction phase. The operations involved can be different for different applications, but this section discusses some of the most common activities. Most of these have already been discussed in the previous sections on media processing. These include the following:

- media-type conversions which might include splitting a color channel into color channels, conversion from color to grayscale or binary, quantization of color or gray shades (Section 1.3)
- color conversions which might include RGB to HSV or Lab color spaces (Section 1.3)
- geometric transformations which might include cropping, scaling, and rotating (Section 1.5)
- tonal correction which might include contrast stretching, gamma adjustments, and histogram equalization (Section 1.6)
- noise filtering (Section 1.6)
- morphological operations (Section 1.6)
- edge detection (Section 1.6)
- segmentation of objects of interest (Section 1.7)
- temporal filtering (Section 2.8) and spectral filtering (Section 2.9)

4.5 Feature Extraction

As already mentioned, feature vectors computed from an image provide a numerical representation of the visual attributes of the image content. One of the first steps in characterizing the image content is to distinguish between the background and the foreground objects. Foreground objects are identified by locating edges, corners, and blobs to demarcate their boundaries. These are collectively termed as **interest points** and usually indicate points in an image where there is a major change in intensity with respect to their surroundings. Since these interest points are identified by analyzing small sections of the image in a repetitive way, these are also called **local features** and the small image sections are termed as **local neighborhood**. Using local features enable these algorithms to better handle scale changes, rotation, and occlusion. The algorithms can be divided into two broad categories: detecting location of the points and generating a descriptor to represent them. Algorithms for detecting interest points include Harris, Minimum eigenvalue, FAST and MSER methods, while algorithms for generating descriptors include the SURF, KAZE, BRISK, LBP, and HOG methods. In some cases, the one detection method can be used to locate the points and then a different descriptor method can be used to build the feature vectors for comparing them between images.

4.5.1 Minimum Eigenvalue Method

An **edge** is a line such that there is a sudden change in intensity perpendicular to it, while a **corner** is a junction of two edges. Edges and corners can be identified when analyzing a small image portion through a window. If the window is placed on a flat region, then shifting the window by a small amount in any direction does not change image intensity by a substantial amount. If the window is on an edge, then there is a large change in intensity perpendicular to the edge line but no change when the window is moved along the edge. If the window is placed on a corner, then there is a significant change in intensity along all directions. If $I(x,y)$ be the image intensity at the original location, (x,y) and $I(x + u, y + v)$ be the intensity at the new location shifted by (u,v), then the change in intensity is given by the sum of squared differences (SSD), where $w(x,y)$ is the window function:

$$E(u,v) = \sum_{x,y} \left[w(x,y) \cdot \left\{ I(x+u, y+v) - I(x,y) \right\}^2 \right]$$

The term $I(x+u, y+v)$ can be approximated by a Taylor series expansion. Named after Brook Taylor (Taylor, 1715), the Taylor series of a function $f(x)$ is an infinite sum of terms expressed in terms of the function's derivatives at a single point $x = a$

$$f(a) + \frac{f'(a)}{1!}(x-a) + \frac{f''(a)}{2!}(x-a)^2 + \cdots = \sum_{n=0}^{\infty} \frac{f^{(n)}(a)}{n!}(x-a)^n$$

In this case, if $Ix = \partial I / \partial x$ and $Iy = \partial I / \partial y$ be the partial derivatives of I, then the Taylor series can be approximated as follows:

$$I(x+u, y+v) \approx I(x,y) + Ix \cdot u + Iy \cdot v$$

$$E(u,v) = \sum_{x,y}\left(Ix.u + Iy.v\right)^2 = (u,v)\cdot M \cdot (u,v)^T$$

Here, M is a 2×2 matrix computed from image derivatives

$$M = \sum_{x,y} w(x,y)\begin{bmatrix} Ix^2 & Ix\cdot Iy \\ Ix\cdot Iy & Iy^2 \end{bmatrix}$$

The **Minimum Eigenvalue Method** (Shi and Tomasi, 1994) proposes that E should be large for small shifts in all directions. To achieve this, the minimum of E should be large over all vectors (u,v). If λ_1 and λ_2 be the eigenvectors of the matrix M, then this minimum value of E is given by the smaller eigenvalue λ_{min}. The algorithm therefore computes all values of λ_{min} which are local maximum or larger than a specified threshold to find corners in an image. The CVST function **detectMinEigenFeatures** detects corners using the minimum eigenvalue algorithm to find feature points in a grayscale image. The following example detects the corners of a checkerboard pattern using the minimum eigenvalue algorithm (Figure 4.4).

Example 4.5: Write a program to detect corners in an image using the Minimum Eigenvalue Method.

```
clear; clc;
I = checkerboard(20);
corners = detectMinEigenFeatures(I);
imshow(I);
hold on;
plot(corners);
hold off;
```

The **eigenvector** v of a square matrix M is a vector which when multiplied with the matrix returns a scaled version of the vector itself. The scalar multiplier is called the **eigenvalue** λ of the matrix i.e.

$$\boldsymbol{M}\cdot\boldsymbol{v} = \lambda\cdot\boldsymbol{v}$$

FIGURE 4.4
Output of Example 4.5.

An example is shown below

$$\begin{bmatrix} 2 & 3 \\ 2 & 1 \end{bmatrix} \cdot \begin{bmatrix} 3 \\ 2 \end{bmatrix} = 4 \cdot \begin{bmatrix} 3 \\ 2 \end{bmatrix}$$

The eigenvalues are obtained from the above equation by finding the roots of $|M - \lambda \cdot I| = 0$, where I is an identity matrix corresponding to M. Substitution of the values of λ in the equation yields the eigenvectors. The BM function **eig** is used to return the eignvectors and eigenvalues. In the following example, the BM function **gallery** is used to invoke a gallery which holds over 50 different test matrix functions used for testing algorithms. The option *circul* returns a circulant matrix that has the property that each row is obtained from the previous one by cyclically permuting the entries one step forward. In the second case, the BM function **magic** is used to generate a matrix of the specified dimension with equal row and column sums.

> **Example 4.6: Write a program to calculate the eigenvalues and eigenvectors of a square matrix .**
>
> ```
> clear; clc;
> M = gallery('circul',4);
> [vec, val] = eig(M)
>
> % verification: should return zero
> D1 = M*vec - vec*val
>
> clear;
> N = magic(5);
> [vec, val] = eig(N)
> D2 = N*vec - vec*val
> ```

4.5.2 Harris Corner Detector

The **Harris Corner Detector** (Harris and Stephens, 1988) is an alternative method which is computationally more efficient than the previous method as it avoids direct computation of the eigenvalues and only computes the determinant and trace of the matrix M. The response measure R of the Harris detector can be represented in terms of the eigenvectors λ_1 and λ_2 of M, where k is an empirical constant having value 0.04–0.06:

$$R = \det(M) - k \cdot (\text{trace}(M))^2$$

$$\det M = \lambda_1 \cdot \lambda_2$$

$$\text{trace } M = \lambda_1 + \lambda_2$$

If both λ_1 and λ_2 are small, then absolute value $|R|$ is small which indicates a flat region in the image. If either λ_1 or λ_2 is large, then $R < 0$ which indicates an edge in the image. If both λ_1 and λ_2 are large, then $R > 0$ which indicates a corner in the image. To find points with a large corner response, values of R larger than a specified threshold should be retained. The CVST function **detectHarrisFeatures** detects corners using the Harris Corner Detector algorithm. In the following example, out of all the candidate corner points, the strongest 50% of the points are retained and plotted on the image (Figure 4.5).

FIGURE 4.5
Output of Example 4.6.

Example 4.7: Write a program to detect corners in an image using the Harris Corner Detector.

```
clear; clc;
I = imread('cameraman.tif');
corners = detectHarrisFeatures(I);
imshow(I); hold on;
nc = size(corners,1);
plot(corners.selectStrongest(nc/2));
```

4.5.3 FAST Algorithm

The FAST algorithm has been proposed for a computationally efficient way to compute interest points (Rosten and Drummond, 2006). To identify a prospective interest point, a circle of 16 pixels around the pixel P undertest is considered. The point P is taken to be an interest point if there exist a set of 12 contiguous pixels in the circle which are all brighter than $I(P) + T$ or all darker than $I(P) - T$, where $I(P)$ is the intensity value of pixel P and T is a specified threshold. To make the computations fast, the intensities of pixels 1, 5, 9, and 13 are first compared with $I(P)$. For P to be qualified as an interest point, at least three of these four pixels should satisfy the threshold criterion, otherwise P is rejected as a candidate for interest point. If the criterion is satisfied, then all 16 pixels are checked. The CVST function **detectFASTFeatures** detects corners using the FAST algorithm. The following example plots the detected FAST feature points over the image (Figure 4.6).

FIGURE 4.6
Output of Example 4.7.

Example 4.8: Write a program to compute interest points from an image using the FAST algorithm.

```
clear; clc;
I = rgb2gray(imread('peppers.png'));
corners = detectFASTFeatures(I);
imshow(I);
hold on;
plot(corners);
hold off;
```

4.5.4 MSER Algorithm

The MSER algorithm has been proposed as a method for blob detection in images (Matas et al., 2002), specifically to find correspondences between image elements from different viewpoints. The algorithm is based on finding regions which stay stable through a wide range of binarization thresholds. For a grayscale image, starting from a binarization threshold of 0 when the entire image appears white, if the thresholds are incremented through all the 256 levels, to the final value of 255, the image will ultimately appear as black. During all the intermediate steps, the image will generally appear as a collection of black and white regions, starting from small black dots to larger black regions as a number of smaller sets combine to grow larger in size. The set of all connected components is referred to as extremal regions. The area of these extremal regions can be plotted as a function of varying threshold value. If the area of some extremal regions exhibits minimal variation over a wide range of threshold changes, then these are considered as maximally stable and hence regions of interest. The CVST function **detectMSERFeatures** can be used to detect objects in a 2-D grayscale image. It returns the region structure (*re*) containing features like object count, location, orientation, a connected component structure (*cc*), and so on. The region structure can be plotted to indicate the position, size, and orientations of the objects found by overlaying ellipses on them. The following example shows objects detected in an image indicated by overlaid ellipses. The example also illustrates how to detect circular objects by thresholding the eccentricity values of the overlaying ellipses. The eccentricity of an ellipse is the ratio of the distance between the focii of the ellipse and its major axis length. The eccentricity of a circle is 0 and that of a line segment is 1. Hence, ellipses with low eccentricity values can be used to detect circles (Figure 4.7).

Detecting blobs
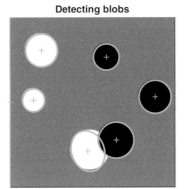

Detecting circles
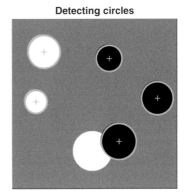

FIGURE 4.7
Output of Example 4.8.

Example 4.9: Write a program to compute regions of interest from an image using the MSER algorithm.

```
clear; clc;
fn = 'circlesBrightDark.png';
I = imread(fn);
[re,cc] = detectMSERFeatures(I);
figure
subplot(121)
imshow(I); hold on; plot(re);
title('Detecting blobs');

stats = regionprops('table',cc,'Eccentricity');
ei = stats.Eccentricity < 0.55;
cr = re(ei);
subplot(122),
imshow(I); hold on; plot(cr);
title('Detecting circles'); hold off;
```

4.5.5 SURF Algorithm

The task of pattern recognition can be divided into three steps: detection, description, and matching. The first step involve detection of interest points. The previous paragraphs discuss methods of detection and detector. A desirable property of a detector is its repeatability i.e. finding the same interest points under different transformations of the image like scaling and rotation. The second step after detection is a description of the feature points. This is done by representing the neighborhood of every interest point by a feature vector. A desirable property of a descriptor is that it should be distinctive, have low dimension, and robust to noise. The third step is to find a method of matching the feature vectors between different images. This is usually performed using a distance or similarity metric. Quick and reliable matching is desirable especially for real-time applications, for which a low-dimensional descriptor is preferred, however reducing the dimensions should not compromise its distinctive property.

One of the major schemes for interest point detection and description is called **Scale Invariant Feature Transform** (SIFT) proposed by David Lowe (Lowe, 2004). The main advantages of SIFT compared to earlier approach were two-fold, first, it not only detects interest points but also generates a descriptor for each point using which images can be compared, and second, the descriptors are robust and invariant to geometric transformations and illumination variations. There are four steps involved in the SIFT scheme: (1) scale space peak selection; (2) keypoint localization; (3) orientation assignment; and (4) keypoint descriptor. The first step involves the concept of **scale space** representation which is a framework for multi-scale signal analysis. It originates from the observation that images can be composed of different structures at different sizes or level of details which can be observed at the proper scale. Since there is no prior information about which scale is appropriate for which image, multiple scales are used to detect features at various details, and this continuum of scales is called the scale space (Witkin, 1983). Each image is convolved with a Laplacian of Gaussian (LoG) kernel both along the horizontal and vertical directions to obtain the scale space images. Multiple scale space representations are obtained by changing the sigma parameter of the Gaussian kernel. To detect interest points, a 3×3 neighborhood around each point in each scale space image is observed along with a similar neighborhood in the scale above it and in the scale below it i.e. a total of 27 points. A point is considered as an interest point if it is the minima or maxima of all these 27 points. The LoG operator when used like this is referred to as the "blob detector" since it gives a large response for blobs of the same scale as the LoG operator. After smoothing the image with appropriate filters with different sigma parameters, the image is then subsampled by a factor of 2 along each direction, resulting in a smaller image and then again repeating the smoothing operation as before. This cycle is repeated several times, each step resulting in smaller and smaller images. The entire multi-scale representation can be visually arranged as an "image pyramid" by placing the original image at the bottom and each cycle's resulting smaller image stacked on top of each other (Burt and Andelson, 1983). All scales must be examined to identify scale-invariant features. An efficient way to compute the scale space representation is to replace the LoG by the Difference of Gaussian (DoG). The images are convolved with standard Gaussian kernels at several scales, and each scale space image is subtracted from the previous image to get the difference image, which simulates the LoG operation. The second step, **keypoint localization** indicates for each interest point what is the values of parameters x,y and σ i.e. coordinates and scale. Outlier rejection is done by accepting only those values of minima/maxima which exceeds a specified threshold. The third step, **orientation assignment** is done by computing the gradient magnitude and directions at the interest points. For a keypoint at (x,y) with intensity $I(x,y)$, the gradient magnitude G and orientation θ are given as follows:

$$Gx = \frac{\partial I(x,y)}{\partial x} = I(x+1,y) - I(x-1,y)$$

$$Gy = \frac{\partial I(x,\ y)}{\partial y} = I(x,y+1) - I(x,y-1)$$

$$G = \sqrt{Gx^2 + Gy^2}$$

$$\theta = \arctan\left(\frac{Gy}{Gx}\right)$$

Around each interest point, a 36-bin orientation histogram (10° angular separation) is created using the orientation angles of the neighboring 4×4 points. Each orientation vector is weighted by the corresponding gradient magnitude. The angular direction indicated by the peak of the histogram is taken as the dominant orientation of the interest point.

In order to achieve rotation invariance, the gradient orientations are rotated relative to the dominant keypoint orientation. The fourth step of generating the **keypoint descriptor** around each interest point takes into account the gradient magnitude and orientation of the 4×4 local neighborhood. The orientations of the pixels within each of these 4×4 sub-regions are used to generate an 8-bit orientation histogram with the length of each orientation vector corresponding to the magnitude of that histogram entry. The descriptor is formed from a vector containing the values of all the 8-bin orientation histogram entries within each of the 4×4 cells leading to a $4 \times 4 \times 8 = 128$ element feature vector for each keypoint.

The **Speeded Up Robust Features** (SURF) algorithm (Bay et al., 2008) has been proposed as an improved version of the SIFT algorithm. SURF uses square shaped box filters instead of Gaussian filters for computing the scale space, which can be done at low computational cost using integral images. The entry of an integral image at point $P(x,y)$ represents the sum of all pixels in the input image I within a rectangular region formed by the origin and point P. The blob detector is based on the Hessian matrix to find points of interest. The Hessian matrix is the set of partial double derivatives:

$$H\big[I(x,y)\big] = \begin{bmatrix} \dfrac{\partial I^2(x,y)}{\partial x^2} & \dfrac{\partial I^2(x,y)}{\partial x \cdot \partial y} \\[2ex] \dfrac{\partial I^2(x,y)}{\partial x \cdot \partial y} & \dfrac{\partial I^2(x,y)}{\partial y^2} \end{bmatrix}$$

For scale invariance, the image is filtered by a Gaussian kernel, so given a point $P(x,y)$, the Hessain matrix $H(P,\sigma)$ at P and scale σ is defined as:

$$H(P,\sigma) = \begin{bmatrix} L_{xx}(P,\sigma) & L_{xy}(P,\sigma) \\[1ex] L_{yx}(P,\sigma) & L_{yy}(P,\sigma) \end{bmatrix}$$

Here, $L_{xx}(P,\sigma)$ is the convolution of the second-order derivative of the Gaussian function with the image I at point P. The Gaussian second-order derivatives are approximated by box filters. In contrast to SIFT which builds the image pyramid scale space by repeatedly sub-sampling the original image, due to the use of the box filters and integral image, SURF applies up-scaled filters on the original image. So for each new octave, the filter size doubles, while the image size remains same. To extract the dominant orientation of the keypoint, SURF uses Haar-wavelet responses in X and Y directions in a circular neighborhood at periodic intervals of $\pi/3$ radians. The direction producing the largest sum of the vertical and horizontal wavelet responses gives the dominant orientation. To generate the descriptor, a square region is constructed centered around the keypoint and oriented along the dominant orientation. The square region is split into 4×4 square sub-regions, and within each sub-region Haar-wavelet responses are computed along horizontal and vertical directions. If dx and dy denote the wavelet responses along the horizontal and vertical directions, then the four-dimensional (4-D) descriptor vector for each sub-region is given by: $F = (\Sigma\, dx, \Sigma\, dy, \Sigma\, |dx|, \Sigma\, |dy|)$. Appending the results for all 4×4 sub-regions generates the $4 \times 4 \times 4 = 64$-dimensional SURF descriptor. The CVST function **detectSURFFeatures** detects corners using the SURF algorithm. The method *selectStrongest* is used to return points with the strongest metrics. The CVST function

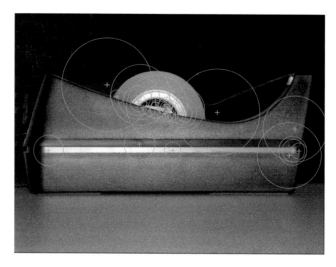

FIGURE 4.8
Output of Example 4.9.

extractFeatures can then be used to generate a descriptor based on the local neighbor-hood of the interest points. In the following example, interest points returned by the SURF algorithm are stored in the variable *sp* which has datatype *SURFPoints*. The fea-ture descriptor *fd* contains the 64-dimensional feature vectors for all the valid points *vp*. The valid points *vp* can be the same or a subset of SURF points *sp* excluding those which lie along an edge or boundary of the image. The strongest 30 points are plotted over the image to show their location and magnitude, the latter indicated by the radius of the cor-responding circle (Figure 4.8).

> **Example 4.10: Write a program to compute interest points from an image using the SURF algorithm and generate a descriptor for the same.**
>
> ```
> clear; clc;
> I = imread('tape.png');
> I = rgb2gray(I);
> sp = detectSURFFeatures(I);
> [fd, vp] = extractFeatures(I, sp, 'Method', 'SURF');
> figure; imshow(I); hold on;
> plot(vp.selectStrongest(30));
> hold off;
> ```

4.5.6 KAZE Algorithm

The KAZE algorithm (Japanese word meaning "wind") (Alcantarilla et al., 2012) takes a similar approach to SIFT to build the scale space, but unlike SIFT, the image is not downs-ampled but processed in its original form. First the image is convolved with a Gaussian kernel to reduce noise, after which the image gradient histogram is computed. A nonlinear diffusion filtering (Perona and Malik, 1990) is applied using a function dependent on the gradient magnitude in order to reduce diffusion at the location of edges. The set of dis-crete scale levels in pixel units is converted to time units as nonlinear diffusion filtering is defined in temporal terms. The set of evolution times is used to build a non-linear scale space. For detecting interest points, the determinant of the Hessian matrix at multiple

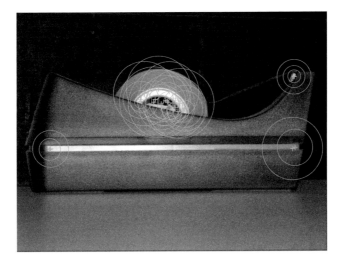

FIGURE 4.9
Output of Example 4.10.

scales is used. Extrema are searched over a rectangular window using Scharr filters of different derivative step sizes. Similar to SURF, the dominant orientation in a local neighborhood of a keypoint is estimated. The derivative responses in a 4×4 sub-region are summed into a descriptor vector $V = (\Sigma\ Lx,\ \Sigma\ Ly,\ \Sigma\ |Lx|,\ \Sigma\ |Ly|)$ computed according to the dominant orientation, generating a 64-element vector. Here, Lx and Ly are first-order derivatives weighted with a Gaussain centered at the interest point. The CVST function **detectKAZEFeatures** can be used to compute interest points using the KAZE algorithm. In the following example, interest points returned by the KAZE algorithm are stored in the variable *kp* which has datatype *KAZEPoints*. The feature descriptor *fd* contains the 64-dimensional feature vectors for all the valid points *vp*. Valid points *vp* can be same or a subset of KAZE points *kp* excluding those which lie along an edge or boundary of the image. The strongest 30 points returned by the method *selectStrongest* are plotted over the image to show their location and magnitude, the latter indicated by the radius of the corresponding circle (Figure 4.9).

> **Example 4.11: Write a program to compute interest points from an image using the KAZE algorithm and generate a descriptor for the same.**
>
> ```
> clear; clc;
> I = imread('tape.png'); I = rgb2gray(I);
> kp = detectKAZEFeatures(I);
> [fd, vp] = extractFeatures(I, kp, 'Method', 'KAZE');
> figure; imshow(I); hold on;
> plot(vp.selectStrongest(30));
> hold off;
> ```

4.5.7 BRISK Algorithm

The BRISK algorithm (Leutenegger et al., 2011) identifies interest points across both image and scale dimensions via a scale space. Keypoints are detected in octave layers of image pyramid as well as layers in-between. The location and scale of each keypoint are obtained via a quadratic function fitting. The keypoint descriptor is generated by a

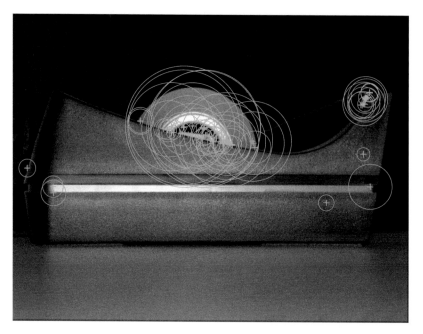

FIGURE 4.10
Output of Example 4.11.

sampling pattern consisting of points lying on appropriately scaled concentric circles. The BRISK descriptor is composed as a binary string by comparing the intensity pairs of points on the pattern and is of length 512 bits. Comparison of two descriptors is performed by computing Hamming distance between them. The CVST function **detectBRISKFeatures** is used to detect interest points using the BRISK algorithm. In the following example, interest points returned by the BRISK algorithm are stored in the variable *bp* which has datatype *BRISKPoints*. The feature descriptor *fd* contains the binary features for all the valid points *vp*. The valid points *vp* can be the same or a subset of BRISK points *bp* excluding those which lie along an edge or boundary of the image. The valid points are plotted over the image to show their location and magnitude, the latter indicated by the radius of the corresponding circle (Figure 4.10).

> **Example 4.12: Write a program to compute interest points from an image using the BRISK algorithm and generate a descriptor for the same**
>
> ```
> clear; clc;
> I = imread('tape.png');
> I = rgb2gray(I);
> bp = detectBRISKFeatures(I);
> [fd, vp] = extractFeatures(I, bp, 'Method', 'BRISK');
> figure; imshow(I); hold on;
> plot(vp);
> ```

4.5.8 LBP Algorithm

The LBP algorithm (Ojala et al., 2002) is used to encode local texture information. For each interest point, the method compares the pixel to each of its eight neighbors in an ordered way. If the center pixel's value is greater than its neighbor, a "0" is written to the output string, else a "1" is written. This generates an 8-bit binary number which is converted to its decimal equivalent value. Next, a histogram of the frequency of each number is generated which produces a 256-dimensional feature vector. The concept of *uniform pattern* can be used to reduce the length of the feature vector. A binary pattern is called uniform if it contains a maximum of two 0-to-1 or 1-to-0 transitions. Recognizing the fact that uniform patterns occur more commonly than non-uniform patterns (containing larger number of transitions), one bin is allotted to each uniform pattern but all the non-uniform patterns are clubbed into a single bin. The 58 uniform binary patterns correspond to the decimal numbers 0, 1, 2, 3, 4, 6, 7, 8, 12, 14, 15, 16, 24, 28, 30, 31, 32, 48, 56, 60, 62, 63, 64, 96, 112, 120, 124, 126, 127, 128, 129, 131, 135, 143, 159, 191, 192, 193, 195, 199, 207, 223, 224, 225, 227, 231, 239, 240, 241, 243, 247, 248, 249, 251, 252, 253, 254, and 255. The 59-th bin is used for all non-uniform patterns, generating a 59-element feature vector. The CVST function **extractLBPFeatures** extracts uniform LBP from a grayscale image, which can be used to classify textures based on the local texture property. The following example extracts LBP features from three kinds of textures and then compares them using squared errors between them. The squared error is visualized as a bar graph to assess difference between the textures across the 59-element feature vector. The *Upright false* option specifies those features which are rotationally invariant to be used in the computations. The plot shows that the differences between the first and second textures which are visually similar (*a vs. b*) are much smaller than those between the first and third textures which are visually dissimilar (*a vs. c*) (Figure 4.11).

Example 4.13: Write a program to compute interest points from an image using the LBP algorithm.

```
clear; clc;
a = imread('bricks.jpg');
b = imread('bricksRotated.jpg');
```

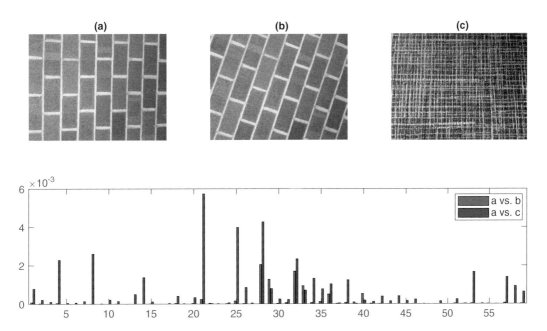

FIGURE 4.11
Output of Example 4.13. Textures (a) bricks (b) rotated bricks (c) carpet (d) plots of a vs. b and a vs. c.

```
c = imread('carpet.jpg');
af = extractLBPFeatures(a);
bf = extractLBPFeatures(b);
cf = extractLBPFeatures(c);
dab = (af - bf).^2;
dac = (af - cf).^2;

figure,
subplot(231), imshow(a); title('a');
subplot(232), imshow(b); title('b');
subplot(233), imshow(c); title('c');
subplot(2,3,[4:6]),
bar([dab ; dac]', 'grouped');
legend('a vs. b', 'a vs. c');
```

4.5.9 HOG Algorithm

The Histogram of Oriented Gradients (HOG) algorithm (Dalal and Triggs, 2005) is based on calculating the magnitude and orientation of image gradients of pixels over the entire image. As discussed in Section 4.5.5, for a pixel at (x,y) having intensity $I(x,y)$, the gradients along the horizontal and vertical directions can be computed as

$$Gx = I(x+1,y) - I(x-1,y)$$

$$Gy = I(x,y+1) - I(x,y-1)$$

This can be implemented by filtering the image with the following horizontal and vertical kernels $[-1,0,1]$ and $[-1,0,1]^T$. The image is split into rectangular blocks of size 8×8, typically with 50% overlap. Within each block gradient, magnitude and directions are calculated. The gradient directions are then quantized into nine bins with 20° angular separation. A 9-bin orientation histogram is constructed with each entry scaled by the gradient magnitude. The 9-bin histograms of all the image blocks are then concatenated to form a single 1-D feature vector for the entire image. The CVST function **extractHOGFeatures** extracts HOG features from images. In the following example, the feature descriptor *fd* stores the gradient histogram and the *Visualization* object *hv* is used to display the scaled orientation gradients for each point of the image (Figure 4.12).

Example 4.14: Write a program to compute interest points from an image Using the HOG algorithm.

```
clear; clc;
I = imread('circlesBrightDark.png');
[fd, hv] = extractHOGFeatures(I);
imshow(I); hold on;
plot(hv); hold off;
```

4.6 Clustering

4.6.1 Similarity Metrics

In a pattern recognition problem, it is required to categorize data into classes. If data can be categorized based on similarity, a query needs to only search within relevant categories. This forms the basis of indexing and retrieval in the database. There are two types of

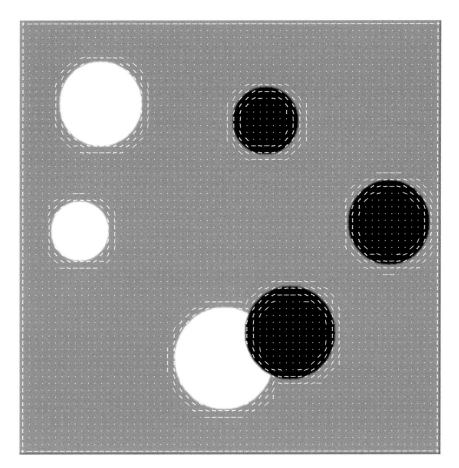

FIGURE 4.12
Output of Example 4.14.

categorization: clustering and classification. In **clustering**, also known as **unsupervised grouping**, categorization is done solely on the basis of the values of the data points and without any other prior information. In **classification**, also known as **supervised grouping**, categorization is done based on data values and prior information about the characteristics of the classes. In both cases, each data point is grouped into a cluster or a class based on its similarity with other data points. Given a query item, two types of searches may be done on similar items: (1) find the X most similar items with respect to the query and (2) find items X % or more similar to the query. So, there needs to be a mathematical definition of similarity using which data points may be compared with each other. In pattern recognition problems, data points are mostly represented using n-dimensional vectors. Given a vector $X = (x_1, x_2,..., x_n)$, its difference can be measured with another vector $Y = (y_1, y_2,..., y_n)$ using the L_p (Minkowski) metric, named after German mathematician Ermann Minkowski:

$$D_p(X,Y) = \left(\sum_{i=1}^{n} |x_i - y_i|^p \right)^{\frac{1}{p}}.$$

Since this is a difference metric, larger the value of D_p, less will be the similarity between the data values or vectors, and vice versa. Physically this represents a measure of the distance between the data points X and Y when plotted in an n-dimensional space (Deza and Deza, 2006). Since the points represent some feature based on which pattern recognition is done, X and Y are often called **feature vectors** and the n-dimensional space is called **feature space**.

If $p = 1$, then the difference is called L_1 metric or cityblock distance or **Manhattan distance**:

$$D_1(X,Y) = \sum_{i=1}^{n} |x_i - y_i|$$

If $p = 2$, then the difference is called L_2 metric or **Euclidean distance**:

$$D_2(X,Y) = \left(\sum_{i=1}^{n} |x_i - y_i|^2 \right)^{\frac{1}{2}}$$

If $p \rightarrow \infty$, then the difference is called L_∞ metric or Chebychev metric or **chessboard distance**:

$$D_\infty(X,Y) = \max_{1 \le i \le n} |x_i - y_i|$$

Apart from the Minkowski metrics, other distance measure are also frequently used. One of these is the **cosine measure** which is equal to {1 − cosine of angle (θ) between two vectors} and • indicates dot product:

$$\cos(\theta) = \frac{X \bullet Y}{|X||Y|} = \frac{\sum_{i=1}^{n} x_i y_i}{\sqrt{\sum_{i=1}^{n} x_i^2} \sqrt{\sum_{i=1}^{n} y_i^2}}$$

The SMLT function **pdist** is used to calculate the distance between pairs of points. The appropriate arguments specify which distance is to be computed. The following example computes pairwise distances between three specified points i.e. P and Q, P and R, and Q and R

Example 4.15: For the three points P = (1, 1), Q = (2, 3), R = (4, 6), find out the L1, L2, L5, L∞, cosine metrics between all pairs of points

```
clear; clc; format compact;

P = [1, 1]; Q = [2, 3]; R = [4, 6];
X = [P ; Q ; R];
fprintf('Cityblock distance\n');        pdist(X, 'cityblock')
fprintf('Euclidean distance\n');        pdist(X, 'euclidean')
fprintf('L5 distance\n'); p=5;          pdist(X, 'minkowski', p)
fprintf('Chebychev distance\n');        pdist(X, 'chebychev')
fprintf('Cosine distance\n');           pdist(X, 'cosine')
```

The program output is:

```
Cityblock distance: 3, 8, 5
Euclidean distance: 2.2361, 5.8310, 3.6056
L5 distance: 2.0123, 5.0754, 3.0752
Chebychev distance: 2, 5, 3
Cosine distance: 0.0194, 0.0194, 0.0000
```

The **Mahalanobis distance** is a measure between a sample point, represented by vector A, and a distribution with mean μ and covariance σ:

$$D_M = \sqrt{(A - \mu).\left(\frac{1}{\sigma}\right).(A - \mu)^T}$$

This distance represents how far A is from the mean in number of standard deviations. In the following example, the SMLT function **mvnrnd** is used to generate a multivariate normal distribution X of 1000 points with a specified mean and specified covariance. Four test points Y are generated which are equidistant from the mean of X in Euclidean distance. The SMLT function **mahal** is used to calculate the squared Mahalanobis distance of each test point from the distribution X, then compared with the corresponding Euclidean distances. It shows that even though the Euclidean distances of all four points are nearly the same, the Mahalanobis distance of the points along the axis of the distribution are much less than those of the points outside the distribution (Figure 4.13).

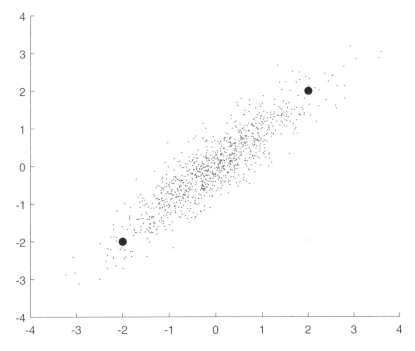

FIGURE 4.13
Output of Example 4.16.

Example 4.16: Write a program to compare Euclidean distance and Mahalanobis distance for a correlated bivariate sample data set

```
clear; clc;
rng('default') % For reproducibility
X = mvnrnd([0;0],[1 .9;.9 1],1000);
Y = 2*[1 1;1 -1;-1 1;-1 -1];
d2_mahal = mahal(Y,X);
d2_Euclidean = sum((Y-mean(X)).^2,2);
scatter(X(:,1),X(:,2),10,'.')
hold on;
scatter(Y(:,1),Y(:,2),50,d2_mahal,'o','filled');
hold off;
```

4.6.2 *k*-means Clustering

The **k-means algorithm** is a frequently used clustering method implemented as per the following steps: (1) Each datum is represented by an n-dimensional feature vector. (2) Let there be m such data points which are to be grouped into k classes. (3) The data are plotted as points in an n-dimensional feature space. (4) An initial estimate of k cluster means (or centroid) is assumed. (5) Distance of all data points from each of the k cluster means computed. (6) Each data point is classified to that group with which it has the least distance. (7) New cluster means (or centroid) are calculated and the process is repeated. (8) The process is assumed to have converged when data points do not change clusters. The SMLT function **kmeans** is used to perform k-means clustering on a data matrix X into k clusters and returns a vector containing the cluster index of each data point, as well as centroid of each cluster. The following example shows five given points being divided into two clusters. The cluster means are marked by crosses (Figure 4.14).

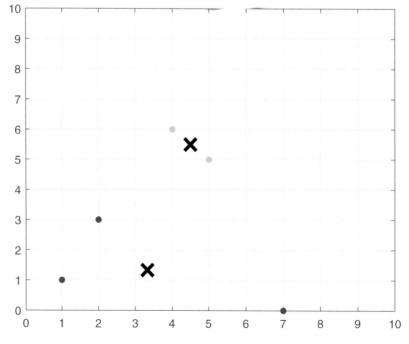

FIGURE 4.14
Output of Example 4.17.

Example 4.17: Write a program to divide the five points A(1,1), B(2,3), C(4,6), D(5,5), E(7,0) into 2 clusters Using k-means clustering and find the cluster means

```
clear; clc;
A = [1, 1]; B = [2, 3]; C = [4, 6]; D = [5, 5]; E = [7, 0];
X = [A ; B ; C ; D ; E]
rng(2);
k = 2; [idx, c] = kmeans(X, k)
plot(c(:,1),c(:,2),'kx','MarkerSize',15,'LineWidth',3);
grid; axis([0,10,0,10]);
hold on;
s = scatter(X(:,1), X(:,2), 30, idx, 'filled');
colormap(cool);
hold off;
```

The program output is show below:

```
X =
     1    1
     2    3
     4    6
     5    5
     7    0
idx =
     2
     2
     1
     1
     2
c =
    4.5000    5.5000
    3.3333    1.3333
```

It shows that two points are assigned to cluster 1 with mean (4.5, 5.5) and remaining three points are assigned to cluster 2 with mean (3.33, 1.33). It is to be noted that this cluster membership can change on subsequent execution of the algorithm as it depends on the initial assignment of the cluster means which the function does randomly. To produce predictable results, the random number generator (*rng*) is seeded with a non-negative integer.

A variation of the k-means algorithm is the **k-medoids algorithm**, which is similar to the k-means clustering in that it also divides the data points into k subsets, but the difference is that while k-means uses the mean of the subsets, also called **centroid**, as the center of the subsets, in the k-medoids algorithm, the center of the subset is a member of the subset, called a **medoid**. This allows you to use the algorithm in situations where the mean of the data does not exist within the data set. The SMLT function **kmedoids** is used to implement the k-medoids clustering algorithm. The following example shows clustering of five points using the k-medoids algorithm. The cluster medoids are marked by circles (Figure 4.15).

Example 4.18: Write a program to divide the five points A(1,1), B(2,3), C(4,6), D(5,5), E(7,0) into 2 clusters using k-medoids clustering and find the cluster medoids

```
clear; clc;
A = [1, 1]; B = [2, 3]; C = [4, 6]; D = [5, 5]; E = [7, 0];
X = [A ; B ; C ; D ; E]
rng(2);
k = 2; [idx, c] = kmedoids(X, k)
plot(c(:,1),c(:,2),'ko','MarkerSize',15,'LineWidth',2);
grid; axis([0,10,0,10]);
hold on;
s = scatter(X(:,1), X(:,2), 30, idx, 'filled');
s.MarkerEdgeColor = [0 0 1];
colormap(autumn);
hold off;
```

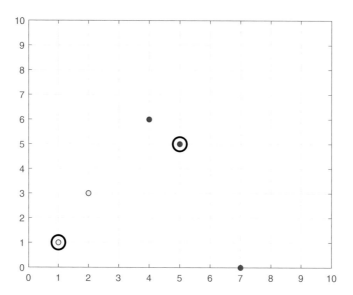

FIGURE 4.15
Output of Example 4.18.

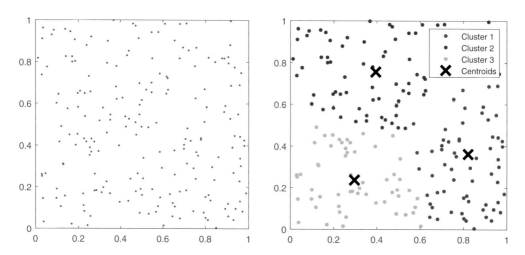

FIGURE 4.16
Output of Example 4.19.

The *k*-means clustering algorithm can also be used for arbitrarily large population of data-points. The following example depicts clustering on 100 data points. The BM function **rand** can be used with random data points using a specified dimension (Figure 4.16).

> **Example 4.19: Write a program to generate 100 2-D random points and divide them into 3 clusters using *k*-means clustering and find the cluster centroids**
>
> ```
> clear; clc;
> rng('default');
> X = [rand(100,2); rand(100,2)];
> figure,
> subplot(121),
> plot(X(:,1),X(:,2),'.');
> ```

```
[idx,C] = kmeans(X,3);
subplot(122),
plot(X(idx == 1,1), X(idx == 1,2),'r.','MarkerSize',12); hold on;
plot(X(idx == 2,1), X(idx == 2,2),'b.','MarkerSize',12);
plot(X(idx == 3,1), X(idx == 3,2),'g.','MarkerSize',12);
plot(C(:,1),C(:,2),'kx','MarkerSize',15,'LineWidth',3);
legend('Cluster 1','Cluster 2','Cluster 3','Centroids');
```

The cluster means calculated from the Fisher-Iris data set are: setosa (1.4620, 0.2460), versicolor (4.2600, 1.3260), virginica (5.5520, 2.0260), as discussed in Section 4.3. This can be considered as *ground truth* since the clusters have been discriminated using the actual species name. In contrast, if the same data set is clustered using the k-means algorithm, then this would generate slightly different centroids. The following example shows that the cluster means computed by k-means are given by: (1.4620, 0.2460), (4.2926, 1.3593), (5.6261, 2.0478). While the first cluster center remains the same as it is non-overlapping with the other two, the second and third cluster means deviate to the extent of around 1%–2% (Figure 4.17).

Example 4.20: Write a program to divide the Fisher-Iris dataset them into 3 clusters using k-means clustering and find the cluster centroids

```
clear; clc;
load fisheriris;
X = meas(:, 3:4);    % petal length & petal width

% ground truth
figure, subplot (121),
c3 = X(1:50,:); c2 = X(51:100,:); c1 = X(101:150,:);
plot(c1(:,1), c1(:,2), 'b.', 'MarkerSize',12); hold on;
plot(c2(:,1), c2(:,2), 'g.', 'MarkerSize',12);
plot(c3(:,1), c3(:,2), 'r.', 'MarkerSize',12);
c1mx = mean(c1(:,1)); c1my = mean(c1(:,2));
c2mx = mean(c2(:,1)); c2my = mean(c2(:,2));
c3mx = mean(c3(:,1)); c3my = mean(c3(:,2));
c = [c1mx, c1my ; c2mx, c2my ; c3mx, c3my];
plot(c(:,1),c(:,2),'kx','MarkerSize',12,'LineWidth',2);
legend('Cluster 1','Cluster 2','Centroids');
title('ground truth'); hold off; axis square;
xlabel('petal length'); ylabel('petal width');
```

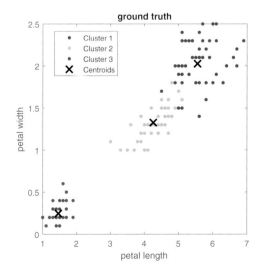 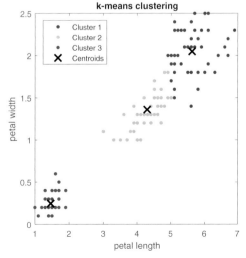

FIGURE 4.17
Output of Example 4.20.

4.6.3 Hierarchical Clustering

Another clustering method known as **hierarchical clustering** is implemented through the following steps: (1) Compute distances between all pairs of data. (2) Place pairs of data corresponding to the shortest distance into a cluster. (3) Replace a cluster by its centroid. (4) Repeat above process until there are k clusters. Hierarchical clustering is visually denoted by a graph called a **dendrogram** which depicts which points are combined into clusters and in which sequence. The SMLT function linkage provides an agglomerative hierarchical cluster tree which displays the minimum distance between pairs of points at each iteration. The SMLT function **cluster** generates clusters from the set of minimum distances between pairs of points. The BM function **scatter** is used to plot points as circles. The SMLT function **dendrogram** is used to generate a dendrogram plot. In the following example, the program output shows that the first four points are grouped into the first cluster and the last point is grouped into the second cluster. The dendrogram depicts that points C and D are first combined into F, next points A and B are combined into G, and finally F and G are combined together into H. The final clusters are H and E (Figure 4.18).

> **Example 4.21: Write a program to divide the five points A(1,1), B(2,3), C(4,6), D(5,5), E(7,0) into 2 clusters using Hierarchical clustering and find cluster means. Also display the corresponding dendrogram**
>
> ```
> clear; clc;
> P = [1, 1]; Q = [2, 3]; R = [4, 6]; S = [5, 5]; T = [7, 0];
> X = [P ; Q ; R ; S ; T]
> Z = linkage (X, 'centroid');
> c = cluster(Z, 'maxclust', 2);
> figure,
> subplot(121),
> scatter(X(:,1), X(:,2), 30, c, 'filled');
> colormap(cool); grid;
> axis([0 10 -1 10]); title('data');
> subplot(122),
> dendrogram(Z); title('dendrogram');
> ```

In some cases, data points can be randomly generated and then clustered using algorithms. The following example shows clustering of 100 random 2-D points into three clusters using hierarchical clustering, with different colors depicting the clusters. The cluster means are calculated and displayed, as well as the dendrogram (Figure 4.19).

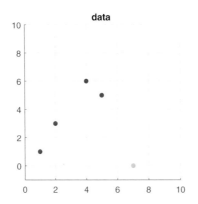
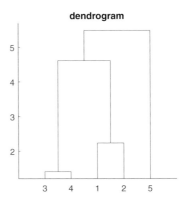

FIGURE 4.18
Output of Example 4.21.

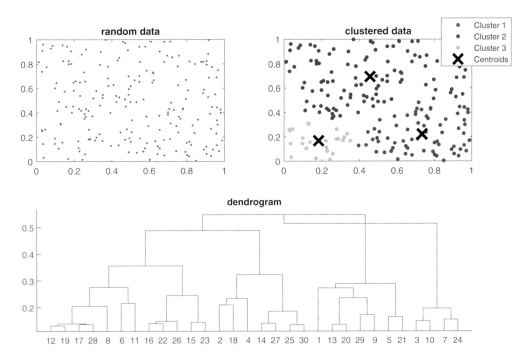

FIGURE 4.19
Output of Example 4.22.

Example 4.22: Write a program to generate 100 2-D random points and divide them into 3 clusters using hierarchical clustering and find the cluster centroids

```
clear; clc;
rng('default');
X = [rand(100,2); rand(100,2)];
figure, subplot(221), plot(X(:,1),X(:,2),'.');
title('random data');

z = linkage(X, 'centroid');
idx = cluster(z, 'maxclust', 3);
c1x = X(idx == 1,1); c1y = X(idx == 1,2); c1m = [mean(c1x), mean(c1y)];
c2x = X(idx == 2,1); c2y = X(idx == 2,2); c2m = [mean(c2x), mean(c2y)];
c3x = X(idx == 3,1); c3y = X(idx == 3,2); c3m = [mean(c3x), mean(c3y)];
subplot(222),
plot(X(idx == 1,1),X(idx == 1,2),'r.','MarkerSize',12); hold on;
plot(X(idx == 2,1),X(idx == 2,2),'b.','MarkerSize',12);
plot(X(idx == 3,1),X(idx == 3,2),'g.','MarkerSize',12);
c = [c1m ; c2m ; c3m];
plot(c(:,1),c(:,2),'kx','MarkerSize',15,'LineWidth',3);
legend('Cluster 1','Cluster 2','Cluster 3','Centroids');
title('clustered data');
subplot(2,2,[3,4]),dendrogram(z); title('dendrogram');
```

The BM function **scatter3** can be used to generate 3-D scatter plots consisting of filled or unfilled circles. The BM function **colormap** can be used to color the data points according to their cluster memberships. The following example shows a random set of 1000 3-D points being clustered into four clusters using hierarchical clustering with colormaps being used to change the color schemes of the displayed clusters (Figure 4.20).

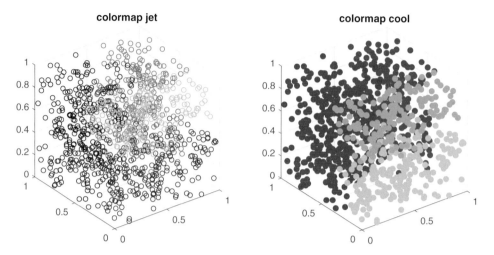

FIGURE 4.20
Output of Example 4.23.

Example 4.23: Write a program for clustering of 1000 3-D random data values into 4 clusters

```
clear; clc;
rng(1);
X = rand(1000, 3);
Z = linkage(X, 'average');
idx = cluster(Z, 'maxclust', 4);

ax1 = subplot(121); map1 = jet(8);
scatter3(X(:,1), X(:,2), X(:,3), 25, idx);
colormap(ax1, map1); title('colormap jet');

ax2 = subplot(122); map2 = cool(8);
scatter3(X(:,1), X(:,2), X(:,3), 40, idx, 'filled');
colormap(ax2, map2); title('colormap cool');
```

The following example shows the Fisher-Iris data set being divided into three clusters using the hierarchical clustering technique. The cluster centroids returned are: (1.4620, 0.2460), (4.1913, 1.3022), and (5.5148, 1.9944) (Figure 4.21), compare with *k*-means results of (1.4620, 0.2460), (4.2926, 1.3593), (5.6261, 2.0478) and the ground truth of (1.4620, 0.2460), (4.2600, 1.3260), (5.5520, 2.0260) shown previously.

Example 4.24: Write a program to divide the Fisher-Iris dataset them into 3 clusters using hierarchical clustering and find the cluster centroids

```
clear; clc;
load fisheriris;
X = meas(:,3:4);
figure, subplot(121),
plot(X(:,1),X(:,2),'k.','MarkerSize',12);
z = linkage(X, 'centroid'); idx = cluster(z, 'maxclust', 3);
title('data'); axis square;
xlabel('petal length'); ylabel('petal width');

subplot (122),
plot(X(idx == 1,1),X(idx == 1,2),'r.','MarkerSize',12); hold on;
plot(X(idx == 2,1),X(idx == 2,2),'b.','MarkerSize',12);
```

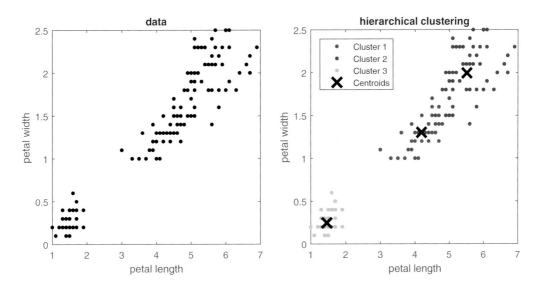

FIGURE 4.21
Output of Example 4.24.

```
plot(X(idx == 3,1),X(idx == 3,2),'g.','MarkerSize',12);

c1x = X(idx == 1,1); c1y = X(idx == 1,2); c1m = [mean(c1x), mean(c1y)];
c2x = X(idx == 2,1); c2y = X(idx == 2,2); c2m = [mean(c2x), mean(c2y)];
c3x = X(idx == 3,1); c3y = X(idx == 3,2); c3m = [mean(c3x), mean(c3y)];

c = [c1m ; c2m ; c3m];
plot(c(:,1),c(:,2),'kx','MarkerSize',15,'LineWidth',3);
legend('Cluster 1','Cluster 2','Cluster 3','Centroids');
title('hierarchical clustering'); hold off; axis square;
xlabel('petal length'); ylabel('petal width');

c   % cluster centroids
```

4.6.4 GMM-Based Clustering

Gaussian Mixture Models (GMM) can be used for clustering by trying to fit multivariate Gaussian components to a given data distribution. The fitting is done using an iterative expectation–maximization (EM) algorithm. It returns the mean and covariance the GMMs that best fit the data. The following example generates 200 random data points by a random number generator by using the BM function **rng** set to default settings for reproducibility, specifies the maximum number of iterations to be 1000 by using the SMLT function **statset**, and then uses the SMLT function **fitgmdist** to fit a specified number of GMMs to fit the data. The function returns the GMM distribution components, each component defined by its mean and covariance and the mixture defined by a vector of mixing proportions. The properties of each GMM component can be displayed by specifying *properties(GMM{j})* for the *j*-th GMM. The example lists out some of the important properties like mean, covariance, component proportions, and number of iterations. The SMLT function **pdf** returns the probability distribution function (PDF) for the GMMs, and the Symbolic Math Toolbox (SMT) function **ezcontour** plots the contour lines of the fitted GMMs over the given data (Figure 4.22).

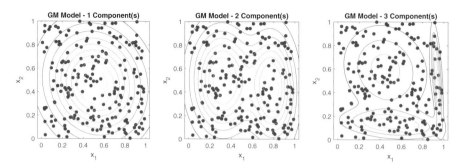

FIGURE 4.22
Output of Example 4.25.

Example 4.25: Write a program to fit a random distribution of points using a specified number of GMMs.

```
clear; clc;
rng('default'); % For reproducibility
X = [rand(100,2); rand(100,2)];
GMM = cell(3,1); % Preallocation
options = statset('MaxIter',1000);
for j = 1:3
    GMM{j} = fitgmdist(X,j,'Options',options);
    fprintf('\n GMM properties for %i Component(s)\n',j);
    Mu = GMM{j}.mu
    Sigma = GMM{j}.Sigma
    CP = GMM{j}.ComponentProportion
    NI = GMM{j}.NumIterations
end;

figure,
for j = 1:3
    subplot(1,3,j),
    gscatter(X(:,1),X(:,2));
    h = gca;
    hold on;
    ezcontour(@(x1,x2)pdf(GMM{j},[x1 x2]),[h.XLim h.YLim],100);
    title(sprintf('GM Model - %i Component(s)',j));
    axis square;
    hold off;
end;
```

The SMLT function **cluster** can be used to cluster the given data using the GMMs generated to fit the data. The following example is used to cluster the petal measurements of the Fisher-Iris data set using Gaussian Mixture Models into three clusters and returns the cluster means. The cluster centroids returned are (1.4620, 0.2460), (4.3066, 1.3375), and (5.5863, 2.0602) (Figure 4.23). Compare with k-means results of (1.4620, 0.2460), (4.2926, 1.3593), and (5.6261, 2.0478) and the ground truth of (1.4620, 0.2460), (4.2600, 1.3260), and (5.5520, 2.0260) shown in Figure 4.17 and hierarchical clustering results (1.4620, 0.2460), (4.1913, 1.3022), and (5.5148, 1.9944) shown in Figure 4.21.

Example 4.26: Write a program to implement clustering on the first two measurements of the Fisher-Iris dataset using Gaussian Mixture Models and return the cluster means

```
clear; clc;

load fisheriris;
```

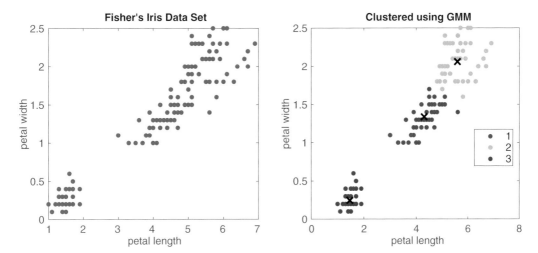

FIGURE 4.23
Output of Example 4.27.

```
X = meas(:,3:4);
rng(3); % For reproducibility

figure; subplot(121),
plot(X(:,1),X(:,2),'.','MarkerSize',15);
title('Fisher''s Iris Data Set');
xlabel('petal length'); ylabel('petal width');
k = 3;
subplot(122),
GMM = fitgmdist(X,k,'CovarianceType','diagonal','SharedCovariance',true);
clusterX = cluster(GMM,X);
h1 = gscatter(X(:,1),X(:,2),clusterX);
hold on;
plot(GMM.mu(:,1),GMM.mu(:,2),'kx','LineWidth',2,'MarkerSize',10);
xlabel('petal length'); ylabel('petal width');
title('Clustered using GMM');
legend(h1,{'1','2','3'});
hold off;
GMM
```

4.7 Classification

In classification, also known as **supervised grouping**, categorization is done based on prior information about classes and all data points are grouped as members of one or other of the classes based on their data values. Accordingly the data set is split into a **training set** and a **testing set**. A training phase is used to inform the system about the discriminative characteristics of each class, while a testing phase is used to utilize the learned characteristics to categorize unknown data into classes. During the training phase, the input values are fed to the classifier and the output values i.e. class labels are also known. These input and output values are used to determine the classifier parameters, e.g. weights in a neural network. However, one problem with this scheme is known as **overfitting** i.e. the classifier parameters are tuned more toward modeling the training set samples. This is not a desirable feature since the testing set is different from the training set, and these parameters would not produce a good classification result during the testing phase. Rather a generalization in modeling, the parameters is desirable which would enable recognition of the testing samples based on the characteristics learned during the training phase. To avoid overfitting, the data set is split into a third part called the **validation set**. The typical proportion of training, validation, and testing samples is 60:20:20. Like the training set, the validation set also has the input and output values known but the training is not done on it. Once the parameters are determined during an iteration of the training phase, these parameters are then applied to the validation set to check whether the output classes can be predicted based on it. Since the class labels of the validation set are already known beforehand, the error between the estimated and the actual classes can be computed. As the training progresses, the error in both the training set and the validation set should ideally reduce. After a certain number of iteration completes, if the error for the training set reduces but that for the validation set increases, it indicates that overfitting has started since the trained parameters are unable to correctly classify the samples in the validation set. The training should be stopped at this point, and the final parameters obtained should be used to classify samples of the test set. This section discusses different classifiers: k-NN, ANN, DT, DA, NB, and SVM.

4.7.1 k-NN Classifiers

One of the frequently used classification algorithms is the **k-nearest neighbour** (k-NN) algorithm. This is a supervised grouping algorithm in which the class membership of a test data point is estimated as belonging to the class for which there are the largest number of data points among its k neighbors (Roussopoulos et al., 1995).The value k is a parameter and specify how many neighbors influence the decision making. Different distance metrics can be used to find the nearest neighbors. The value of k can be tuned to produce the best results for a particular problem. In the following example, the k-NN algorithm is illustrated using the petal measurements of the Fisher-Iris Data set. The query point is designated by the *newpoint* vector. The BM function **line** represents the query point using an X marker of the specified size and appearance. The SMLT function **knnsearch** is used to find the nearest neighbors to the query point specifying the value of k to be 5. It returns the index values of the nearest neighbors and their distances from the query point. The index values are used to print out the names of the corresponding species. The nearest neighbors are indicated by gray circles of the specified size. The BM function **legend** adds legends to the figure to designate the three classes of flowers, the query point and the nearest neighbors found. Finally, the SMLT function **tabulate** is used to generate a frequency table of the species names found among the five nearest neighbors. The output of the program shows that out of the five nearest neighbors found, three are of species *setosa* and two are of species *versicolor*. The decision taken by the system is that the query point belongs to the class *setosa* since the majority of the neighbors belong to that species (Figure 4.24).

> **Example 4.27: Write a program to classify the point (2.5, 0.7) to the Fisher Iris dataset using 5 nearest neighbours**
>
> ```
> clear; clc;
>
> load fisheriris;
> x = meas(:, 3:4);
> gscatter (x(:,1), x(:,2), species);
>
> newpoint = [2.5, 0.7];
> ```

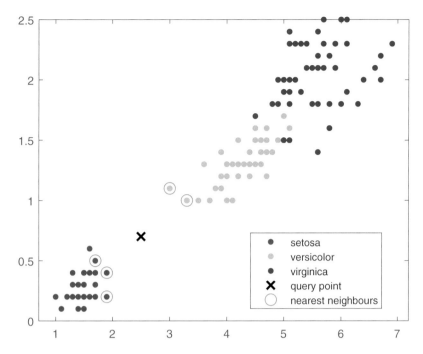

FIGURE 4.24
Output of Example 4.26.

```
line(newpoint(1), newpoint(2), ...
    'marker', 'x', 'color', 'k', 'markersize', 10, 'linewidth', 2, 'Linestyle','none');

[n, d] = knnsearch(x, newpoint, 'k', 5);
line(x(n,1), x(n,2), 'color', [0.5, 0.5, 0.5], ...
    'marker', 'o', 'linestyle', 'none', 'markersize', 10);

legend('setosa','versicolor','virginica','query point','nearest neighbours');

n, d, species(n)
tabulate(species(n))
```

4.7.2 Artificial Neural Network (ANN) classifiers

Supervised classification algorithms can broadly be divided into two types: lazy and eager algorithms. The k-NN algorithm discussed above is a lazy classification algorithm because all the system uses all the training samples in the last step to generate a decision regarding the classification of the query point. The other type i.e. the eager classification algorithm generates a separate data model from the training samples and uses that model to make a decision instead of the actual data. An example of eager algorithm is the neural network. The basic structural unit of an **artificial neural network** (ANN) is a **neuron** which has multiple signal input lines, a single output line, and a threshold (Hopfield, 1982). If sum of input signals equals or exceeds the threshold, the neuron produces a non-zero output, otherwise a zero output. A neuron with two input lines with binary data can be used to simulate a logical OR gate using a threshold of 1 and a logical AND gate using a threshold of 2. A **perceptron** is a type of neuron whose inputs are associated with weights (Rosenblatt, 1958). If the input vector is X, and the weights are inserted into a vector W, then additionally the combined input is I is given by the following relation, where b is the weight associated with a bias line with a signal of logical 1:

$$X = (x_1, x_2, \ldots, x_n)$$

$$W = (w_1, w_2, \ldots, w_n)$$

$$I = b + W \cdot X^T = b + \sum_{i=1}^{n} w_i x_i$$

The output y is produced by a transfer function f which triggers an output value based on the combined input I. In the simplest case, $y = f(x)$ is a simple step function of the combined input:

$$y = 0 \text{ if } x < 0$$

$$y = 1 \text{ if } x \geq 0$$

In most practical cases however instead of step functions sigmoid functions are used, so called because of their S-shaped curves. There are two types of sigmoid functions log-sigmoid (y_1) and tan-sigmoid (y_2):

$$y_1 = \frac{1}{1 + e^{-x}}$$

$$y_2 = \frac{e^x - e^{-x}}{e^x + e^{-x}}$$

Perceptrons are associated with a training phase and a testing phase, for which the entire data set is split into two parts: the training set and testing set. For the training set, the class information of the data points are known which are referred to as target values. Using the known input values and known target (class) values, the weights are adjusted in an iterative manner so that the error between the target and actual output values falls below a specified minimum threshold. The weights at the end of the training phase are the final weights which are used during the testing phase to estimate the class of the test data points. The training phase continues until one of two conditions is satisfied: either the error falls below the specified threshold or a maximum number of specified iterations have been completed. The Neural Network Toolbox (NNT) function **perceptron** is used to implement a simple perceptron which can be used to simulate logic gates. The NNT function **train** is used to train the perceptron so that it produces the correct output for a given set of inputs to simulate the logic gates. The NNT function **view** displays a graphical diagram of the perceptron along with the input and output. In the following example, perceptrons p_1 and p_2 are used to simulate an OR gate and an AND gate, respectively.

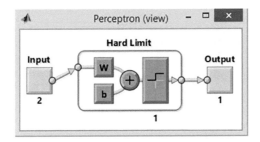

FIGURE 4.25
Output of Example 4.28.

After the perceptrons have been trained, they will produce correct outputs for a given set of inputs e.g. $= p_1 (1,0) = 1, y = p_2 (1,0) = 0$, and so on (Figure 4.25).

> **Example 4.28: Write a program to simulate the OR gate and AND gate using perceptrons**
>
> ```
> clear; clc;
>
> 5x = [0 0 1 1; 0 1 0 1];
> t = [0 1 1 1];
> p1 = perceptron;
> p1 = train(p1,x,t);
> view(p1)
> y = p1(x);   % OR gate
>
> x = [0 0 1 1; 0 1 0 1];
> t = [0 0 0 1];
> p2 = perceptron;
> p2 = train(p2,x,t);
> view(p2)
> y = p2(x);   % AND gate
> ```

Until now we have been seeing that the training and testing data sets are the same. In pattern recognition applications however, the training and testing sets are usually similar but not exactly identical. This is because in practical situations the training set is acquired during designing the system but the test set can be obtained at any point of time after the system design from different field conditions. Even though the training is done on the training set, the recognition system should be robust enough to correctly recognize samples of the testing set, which is typically different in some ways from the training set. In the following example, three characters A, B, C created from a grid of binary numbers, are trained using a perceptron based recognition system, which is then used to recognize three degraded versions of the same characters. Each character is made up of a grid of seven columns and nine rows, the background is represented by white and the characters by black. Since a perceptron has only a single output, the system is designed to recognize only a single character, in this case A. When feeding into the system input, the characters are converted into a 63-element vector, for which there should be 63 input elements to the perceptron. The input matrix consist of 63 rows and three columns for the three characters, while the output vector is [1, 0, 0] for recognizing character A. After the system is trained the test set is fed again in the form of 63 × 3 matrix. If the system correctly recognizes the characters, then the expected output should be [1, 0, 0]. The transfer function of the perceptron is a step function by default (Figure 4.26).

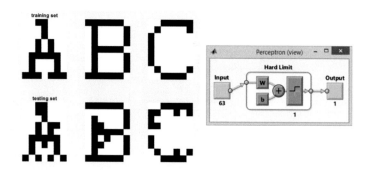

FIGURE 4.26
Output of Example 4.29.

Example 4.29: Write a program to implement character recognition using a perceptron.

```
clear; clc;

% training set
xa = [  0, 0, 1, 1, 0, 0, 0 ; 0, 0, 0, 1, 0, 0, 0 ; 0, 0, 0, 1, 0, 0, 0 ; ...
        0, 0, 1, 0, 1, 0, 0 ; 0, 0, 1, 0, 1, 0, 0 ; 0, 1, 1, 1, 1, 1, 0 ; ...
        0, 1, 0, 0, 0, 1, 0 ; 0, 1, 0, 0, 0, 1, 0 ; 1, 1, 1, 0, 1, 1, 1 ];

xb = [  1, 1, 1, 1, 1, 1, 0 ; 0, 1, 0, 0, 0, 0, 1 ; 0, 1, 0, 0, 0, 0, 1 ; ...
        0, 1, 0, 0, 0, 0, 1 ; 0, 1, 1, 1, 1, 1, 0 ; 0, 1, 0, 0, 0, 0, 1 ; ...
        0, 1, 0, 0, 0, 0, 1 ; 0, 1, 0, 0, 0, 0, 1 ; 1, 1, 1, 1, 1, 1, 0 ];

xc = [  0, 0, 1, 1, 1, 1, 1 ; 0, 1, 0, 0, 0, 0, 1 ; 1, 0, 0, 0, 0, 0, 0 ; ...
        1, 0, 0, 0, 0, 0, 0 ; 1, 0, 0, 0, 0, 0, 0 ; 1, 0, 0, 0, 0, 0, 0 ; ...
        1, 0, 0, 0, 0, 0, 0 ; 0, 1, 0, 0, 0, 0, 1 ; 0, 0, 1, 1, 1, 1, 0 ];

% testing set
xA = [  0, 0, 1, 1, 0, 0, 0 ; 0, 0, 1, 1, 0, 0, 0 ; 0, 0, 0, 1, 0, 0, 0 ; ...
        0, 0, 1, 0, 1, 0, 0 ; 0, 0, 1, 0, 1, 0, 0 ; 0, 1, 1, 1, 1, 1, 0 ; ...
        0, 1, 0, 1, 0, 1, 0 ; 0, 1, 0, 0, 0, 1, 0 ; 1, 0, 1, 0, 1, 0, 1];

xB = [  1, 1, 1, 1, 1, 1, 0 ; 0, 0, 0, 0, 0, 0, 1 ; 0, 1, 0, 0, 0, 0, 1 ; ...
        0, 1, 1, 0, 0, 0, 1 ; 0, 1, 1, 1, 1, 1, 0 ; 0, 1, 0, 1, 0, 0, 1 ; ...
        0, 1, 1, 0, 0, 0, 1 ; 0, 1, 0, 0, 0, 0, 1 ; 1, 1, 1, 1, 1, 1, 0];

xC = [  0, 0, 1, 1, 1, 1, 1 ; 1, 1, 0, 0, 1, 0, 1 ; 1, 0, 0, 0, 0, 0, 0 ; ...
        1, 1, 0, 0, 0, 0, 0 ; 0, 0, 0, 0, 0, 0, 0 ; 1, 0, 0, 0, 0, 0, 0 ; ...
        1, 0, 0, 0, 0, 0, 0 ; 0, 1, 0, 0, 1, 0, 1 ; 0, 0, 1, 1, 1, 1, 0];

subplot(231), imshow(~xa); subplot(232), imshow(~xb); subplot(233), imshow(~xc);
subplot(234), imshow(~xA); subplot(235), imshow(~xB); subplot(236), imshow(~xC);

% training phase

xa = xa(:); xb = xb(:); xc = xc(:);      % converting to vector
x = [xa xb xc];
t = [1 0 0];
p = perceptron;
p = train(p, x, t);
view(p)
y = p(x)

% testing phase

xA = xA(:); xB = xB(:); xC = xC(:);      % converting to vector
X = [xA xB xC];
Y = p(X)
```

Since a perceptron has a single output, to check the results for all three classes in the previous example, a neural network has to be formed by connecting three perceptrons together. Such a structure is called **single layer perceptron** (SLP) and consists of multiple inputs and multiple outputs but a single layer of perceptrons between the input and output. In the following example, the three characters are recognized using a SLP with 63 inputs and three outputs. The target is modified to reflect the three output lines. When the first column of input signals are fed to the SLP, the target is designated as [1, 0, 0] which indicates class 1 is true, when the second column is fed to the input the target is designated as [0 1 0] which indicates class 2 is true, and for the third column of inputs, the target of [0 0 1] indicates class 3 is true. The last line saves the train and test workspace variables x and X containing the three characters A, B, C to a MAT file *abc.mat* using the BM function **save** so that in a subsequent program we do not need to define the character arrays again, we can simply load the variables from the file into the workspace (Figure 4.27).

Example 4.30: Write a program to implement character recognition using a single layer neural network.

```
clear; clc;

% training set
xa = [   0, 0, 1, 1, 0, 0, 0 ; 0, 0, 0, 1, 0, 0, 0 ; 0, 0, 0, 1, 0, 0, 0 ; ...
         0, 0, 1, 0, 1, 0, 0 ; 0, 0, 1, 0, 1, 0, 0 ; 0, 1, 1, 1, 1, 1, 0 ; ...
         0, 1, 0, 0, 0, 1, 0 ; 0, 1, 0, 0, 0, 1, 0 ; 1, 1, 1, 0, 1, 1, 1 ];

xb = [   1, 1, 1, 1, 1, 1, 0 ; 0, 1, 0, 0, 0, 0, 1 ; 0, 1, 0, 0, 0, 0, 1 ; ...
         0, 1, 0, 0, 0, 0, 1 ; 0, 1, 1, 1, 1, 1, 0 ; 0, 1, 0, 0, 0, 0, 1 ; ...
         0, 1, 0, 0, 0, 0, 1 ; 0, 1, 0, 0, 0, 0, 1 ; 1, 1, 1, 1, 1, 1, 0 ];

xc = [   0, 0, 1, 1, 1, 1, 1 ; 0, 1, 0, 0, 0, 0, 1 ; 1, 0, 0, 0, 0, 0, 0 ; ...
         1, 0, 0, 0, 0, 0, 0 ; 1, 0, 0, 0, 0, 0, 0 ; 1, 0, 0, 0, 0, 0, 0 ; ...
         1, 0, 0, 0, 0, 0, 0 ; 0, 1, 0, 0, 0, 0, 1 ; 0, 0, 1, 1, 1, 1, 0 ];

% testing set
xA = [   0, 0, 1, 1, 0, 0, 0 ; 0, 0, 1, 1, 0, 0, 0 ; 0, 0, 0, 1, 0, 0, 0 ; ...
         0, 0, 1, 0, 1, 1, 0 ; 0, 0, 1, 0, 1, 0, 0 ; 0, 1, 1, 1, 1, 1, 0 ; ...
         0, 1, 0, 1, 0, 1, 0 ; 0, 1, 0, 0, 0, 1, 0 ; 1, 0, 1, 0, 1, 0, 1];

xB = [   1, 1, 1, 1, 1, 0, 0 ; 0, 0, 0, 0, 0, 0, 1 ; 0, 1, 0, 0, 0, 0, 1 ; ...
         0, 1, 1, 0, 0, 0, 1 ; 0, 1, 1, 1, 1, 1, 0 ; 0, 1, 0, 1, 0, 0, 1 ; ...
         0, 1, 1, 0, 0, 0, 1 ; 0, 1, 0, 0, 0, 0, 1 ; 1, 1, 1, 1, 1, 1, 0];

xC = [   0, 0, 1, 1, 1, 1, 1 ; 1, 1, 0, 0, 1, 0, 1 ; 1, 0, 0, 0, 0, 0, 0 ; ...
         1, 1, 0, 0, 0, 0, 0 ; 0, 0, 0, 0, 0, 0, 0 ; 1, 0, 0, 0, 0, 0, 0 ; ...
         1, 0, 0, 0, 0, 0, 0 ; 0, 1, 0, 0, 1, 0, 1 ; 0, 0, 1, 1, 1, 1, 0];
```

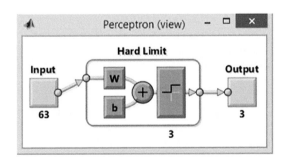

FIGURE 4.27
Output of Example 4.30.

```
subplot(231), imshow(~xa); subplot(232), imshow(~xb); subplot(233), imshow(~xc);
subplot(234), imshow(~xA); subplot(235), imshow(~xB); subplot(236), imshow(~xC);

% training phase

xa = xa(:); xb = xb(:); xc = xc(:);        % converting to vector
x = [xa xb xc];
t = [1 0 0 ; 0 1 0 ; 0 0 1];
p = perceptron;
p = train(p, x, t);
view(p)
y = p(x)

% testing phase

xA = xA(:); xB = xB(:); xC = xC(:);        % converting to vector
X = [xA xB xC];
Y = p(X)
save abc.mat x X
```

These networks studied above are conceptually simple with a single layer connecting both the input and output and can solve only linearly separable problems i.e. where the input space can be separated using a straight line into two zones for which the outputs would be different. For more complex problems requiring non-linear separation, **multi-layered perceptrons** (MLP) are required where there are one or more hidden layers in between the input and output (Bishop, 2005). An MLP is also referred to as **feedforward** neural network (FNN) since the signal propagates in the forward direction from the input layer through the hidden layers to the output layer. A feedforward network with one hidden layer and enough neurons in the hidden layers, can fit any finite input-output mapping problem. An FNN is generally used to solve non-linear separation problems as opposed to perceptrons which are used for linear separation problems. One of the most simplest and frequently used non-linear mapping problem is the XOR logic gate. The following example shows how an FNN can be used to implement a XOR gate. The NNT function **feedforwardnet** is used to implement an FNN and returns a multi-layered trained network. The function argument of two denotes two neurons in the hidden layer and *trainlm* is a network training function that updates weight and bias values according to an optimization algorithm proposed by Levenberg-Marquardt (Marquardt, 1963; Hagan and Menhaj, 1994). The network property of *dividetrain* specifies to assign all targets to the training set and no targets to the validation or test sets since there are only four samples in the training set. The NNT function **train** is used to train the network which updates its weights so that the input produces the target values. After the perceptron has been trained it will produce the correct output for a given set of inputs i.e. $y = p(0,0) = 0$, $y = p(0,1) = 1$, $y = p(1,0) = 1$, $y = p(1,1) = 0$ (Figure 4.28).

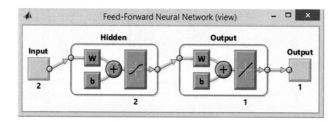

FIGURE 4.28
Output of Example 4.31.

Example 4.31: Write a program to implement a XOR gate using a Feedforward Neural Network

```
clear; clc;
x = [0 1 0 1 ; 0 0 1 1];     % input
t = [0 1 1 0];               % target
p = feedforwardnet(2,'trainlm');
p.divideFcn = 'dividetrain';
p = train(p,x,t);
view(p);
y = round(p(x))
```

To solve the above character recognition problem using an MLP, the NNT function **patternnet** is used to implement a pattern recognition network which are essentially feedforward backpropagation networks that can be trained to classify inputs according to target classes. The term **backpropagation** implies that the errors calculated at each of the output lines flow backward from the output toward the input side and in the process update the weights to new values such that the errors reduce in the next iteration. In the following example, the three character recognition solution is implemented using a MLP with 63 inputs, 3 outputs and a hidden layer with 10 neurons. The target is modified to reflect the three output lines. The trained network correctly recognizes the three test characters (Figure 4.29). The character definitions are loaded into workspace from the previously saved *abc.mat* file by using the BM function **load**.

Example 4.32: Write a program to implement character recognition using a multi layer neural network.

```
clear; clc;
load abc
T = [1 0 0 ; 0 1 0 ; 0 0 1];
nn = patternnet(10); % hidden layer with 10 neurons
nn = train(nn,x,T);
view(nn)
y = nn(x);           % output from training set
Y = round(nn(X))     % output from testing set
```

The following example shows the use of pattern recognition neural network to classify the query point *q* using the Fisher-Iris data set. The third and fourth columns of the measurements are used i.e. petal length and width. The output *y* produced from the training set shows that the first 50 samples belongs to class 1, the next 50 samples belongs to class 2, and the remaining 50 samples belongs to class 3. The output *Y* produced in response to the query classifies it as a member of the second class i.e. *versicolor*, which is plotted on the data set (Figure 4.30).

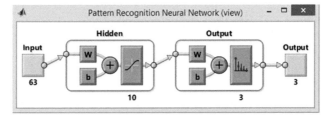

FIGURE 4.29
Output of Example 4.32.

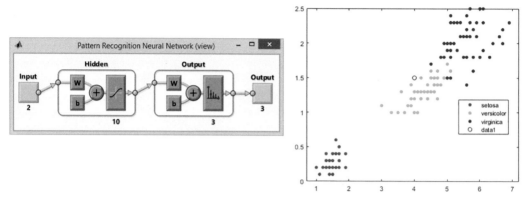

FIGURE 4.30
Output of Example 4.33.

Example 4.33: Write a program to implement pattern recognition using a multi-layer neural network.

```
clear; clc;
[m,t] = iris_dataset;
x = m(3:4,:);          % petal length and width
nn = patternnet(10);
nn = train(nn,x,t);
view(nn)
q = [4, 1.5];          % query point
y = round(nn(x))       % output from training set
Y = round(nn(q'))      % output from query point

x = x';
load fisheriris;
gscatter (x(:,1), x(:,2), species);
hold on;
plot(q(1), q(2), 'ko');
```

4.7.3 Decision Tree Classifiers

A **decision tree** classifier is a predictive model which associates a set of observations about a data item to the item's possible class value (Coppersmith et al., 1999). The branches of the tree represent attribute values, while the leaf nodes represent class labels. The process of constructing a decision tree is usually a top-down approach and starts from the root node which represents the total data set. The data set is subsequently split into subsets in a flow-chart like structure where each internal node, called the decision node, represents a test conducted on an attribute value. The tests encapsulate a series of questions about some attributes of the data. Each time it receives an answer, a follow-up question is asked until a conclusion about the class label of the record is reached. Each branch represents the outcome of a test and each leaf, called terminal node, embodies a class label. Constructing a decision tree is dependent on finding an attribute that returns the highest information gain (i.e. the most homogeneous branches). For the classification of an instance, the tree starts from the root node and moves along the branches until a leaf node is reached which provides the class name of the instance (Breiman et al., 1984). The SMLT function fitctree is used to generate a binary classification decision tree based on the input variables, also called predictors, features or attributes. The following example shows a decision tree based on the petal length and petal width parameters of the Fisher-Iris data set. The test is item

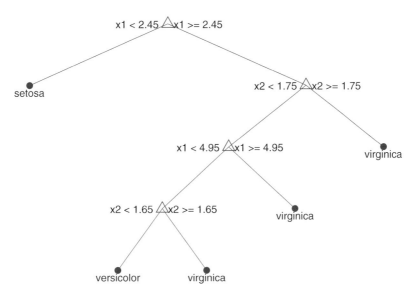

FIGURE 4.31
Output of Example 4.34.

computed from the mean of the data points for the two features mentioned. The feature values of the test item t are: $x_1 = 3.75$, $x_2 = 1.20$. The classification process starts at the top node and branches to the right since $x_1 > 2.45$, at the second node it branches to the left since $x_2 < 1.75$, at the third node it branches to the left since $x_1 < 4.95$, at the fourth node it branches to the left since $x_2 < 1.65$ thereby returning the predicted class label as *versicolor* (Figure 4.31).

Example 4.34: Write a program to implement a decision tree for the Fisher-Iris dataset.

```
clear; clc;
load fisheriris
X = meas(:,3:4);
C = fitctree(X, species);
view(C, 'Mode','graph');
t = mean(X);     % test point
p = predict(C, t)
```

To calculate the classification error or loss, which is the proportion of the observations that the classifier mis-classifies, the *k*-fold cross-validation should be enabled. The following example specifies an 8-fold cross-validation which divides the data set into eight disjoint subsamples or folds chosen randomly but with roughly equal size. The decision trees for the first and last validation cycles are displayed. The individual and average loss due to mis-classification are also computed (Figure 4.32).

Example 4.35: Write a program to implement decision trees using *k*-fold cross-validation for the Fisher-Iris dataset.

```
clear; clc;
load fisheriris;
petal_length = meas(:,3);
petal_width = meas(:,4);
```

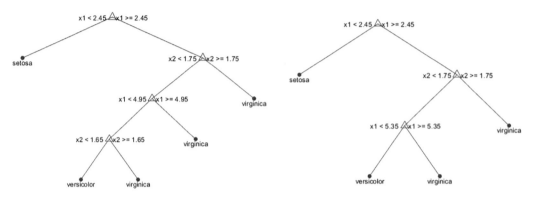

FIGURE 4.32
Output of Example 4.35.

```
rng(1);                    % For reproducibility
dt = fitctree([petal_length, petal_width],species, 'CrossVal','on', 'KFold',8);
view(dt.Trained{1}, 'Mode','graph')
view(dt.Trained{8}, 'Mode','graph')

Lavg = kfoldLoss(dt)
L = kfoldLoss(dt, 'Mode','individual')
```

4.7.4 Discriminant Analysis Classifiers

Discriminant analysis is a classification method that assumes that different classes generate data based on different Gaussian distributions. To train a classifier, a fitting function estimates the parameters of a Gaussian distribution for each class. The function returns the mean values of the distributions and also parameters for class boundaries. The parameters are generated by a linear or quadratic combination of the features considered for classification (Fisher, 1936). The SMLT function **fitcdiscr** returns a fitted discriminant analysis model based on the input variables. In the following example, a **linear discriminant analysis** (LDA) classifier is trained on two features, petal length and petal width, of the Fisher-Iris data set and returns the means of the Gaussian distributions as the centroids of the three classes (setosa, versicolor, virginica) as also the line coefficients k, L for demarcating the class boundaries. A linear discriminant classifier assumes that the Gaussian distributions have the same covariance matrix for each class. For handling the three classes, the classifier is invoked twice, once for discriminating between classes 1 and 2, and next time for discriminating between classes 2 and 3 (Figure 4.33). In the figure, x_1 and x_2 corresponds to *petal length* and *petal width*, respectively. The three centroids of the classes are marked by circles. The class boundaries are defined by the relation:

$$f = k + \begin{bmatrix} x_1 & x_2 \end{bmatrix} \begin{bmatrix} l_1 \\ l_2 \end{bmatrix}$$

Example 4.36: Write a program to implement an LDA classifier on Fisher-Iris dataset.

```
clear; clc;

load fisheriris;
X = meas(:,3:4);
```

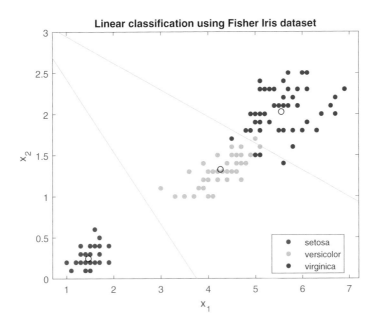

FIGURE 4.33
Output of Example 4.36.

```
C = fitcdiscr(X, species, 'DiscrimType','linear');
gscatter (X(:,1), X(:,2), species);
hold on;
plot(C.Mu(:,1), C.Mu(:,2), 'ko');    % class centroids

K = C.Coeffs(1,2).Const;
L = C.Coeffs(1,2).Linear;
f = @(x1,x2) K + L(1)*x1 + L(2)*x2;
ezplot(f, [0 10 0 3]);

K = C.Coeffs(2,3).Const;
L = C.Coeffs(2,3).Linear;
f = @(x1,x2) K + L(1)*x1 + L(2)*x2;
ezplot(f, [0 10 0 3]);
legend('setosa', 'versicolor', 'virginica');
title('Linear classification using Fisher Iris dataset');
hold off;
```

In complex cases the class boundaries cannot be suitable modeled using straight lines, in which case the discriminant classifier can be used to generate non-linear class boundaries. A **quadratic discriminant analysis** (QDA) classifier assumes that the Gaussian distributions have different covariance matrix for each class and returns the coefficients k, L, Q for generating the class boundaries. The following example shows quadratic boundaries between classes of the Fisher-Iris data set (Figure 4.34). The class boundaries are given by the following:

$$f = k + \begin{bmatrix} x_1 & x_2 \end{bmatrix} \begin{bmatrix} l_1 \\ l_2 \end{bmatrix} + \begin{bmatrix} x_1 & x_2 \end{bmatrix} \begin{bmatrix} q_{11} & q_{12} \\ q_{21} & q_{22} \end{bmatrix} \begin{bmatrix} x_1 \\ x_2 \end{bmatrix} = 0$$

The class boundaries are plotted over the data using the SMT function **ezplot**.

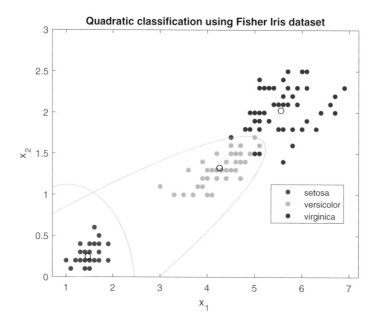

FIGURE 4.34
Output of Example 4.37.

Example 4.37: Write a program to implement quadratic discriminant analysis classifier on Fisher-Iris dataset.

```
clear; clc;

load fisheriris;
X = meas(:,3:4);
C = fitcdiscr(X, species, 'DiscrimType', 'quadratic');
gscatter (X(:,1), X(:,2), species);
hold on;
plot(C.Mu(:,1), C.Mu(:,2), 'ko');

K = C.Coeffs(1,2).Const;
L = C.Coeffs(1,2).Linear;
Q = C.Coeffs(1,2).Quadratic;

f = @(x1,x2) K + L(1)*x1 + L(2)*x2 + Q(1,1)*x1.^2 + ...
    (Q(1,2) + Q(2,1))*x1.*x2 + Q(2,2)*x2.^2;
ezplot(f, [0 10 0 3]);

K = C.Coeffs(2,3).Const;
L = C.Coeffs(2,3).Linear;
Q = C.Coeffs(2,3).Quadratic;

f = @(x1,x2) K + L(1)*x1 + L(2)*x2 + Q(1,1)*x1.^2 + ...
    (Q(1,2) + Q(2,1))*x1.*x2 + Q(2,2)*x2.^2;
ezplot(f, [0 10 0 3]);

legend('setosa', 'versicolor', 'virginica');
title('Quadratic classification using Fisher Iris dataset');
```

To predict the classes of test data, the trained classifier finds the class with the smallest misclassification cost. Parameters used for prediction includes posterior probability, prior probability, and cost. The prior probability $P(k)$ of a class k is the existing probability of a class before the classification process started. The posterior probability $P(x|k)$ that a point

x belongs to class k is the product of the prior probability and the density function of the multivariate Gaussian distribution with mean μ_k and covariance Σ_k at a point x:

$$P(x|k) = \frac{1}{\sqrt{2\pi\Sigma_k}} \exp\left\{-\frac{1}{2}(x-\mu_k)^T \cdot \Sigma_k^{-1} \cdot (x-\mu_k)\right\}$$

Here, $|\Sigma k|$ is the determinant and Σ_k^{-1} is the inverse of the matrix Σk. The posterior probability $P(k|x)$ is then given by:

$$P(k|x) = \frac{P(x|k) \cdot P(k)}{P(x)}$$

The cost(i,j) is the cost of classifying an observation into class j if its true class is i. Typically cost$(i,j) = 1$ when $i \neq j$ and cost$(i,j) = 0$ when $i = j$. The SMLT function **predict** returns predicted class labels for test data based on a trained discriminant analysis classification model. The classification is estimated by minimizing the expected classification cost given below, where y is the predicted classification, K is the number of classes, and $P(k|x)$ is the posterior probability of class k for observation x, $C(y|k)$ is the cost of classifying an observation as y when its true class is k

$$y = \arg \min_{1 \leq y \leq K} \sum_{k=1}^{K} P(k|x) \cdot C(y|k)$$

In the following example, an iris flower with average measurements is classified based on a trained linear discriminant model. The output label is returned as *versicolor* (Figure 4.35).

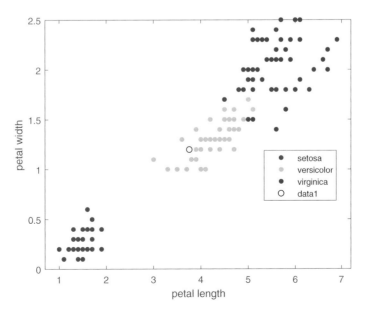

FIGURE 4.35
Output of Example 4.38.

Example 4.38: Write a program to predict a test item using a discriminant analysis classifier trained on Fisher-Iris dataset.

```
clear; clc;
load fisheriris;
X = meas(:,3:4);
C = fitcdiscr(X, species, 'DiscrimType', 'linear');
gscatter (X(:,1), X(:,2), species);
hold on;
t = mean(X);              % test data
plot(t(1), t(2), 'ko');
p = predict(C, t)         % predicted label
xlabel('petal length');
ylabel('petal width');
```

Instead of manually splitting the data set into training and testing portions and to avoid a possible selection bias in doing so, an alternative scheme is called **cross-validation** is also a common practice. In a k-fold cross-validation scheme, the data set is partitioned into k equal sized subsamples. Out of that $(k–1)$ partitions are used as training set and the remaining one as the test set. This process is then repeated k times with each of the k partitions used once as the test set. All the k results are subsequently averaged to produce a single estimation. The advantage of this scheme is that all the samples are used both for training and testing. Typically, 8-fold or 10-fold cross-validation schemes are most frequently used. The SMLT function **cvpartition** creates a cross-validation partition for data. The partition divides the observations into k disjoint subsamples or folds chosen randomly but with roughly equal size. The default value of k is 10. The following example shows a 5 fold partition with stratification enabled, which ensures that all classes occur in equal proportion for each fold of the data set. Alternatively, instead of a k-fold specification, a holdout specification ensures that the data set is partition into a specified number of training and testing (holdout) samples. In the following example, out of the total 150 datapoints of the Fisher-Iris data set, *cv*1 returns five subsets (folds) each containing 120 train samples and 30 test samples, all three classes occurring in equal proportions in each subset, and *cv*2 returns a single set with 135 train samples and 15 test samples (10% holdout). The SMLT function **training** identifies the training samples within each fold, while the SMLT function **test** identifies the test samples (Figure 4.36). The figure depicts that out of 150 samples of the third subset (fold), 120 are training samples and 30 are test samples.

Example 4.39: Write a program to implement k-fold cross-validation on a dataset.

```
clear; clc;
load fisheriris;
[C,~,idx] = unique(species);   % find instances of each class
n = accumarray(idx(:),1)       % Number of instances for each class in species
cv1 = cvpartition(species,'KFold',5,'Stratify',true)
```

FIGURE 4.36
Output of Example 4.39.

```
cv2 = cvpartition(species,'Holdout', 0.1,'Stratify',false)

t = training(cv1,3); t'
s = test(cv1,3); s'
bar(t); hold on; bar(s); legend('train', 'test');
```

4.7.5 Naive Bayes Classifiers

It has already been mentioned that a discriminant analysis classifier assumes that different classes generate data based on different Gaussian distributions. If the covariance matrix of the classes are different, then a quadratic classifier can be used to draw the class boundaries, while a linear classifier assumes that all classes have a common covariance matrix. A Naive Bayes classifier further assumes that the features comprising each class are independent of each other implying that the covariance terms are zero which reduces the covariance matrix of each class to a diagonal matrix (Hastie et al., 2008). During the training step, the method estimates the parameters of a probability distribution. For the prediction step, the method computes the posterior probability of that sample belonging to each class. The SMLT function **fitcnb** is used to implement a Naive Bayes classifier and returns the means and the diagonal covariance matrix of each class. The following example shows the Naive Bayes classifier being applied to two measurements (petal length and petal width) of the Fisher-Iris data set and a test item generated from the mean of all values is predicted to be an estimated class. The BM function **cell2mat** converts a cell array returned from the classifier to an ordinary array from which the cluster means are extracted. The figure shows the centroids of the classes as circles and the boundaries of the individual classes by contour lines estimated from the means and covariance matrices during the training step. The test item is shown as a cross and classified as class *versicolor* (Figure 4.37).

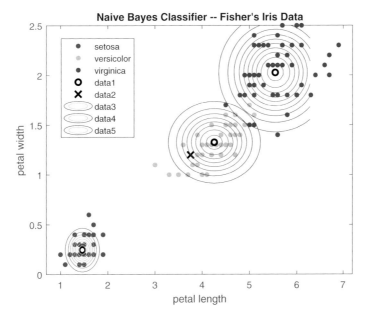

FIGURE 4.37
Output of Example 4.40.

Example 4.40: Write a program to predict a test item using a naive Bayes classifier trained on Fisher-Iris dataset.

```
clear; clc;

load fisheriris;
X = meas(:,3:4);
C = fitcnb(X, species);
p = cell2mat(C.DistributionParameters);
m = p(2*(1:3)-1,1:2);    % Extract the means
s = zeros(2,2,3);        % sigma
for j = 1:3
    s(:,:,j) = diag(p(2*j,:)).^2; % Create diagonal covariance matrix
end
m
s

figure,
gscatter (X(:,1), X(:,2), species);
h = gca;
hold on;
plot(m(:,1), m(:,2), 'ko', 'LineWidth', 2);

t = mean(X);             % test data
plot(t(1), t(2), 'kx', 'LineWidth', 2, 'MarkerSize',10);
p = predict(C, t)        % predicted label

cxlim = h.XLim;
cylim = h.YLim;

ezcontour(@(x1,x2)mvnpdf([x1,x2],m(1,:),s(:,:,1)));
ezcontour(@(x1,x2)mvnpdf([x1,x2],m(2,:),s(:,:,2)));
ezcontour(@(x1,x2)mvnpdf([x1,x2],m(3,:),s(:,:,3)));

h.XLim = cxlim;
h.YLim = cylim;
xlabel('petal length');
ylabel('petal width');
title('Naive Bayes Classifier -- Fisher''s Iris Data');
hold off;
```

4.7.6 Support Vector Machine (SVM) Classifiers

The SVM is a binary classification algorithm which searches for an optimal hyperplane that separates the data into two classes (Christianini and Shawe-Taylor, 2000). The optimal hyperplane maximizes a margin (space that does not contain any observations) surrounding itself, which creates boundaries for the positive and negative classes. The support vectors are the data points that are closest to the separating hyperplane; these points are on the boundary of the margin. The SMLT function **fitcsvm** trains or cross-validates an SVM for binary classification. In the following example, from the Fisher-Iris data set, the *setosa* class is removed, and from the remaining two classes, the first class (*versicolor*) is the negative class, and the second (*virginica*) is the positive class. The support vectors, which are observations that occur on or beyond their estimated class boundaries, are identified and plotted (Figure 4.38). A test item computed from the mean of the datapoints of the two classes submitted to the trained classifier, and the predicted class is returned as *virginica* using the SMLT function **predict**.

Example 4.41: Write a program to predict a test item using a binary SVM classifier trained on Fisher-Iris dataset.

```
clear; clc;
load fisheriris;
species_n = ~strcmp(species,'setosa');  % remove one class
```

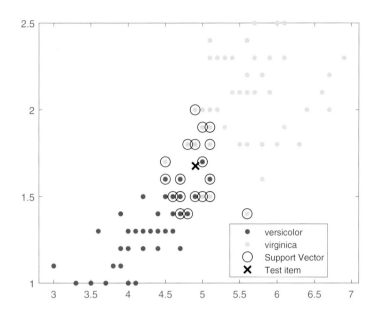

FIGURE 4.38
Output of Example 4.41.

```
X = meas(species_n, 3:4);
y = species(species_n);
C = fitcsvm(X,y);
sv = C.SupportVectors;
t = mean(X);          % test item
p = predict(C, t)     % estimated class

figure,
gscatter(X(:,1),X(:,2),y);
hold on;
plot(sv(:,1),sv(:,2),'ko','MarkerSize',10);
plot(t(1), t(2), 'kx', 'MarkerSize', 10, 'LineWidth', 2);
legend('versicolor','virginica','Support Vector', 'Test item');
hold off;
```

The following example calculates multiple class boundaries using a binary SVM for the three classes of the Fisher-Iris data set. For each class: (1) Create a logical vector (*idx*) indicating whether an observation is a member of the class. (2) Train an SVM classifier using the predictor data and *idx*. (3) Store the classifier in a cell of a cell array. The SVM model is a 3-by-1 cell array, with each cell containing a *ClassificationSVM* classifier. For each cell, the positive class is setosa, versicolor, and virginica, respectively. The SVM classifier is trained using a **radial basis function** (RBF) kernel. For nonlinear SVM, the algorithm forms a Gram matrix using the rows of the predictor data X. The dual formalization replaces the inner product of the observations in X with corresponding elements of the resulting Gram matrix (called the "kernel trick"). Consequently, nonlinear SVM operates in the transformed predictor space to find a separating hyperplane. The Gram matrix of a set of n vectors $\{x_1, x_2, \ldots, x_n\}$ is an n-by-n matrix with element (j,k) defined as $G(x_j, x_k) = < \phi(x_j), \phi(x_k) >$, an inner product of the transformed predictors using the kernel function ϕ. For the Gaussian or RBF kernel, $(x_j, x_k) = \exp(-\|x_1 - x_2\|^2)$. The SMLT function **predict** can be used to predict the labels of the SVM classification tree. The three SVM classifiers can be used to identify the appropriate classes i.e. predict(SVM{1}, X) identifies class 1, predict(SVM{2}, X) identifies class 2, predict(SVM{3}, X) identifies class 3.

Example 4.42: Write a program to predict a test item using multiple classes with an SVM classifier trained on Fisher-Iris dataset.

```
clear; clc;
load fisheriris;
X = meas(:,3:4);
C = cell(3,1);
classes = unique(species);

for j = 1:numel(classes)
    idx = strcmp(species, classes(j)); % Create binary classes for each classifier
    C{j} = fitcsvm(X, idx, 'KernelFunction', 'rbf');
end

t = mean(X);   % Test item
predict(C{1},t)
predict(C{2},t)
predict(C{3},t)
```

4.7.7 Classification Learner App

The Classification Learner App is an interactive and graphical utility to train, validate, and tune classification models. A number of different classification algorithms can be trained simultaneously and their validation errors can be compared side by side before choosing the best model. The best model can then be exported to the workspace to classify new data. On the Apps tab, in the Machine Learning group, select Classification Learner (Figure 4.39).

Click New Session and select data from the workspace or from file. Specify variables to use as predictors (features) and response variables that needs to be classified (Figure 4.40).

Click Start Session. This generates a scatter plot of the data which can be used to investigate which variables are useful for predicting the response. Select different options on the X and Y lists under Predictors to visualize the distribution of species and measurements. Observe which variables separate the species colors most clearly. If necessary click on Feature Selection and add or remove features (Figure 4.41).

In the Model Type section, click All Quick-To-Train. This option will train all the model presets available for your data set that are fast to fit. Click Train. A selection of model types

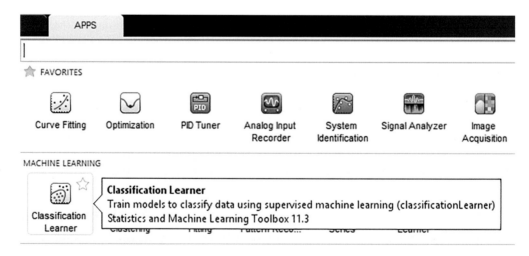

FIGURE 4.39
Classification learner App interface: workspace.

FIGURE 4.40
Classification learner App interface: data set selection.

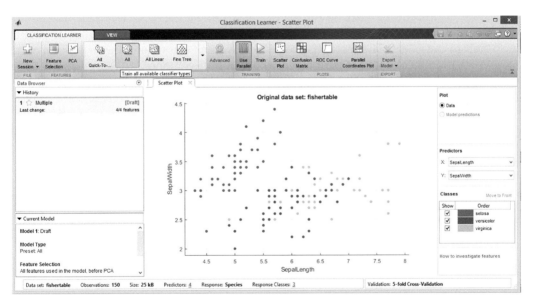

FIGURE 4.41
Classification learner App interface: classifier selection.

appears in the History list. When they finish training, the best percentage Accuracy score is highlighted in a box. Click models in the history list to explore results in the plots. To try all the classifier model presets available for your data set, click All, then click Train. The default validation option is 5-fold cross-validation, which protects against overfitting. For a *k*-fold *cross-validation* scheme, the entire data are partitioned using *k* disjoint sets or folds, and for each fold the model is trained using out-of-fold observations and then its performance is assessed using in-fold data. Finally it calculates average test error over all folds. This scheme is recommended for small data sets. Alternatively, for a *holdout valida-tion* scheme, a specified percentage of data is used as test set, while the remaining data are used as training set. The model is trained on the training set and its performance is assessed with the test set. The score is the accuracy on the held-out observations. This scheme is recommended for large data sets. After a model is trained the scatter plot switches from displaying data to displaying model predictions. Misclassified samples are indicated by crosses (Figure 4.42).

For a specific classifier selected from the History List click on Confusion Matrix to display the results. The rows indicate the true classes and the columns show predicted classes. For holdout or cross-validation schemes, the confusion matrix is calculated on the held-out observations. The diagonal cells show where the true class and predicted class match. If these cells are green, then the classifier has performed well. On selecting True Positive False Negative option, the results for each class are displayed. The ROC (receiver operating characteristic) curve shows true-positive rate (TPR) and false positive rate (FPR): for example a TPR of 0.8 indicates that 80% of the observations have been correctly assigned to the positive class, a FPR of 0.3 indicates that 30% of the observations have been incorrectly assigned to the positive class. AOC (area under curve) is a measure of the overall quality of the classifier using which different classifiers can be compared, larger AOC means better performance (Figure 4.43).

Alternatively, to select specific classifiers manually, click the arrow on the far right of the Model Type section to expand the list of classifiers (Figure 4.44). From the list select a classifier and see the parameters that can be tuned. After finalizing click Train. Repeat to try different classifiers. The score displayed against each classifier is the validation accuracy which estimates a model's performance on new data compared to training data. After selecting the classifier with the best results, click on Export Model. If the training data are included, then the trained model is exported as a structure containing a classification object e.g. ClassificationTree, ClassificationDiscriminant, ClassificationSVM, ClassificationKNN, and so on. On selecting the Export Compact Model option, the trained model is exported without the training data as a compact classification object e.g. CompactClassificationTree.

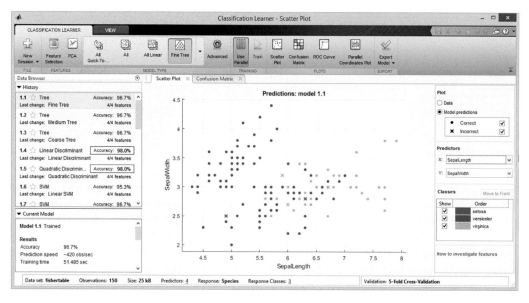

FIGURE 4.42
Classification learner App interface: classification results.

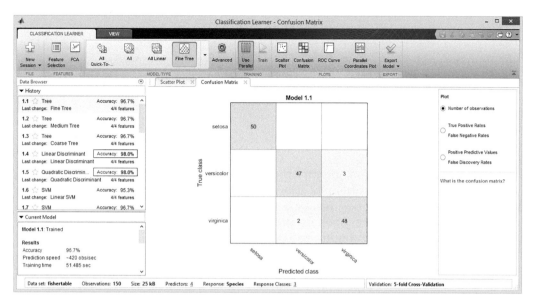

FIGURE 4.43
Classification learner App interface: confusion matrix.

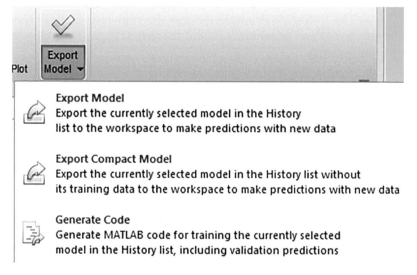

FIGURE 4.44
Classification learner App interface: export.

To use the exported classifier to make predictions for new data, S, use the form: *yfit* = trainedModel.predictFcn(S). The test data S has to be in the same form as the original training data used for training the model i.e. as table, matrix or vector, where the predictor columns or rows occurs in the same order and format. The output *yfit* contains a class prediction for each data point in S. For help making predictions use: trainedModel.HowToPredict.

During export, you can generate MATLAB code which can be used with the new data. The exported code is displayed in the MATLAB Editor from where it can be saved as a file. To make predictions on new data, use a function for loading the trained model and predicting classes of new data:

```
function label = classifyX (X)
    C = loadCompactModel('myModel');
    label = predict(C,X);
end;
```

To optimize feature selection, identify predictors that separate classes well by plotting pairs of predictors on the scatter plot. The scatter plot displays input training data before training and model prediction results after training. Choose features to plot using X and Y lists under Predictors. Look for predictors that separate classes well. Investigate finer details of the plot by zooming and panning controls on the top-right of the plot. If predictors not useful for separating out the classes can be identified, then use Feature Selection control to remove them and train classifiers including only the most useful predictors. Use PCA to reduce the dimensionality of the predictor space (Figure 4.45). PCA linearly transforms predictors in order to remove redundant dimensions, and generates a new set of variables called principal components. In the Features section, select PCA and Enable PCA options. When next clicking on Train, the PCA transforms the selected features before training the classifier. By default PCA keeps only the components that account for 95% of the variance. The SMLT function **pca** is used to compute PCA of raw data. The following example reduces the dimensionality of the Fisher-Iris data set from 4-D to 2-D and generates a plot of the three classes by taking into account all the four measurements.

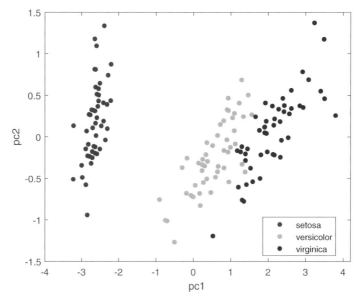

FIGURE 4.45
Output of Example 4.43.

Example 4.43: Write a program to reduce the dimensionality of the Fisher iris dataset from 4 to 2 and generate a 2-D plot of the 3 classes.

```
clear; clc;
load fisheriris;
X = meas;
[coeff,score,latent] = pca(X);
gscatter(score(:,1), score(:,2), species);
xlabel('pc1'); ylabel('pc2');
```

The following example uses the Fisher-Iris data set to train all applicable classifiers and display their accuracy results and confusion matrices.

Example 4.44: Use the Fisher-Iris dataset to train all applicable classifiers and display their accuracy results and confusion matrices using the Classification Learner App.

Fisher's 1936 iris data contains measurements of flowers: sepal length, sepal width, petal length, petal width, for specimens from three species of flowers: setosa, versicolor, virginica.

- Step 1: Create a table of measurement predictors (or features) using variables from the data set to use for a classification.

  ```
  rng('default');
  fishertable = readtable('fisheriris.csv');
  ```

  Alternatively, a table can be created by specifying matrix and vector names

  ```
  rng('default');
  load fisheriris;
  iristable = table(meas, species)
  ```

- Step 2: Open Apps>Machine Learning>Classification Learner
- Step 3: New Session>From Workspace
- Step 4: Start Session
- Step 5: Train All Classifiers>Train
- Step 6: Select best results from History list>Confusion Matrix

The following lists the best results and confusion matrix metrics based on three feature options:

- Case 1:
  - Features: Sepal Length and Sepal Width
  - Best results: Classifier: Cubic SVM, Accuracy: 82% (Figure 4.46)
  - TP (true positive) = (T1.P1 + T2.P2 + T3.P3)/3 = (100 + 72 + 74)/3 = 82%
  - TN (true negative) = {(T2+T3)/2+(T1+T3)/2+(T1+T2)/2}/3 = (100+100+100)/3 = 100%
  - FP (false positive) = {(T2.P1 + T3.P1)/2 + (T1.P2 + T3.P2)/2 + (T1.P3 + T2.P3)/2} = {(0+0)/2 + (0+26)/2 + (0+28)/2} = 27%
  - FN (false negative) = (T1.P2 + T1.P3)/2 + (T2.P1 + T2.P3)/2 + (T3.P1 + T3.P2)/2 = 0+0 + 0+28/2 + 0+26/2 = 27%

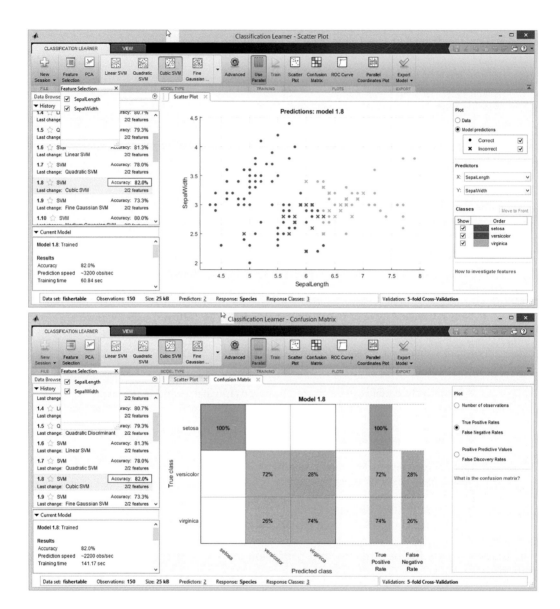

FIGURE 4.46
Output of Example 4.44 Case 1.

- Case 2:
    - Features: Petal Length and Petal Width
    - Best results: Classifier: Quadratic Discriminant, Accuracy: 97.3% (Figure 4.47)
    - TP = (T1.P1 + T2.P2 + T3.P3)/3 = (100+96+96)/3 = 97.3%
    - TN = {(T2 + T3)/2 + (T1 + T3)/2 + (T1 + T2)/2}/3 = (100 + 100 + 100)/3 = 100%
    - FP = {(T2.P1 + T3.P1)/2 + (T1.P2 + T3.P2)/2 + (T1.P3 + T2.P3)/2} = {(0+0)/2 + (0+4)/2 + (0+4)/2} = 4%

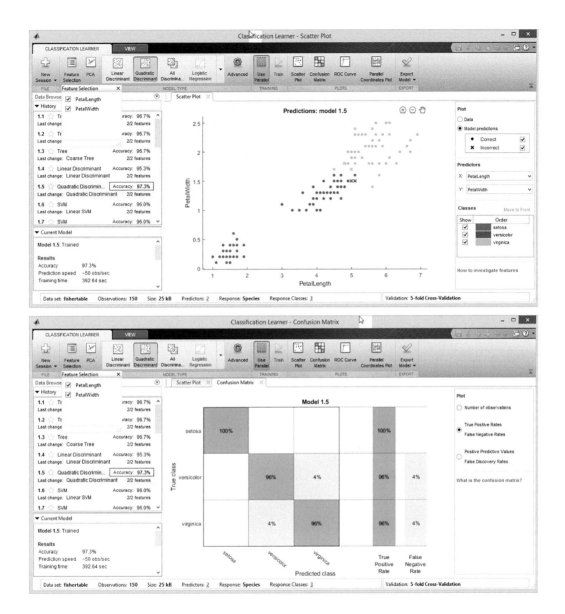

FIGURE 4.47
Output of Example 4.44 Case 2.

- FN = (T1.P2 + T1.P3)/2 + (T2.P1 + T2.P3)/2 + (T3.P1 + T3.P2)/2 = 0+0 + 0+4/2 + 0+4/2 = 4%
- Case 3:
  - Features: Sepal Length, Sepal Width, Petal Length, and Petal Width
  - Best results: Classifier: Linear Discriminant, Accuracy: 98% (Figure 4.48)

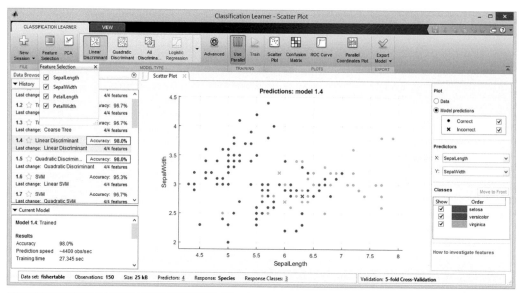

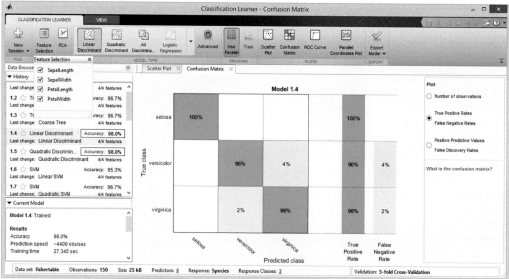

FIGURE 4.48
Output of Example 4.44 Case 3.

- TP = (T1.P1 + T2.P2 + T3.P3)/3 = (100+96+98)/3 = 98%
- TN = {(T2 + T3)/2 + (T1 + T3)/2 + (T1 + T2)/2}/3 = (100 + 100 + 100)/3 = 100%
- FP = {(T2.P1 + T3.P1)/2 + (T1.P2 + T3.P2)/2 + (T1.P3 + T2.P3)/2} = {(0+0)/2 + (0+4)/2 + (0+2)/2} = 3%
- FN = (T1.P2 + T1.P3)/2 + (T2.P1 + T2.P3)/2 + (T3.P1 + T3.P2)/2 = 0+0 + 0+4/2 + 0+2/2 = 3%

4.8 Performance Evaluation

Once clustering or classification is done, the question of performance evaluation arises next to analyze how efficient or effective the recognition system has been. This is especially important in a research scenario where issues of the comparative analysis of the current or proposed methodology vis-à-vis the earlier techniques as surveyed in extant literature is critical. Clustering results are evaluated based on cluster statistics. Clustering is considered more effective when clusters are smaller and farther apart i.e. intra-cluster distances are small and inter-cluster distances are large. One such index for clustering is called the **silhouette value** of a point, which is a measure of how similar that point is to points in its own cluster, when compared to points in other clusters and is the given by the following, where S_i is the silhouette value of the i-th point in a cluster, D_i is the minimum average distance from the i-th point to points in a different cluster minimized over clusters, and d_i is the average distance from the i-th point to other points in the same cluster (Kaufman and Rousseeuw, 1990).

$$S_i = \frac{D_i - d_i}{\max\left(D_i, d_i\right)}$$

The silhouette value ranges from –1 to +1. A high silhouette value indicates that i is well-matched to its own cluster, and poorly matched to neighboring clusters. If most points have a high silhouette value, then the clustering solution is appropriate. If many points have a low or negative silhouette value, then the clustering solution may have either too many or too few clusters. The silhouette clustering evaluation criterion can be used with any distance metric. The SMLT function **silhouette** calculates the silhouette value and generates the silhouette plot. The following example clusters a random set of 20 points into two groups and calculates their silhouette values which are displayed in a scatter plot (Figure 4.49). By default, silhouette uses the squared Euclidean distance between

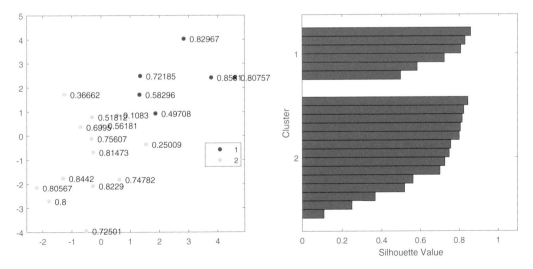

FIGURE 4.49
Output of Example 4.45.

points in X. A silhouette plot is also shown which looks like a bar graph with the length of the bars equal to the silhouette value of the data points.

Example 4.45: Write a program to generate a silhouette plot for performance evaluation.

```
clear; clc;
rng default;  % For reproducibility
X = [randn(10,2)+ones(10,2);randn(10,2)-ones(10,2)];
cidx = kmeans(X,2,'distance','sqeuclidean');
s = silhouette(X,cidx,'sqeuclidean');
d = 0.2;
gscatter(X(:,1), X(:,2), cidx); hold on;
for i=1:20
    text(X(i,1)+d, X(i,2), num2str(s(i)))
end;
figure,
[s, h] = silhouette(X,cidx,'sqeuclidean');
```

The SMLT function **evalclusters** is used to evaluate a clustering distribution and return the optimum number of clusters, using the specified maximum number of clusters. In the following example, the clustering solution is evaluated using Calinski-Harabasz Criterion (Calinski and Harabasz, 1974) and returns the optimum number of clusters, which in this case is 4. This optimum number is then used to execute *k*-means clustering algorithm and generate a plot. The **Calinski-Harabasz** (C-H) criterion is sometimes called the variance ratio criterion (VRC). The Calinski-Harabasz index is defined as:

$$\mathrm{VRC}(k) = \frac{D}{d} \times \frac{N-k}{k-1}$$

Here, D is the overall between-cluster variance and d is the overall within-cluster variance, k is the number of clusters, and N is the number of observations. The following example shows the C-H scores for clustering the Fisher-Iris data set into number of clusters from 1 to 6 and indicates that the optimal value is 3, as shown in the left side figure, while the right side figure shows the grouped scatter plot, as suggested by the C-H criterion (Figure 4.50).

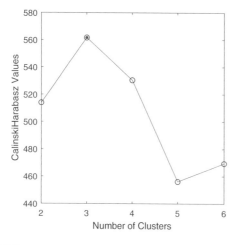
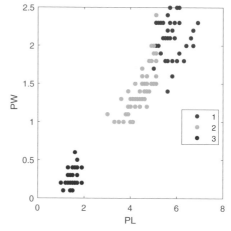

FIGURE 4.50
Output of Example 4.46.

Example 4.46: Write a program to evaluate a clustering distribution and return the optimum number of clusters.

```
clear; clc;
load fisheriris;
rng('default');
X = meas;
eva = evalclusters(X,'kmeans','CalinskiHarabasz','KList',[1:6]);
opc = eva.OptimalK;
subplot(121),
plot(eva)
PL = X(:,3);
PW = X(:,4);
gr = eva.OptimalY;
subplot(122),
gscatter(PL, PW, gr);
```

Classification performance can be evaluated using a confusion matrix. A **confusion matrix** is a grid which shows a comparison between the actual classes and the predicted classes. Suppose in a box there are some rectangular objects (R) and some triangular objects (T). An object is taken out at random from the box and the computer is asked to predict its type (class). Four situations can arise for each class, say R:

- Actually R and predicted as R: true positive (correct classification)
- Actually not R and predicted as not R: true negative (correct classification)
- Actually R but predicted as not R: false negative (incorrect classification)
- Actually not R but predicted as R: false positive (incorrect classification)

The SMLT function **confusionmat** returns a confusion matrix determined by specified known and predicted groups. The following example shows the confusion matrix generated for classification of the Fisher-Iris data using an LDA. The data set is partitioned into training and test sets with a 20% holdout for testing. The BM function **array2table** converts the array to a table structure with named columns. This is done so that an additional column to the right containing the species name can be added based on which the classifier is used to group the data into three classes. The example shows that the classifier misclassifies one *versicolor* sample as *virginica* in the test set (Figure 4.51).

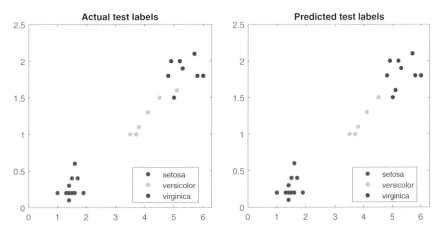

FIGURE 4.51
Output of Example 4.47.

Example 4.47: Write a program to partition 80% of the Fisher-iris dataset for training and remaining 20% for testing and generate the confusion matrix after classification using the linear discriminant classifier

```
clear; clc;
load fisheriris
n = size(meas,1);
rng(1);                              % For reproducibility
cvp = cvpartition(n, 'Holdout', 0.2);
tr = training(cvp);                  % Training set indices
te = test(cvp);                      % Test set indices
trtb = array2table(meas(tr,:));
trtb.e = species(tr);
clf = fitcdiscr(trtb, 'e');          % classifier model
tepl = predict(clf, meas(te,:));     % test set predicted labels
confusionmat(species(te), tepl)

tef = meas(te, 3:4);                 % test set features
teal = species(te);                  % test set actual labels
subplot(121), gscatter(tef(:,1), tef(:,2), teal);
title('Actual test labels');
subplot(122), gscatter(tef(:,1), tef(:,2), tepl)
title('Predicted test labels');
```

The confusion matrix generated by the program is shown below, where each row is the actual class and each column is the predicted class.

```
    15      0      0
     0      6      1
     0      0      8
```

The confusion matrix shows that the measurements belonging to *setosa* and *virginica* are classified correctly, while one measurement belonging to *versicolor* is misclassified as *virginica*. This is indicated in the figure above where green dot pertaining to actual test labels plot is shown as blue in the predicted test labels plot.

The following example shows the Fisher-Iris data set randomly split into two equal halves, one half being trained using three different classifiers i.e. decision tree, linear discriminant, naive bayes, while the other half being treated as the test set whose classes are predicted and the corresponding confusion matrix generated. For the first classifier, percentage of recognition accuracy is $71/75 = 94.67\%$, for the second classifier accuracy is $73/75 = 97.33\%$ and for the third classifier accuracy is $72/75 = 96\%$.

Example 4.48: Write a program to randomly split the Fisher-iris dataset into two equal portions, and compare the performance of three classifiers by generating their confusion matrix.

Case – 1: Decision Tree Classifier

```
clear; clc;
load fisheriris;
rng(0);                  % For reproducibility
n = length(species);%    length of dataset
p = randperm(n);         % random permutations
x = meas(p,:);           % dataset
lb = species(p);         % label
h = floor(n/2);          % half length of dataset
t = x(1:h,:);            % training set
tl = lb(1:h);            % training set labels
m = fitctree(t,tl);      % decision tree classifier
s = x(h+1:end,:);        % testing set
```

```
r = predict(m,s);    % predicted labels
sl = lb(h+1:end);    % actual labels
[c, o] = confusionmat(sl,r, 'Order',{'setosa','versicolor','virginica'})
Program output :

c =
    29     0     0
     0    20     4
     0     0    22
```

Case – 2: Linear Discriminant Classifier

```
clear; clc;
load fisheriris
rng(0);                    % For reproducibility
n = length(species);
p = randperm(n);
x = meas(p,:);
lb = species(p);
h = floor(n/2);
t = x(1:h,:);
tl = lb(1:h);
m = fitcdiscr(t,tl);       % linear discriminant classifier
s = x(h+1:end,:);
r = predict(m,s);
sl = lb(h+1:end);
[c, o] = confusionmat(sl,r, 'Order',{'setosa','versicolor','virginica'})
Program Output :
c =
    29     0     0
     0    22     2
     0     0    22
```

Case – 3: Naive Bayes Classifier

```
clear; clc;
load fisheriris
rng(0);                     % For reproducibility
n = length(species);
p = randperm(n);
x = meas(p,:);
lb = species(p);
h = floor(n/2);
t = x(1:h,:);
tl = lb(1:h);
m = fitcnb(t,tl);        % naive bayes classifier
s = x(h+1:end,:);
r = predict(m,s);
sl = lb(h+1:end);
[c, o] = confusionmat(sl,r, 'Order',{'setosa','versicolor','virginica'})
Program Output :
c =
    29     0     0
     0    21     3
     0     0    22
```

Review Questions

1. Differentiate between supervised and unsupervised grouping.
2. Explain the terms clustering, classification, training phase, testing phase.
3. What is meant by feature vector and feature space?

4. How is data stored in and retrieved from a MAT file?
5. Discuss about the Fisher-Iris data set.
6. How can multiple media files be read and stored for pattern recognition?
7. What is the utility of pre-processing?
8. How can feature points be extracted using the minimum eigenvalue method?
9. How are corners in images be detected using the Harris corner detector?
10. How can feature points be detected using the FAST algorithm?
11. How can feature points be detected using the MSER algorithm?
12. How does the SURF algorithm detect interest points and generate a descriptor for the same?
13. How does the KAZE algorithm detect interest points and generate a descriptor for the same?
14. How does the BRISK algorithm detect interest points and generate a descriptor for the same?
15. How can interest points be computed using the LBP algorithm?
16. How can interest points be computed using the HOG algorithm?
17. Differentiate between the k-means and hierarchical clustering schemes?
18. How can GMMs be used to fit a random distribution of points?
19. How can classification of data be done using the k-NN algorithm?
20. What is an ANN? Differentiate between SLP and MLP?
21. How can decision tree classifiers be used for data classification?
22. How can discriminant analysis classifiers be used for data classification?
23. How can a k-fold cross-validation be done? What is its utility?
24. How can data be classified using the Naïve Bayes classifier?
25. How can a test item be predicted using multiple classes with an SVM classifier?

Function Summary

Basic MATLAB® (BM) Functions

%: comments
:: range of values or all current values
[]: map current values to full range of values
': transpose of a vector or matrix
abs: absolute value
area: filled area 2-D plot
array2table: converts an array to a table
atan: inverse tangent in radians
audiodevinfo: information about audio device
audioinfo: information about audio files
audioplayer: create object for playing audio
audioread: read audio files
audiorecorder: create object for recording audio
audiowrite: write audio files
axes: specify axes appearance and behavior
axis: set axis limits and aspect ratios
bar, bar3: bar graph, 3-D Bar graph
boundary: return coordinates of boundary of a set of points
ceil: round towards positive infinity
cell: create cell array
cell2mat: convert cell array to ordinary array
clc: clear command window of previous text
clear: clear items from workspace memory
colorbar: displays color scalebar in colormap
colormap: Color look-up table
comet: 2-D animated plot
compass: plot arrows emanating from origin
continue: pass control to next iteration of loop
contour, contourf: contour plot of a matrix
cos: cosine of argument in radians
cov: calculate covariance
cylinder: 3-D cylinder
datetick: date formatted tick labels
dir: list folder contents
disp: display value of variable
double: convert to double precision data type
eig: calculate eigenvectors and eigenvalues

ellipsoid: generate 3-D ellipsoid
end: terminate block of code
eps: epsilon, a very small value equal to 2^{-52} or 2.22×10^{-16}
errorbar: line plot with error bars
exp: exponential
ezplot: plots symbolic expressions
feather: plot velocity vectors
fft, fft2: fast Fourier transform, 2-D fast Fourier transform
fftshift: shift zero-frequency component to center of spectrum
figure: create a figure window
fill: fill 2-D polygons
filter, filter2: 1-D digital filter, 2-D digital filter
find: find indices and values of nonzero elements
fliplr: flip matrix left to right
flipud: flip matrix up to down
floor: round toward negative infinity
fmesh: create 3-D mesh
for ... end: repeat statements specified number of times
format: set display format in command window
frame2im: return image data associated with movie frame
fsurf: create 3-D surface
fullfile: build full file name from parts
function: create user-defined function
gallery: invoke gallery of matrix functions for testing algorithms
gca: get current axis for modifying axes properties
gcf: get current figure handle for modifying figure properties
getaudiodata: return recorded audio from array
getFrame: capture axes or figure as movie frame
grid: display grid lines in plot
hasFrame: determine if frame is available to read
hex2dec: convert hexadecimal to decimal
histogram: histogram plot
hold: hold on the current plot
hot: hot colormap array
hsv2rgb: convert colors from HSV space to RGB space
if ... end: execute statements if condition is true
if ... elseif ... else: execute statements if condition is true else other statements
im2double: convert image to double precision
im2frame: convert image to movie frame
imag: imaginary part of complex number
image, imagesc: display image, display scaled image from array
imageDatastore: create datastore for image data
imapprox: approximate indexed image by reducing number of colors
imfinfo: display information about file
imread: read image from file
imresize: resize image
imshow: display image
imwrite: write image to file
ind2rgb: convert indexed image to RGB image

Inf: infinity
jet: jet colormap array
legend: add legend to axes
length: length of largest array dimension
line: create line
linspace: generate vector of evenly spaced values
load: load variables from file into workspace
log: natural logarithm
magic: magic square with equal row and column sums
max: maximum value
mean: average value
mesh, meshc: mesh, mesh with contour plot
meshgrid: generates 2-D grid coordinates
min: minimum value
mmfileinfo: information about multimedia file
movie: play recorded movie frames
NaN: not a number, output from operations which have undefined numerical results
num2str: convert numbers to strings
numel: number of array elements
ones: create array of all ones
pause: stop MATLAB execution temporarily
peaks: sample function of two variables
pie, pie3: pie chart, 3-D pie chart
play: play audio
plot, plot3: 2-D line plot, 3-D line plot
plotyy: plot using two *y*-axis labeling
polar: polar plot
polarhistogram: polar histogram
polyshape: create a polygon defined by 2-D vertices
prod: product of array elements
quiver: arrow plot
rand: uniformly distributed random numbers
randi: uniformly distributed random integers
randperm: random permutation
readFrame: read video frame from video file
real: real part of complex number
recordblocking: record audio holding control until recording completes
rectangle: create rectangle with sharp or curved corners
release: release resources
rgb2gray: convert RGB image to grayscale
rgb2hsv: convert colors from RGB space to HSV space
rgb2ind: convert RGB image to indexed image
rgbplot: plot colormap
rng: control random number generation
round: round to nearest integer
save: save workspace variables to MAT-file
scatter, scatter3: 2-D scatter plot, 3-D scatter plot
set: set graphics object properties
sin: sine of argument in radians

size: return array size
sort: sort array elements
sound: convert matrix data to sound
sphere: generate sphere
spring: spring colormap array
sprintf: format data into string
sqrt: square root
stairs: stairs plot
stem: stem plot
step: run System object algorithm
strcat: concatenate strings horizontally
strcmp: compare strings
struct: create a structure array
subplot: multiple plots in a single figure window
sum: sum of elements
summer: summer colormap array
surf: create surface plot
tan: tangent of argument in radians
text: insert text descriptions in graphical plots
tic: start stopwatch timer
table: create table array with named columns
title: insert title in graphical plots:
toc: stop timer and read elapsed time from stopwatch
uint8: unsigned integer 8-bit (0–255)
unique: unique values in array
ver: displays MATLAB version
VideoReader: read video files
VideoWriter: write video files
view: viewpoint specification
while ... end: execute statements while condition is true
whos: list variables in workspace, with sizes and types
xlabel: label *x*-axis
ylabel: label *y*-axis
zeros: create array of all zeros
zlabel: label *z*-axis

Image Processing Toolbox (IPT) Functions

affine2d: 2-D affine geometric transformation
applycform: apply device-independent color space transformation
blockproc: distinct block processing for image
boundarymask: find region boundaries of segmentation
bwarea: area of objects in binary image
bwareafilt: extract objects from binary image by size
bwboundaries: trace region boundaries in binary image,

bwconncomp: find connected components in binary image
bwconvhull: generate convex hull image from binary image
bwdist: distance transform of binary image
bweuler: euler number of binary image
bwlabel: label connected components in 2-D binary image
bwmorph: morphological operations on binary images
bwperim: find perimeter of objects in binary image
bwtraceboundary: trace object in binary image
checkerboard: create checkerboard image
col2im: rearrange matrix columns into blocks
corr2: 2-D correlation coefficient
dct2: 2-D discrete cosine transform
dctmtx: discrete cosine transformation matrix
deconvblind: deblur image using blind deconvolution
deconvlucy: deblur image using Lucy–Richardson deconvolution
deconvwnr: deblur image using Wiener deconvolution
edge: find edges in intensity image
entropy: calculate entropy of grayscale image
entropyfilt: filter using local entropy of grayscale image
fspecial: create predefined 2-D filter
gabor: create Gabor filter
gray2ind: convert grayscale or binary image to indexed image
graycomatrix: create gray-level co-occurrence matrix from image
grayconnected: select contiguous image region with similar gray values
graycoprops: properties of gray-level co-occurrence matrix
graythresh: global image threshold using Otsu's method
histeq: enhance contrast using histogram equalization
hough: hough transform
houghlines: extract line segments based on Hough transform
houghpeaks: identify peaks in Hough transform
idct2: 2-D inverse discrete cosine transform
ifft2: 2-D discrete Fourier transform
im2bw: convert image to binary
im2col: rearrange image blocks into columns
imabsdiff: absolute difference of two images
imadd: add two images or add constant to image
imadjust: adjust image intensity values or colormap
imageinfo: image information tool
imbinarize: convert grayscale image to binary image
imbothat: bottom-hat filtering
imclose: morphologically close image
imcolormaptool: choose Colormap tool
imcomplement: complement image
imcontrast: adjust Contrast tool
imcrop: crop image
imdilate: morphologically dilate image
imdistline: distance tool
imdivide: divide one image into another

imerode: morphologically erode image
imfill: fill image regions and holes
imfilter: multidimensional filtering of images
imfindcircles: find circles using circular Hough transform
imfuse: composite of two images
imgaborfilt: apply Gabor filter to 2-D image
imgradient: gradient magnitude and direction of an image
imgradientxy: directional gradients of an image
imhist: histogram of image data
imlincomb: linear combination of images
immagbox: magnification box to the figure window
immovie: make movie from multi-frame image
immse: mean-squared error
immultiply: multiply two images
imnoise: add noise to image
imopen: morphological open operation on image
imoverlay: burn binary mask into 2-D image
imoverview: overview tool for displayed image
impixelinfo: pixel information tool
impixelregion: pixel region tool
implay: play movies, videos, or image sequences
imquantize: quantize image using specified quantization levels
imrect: create draggable rectangle
imregconfig: configurations for intensity-based registration
imregister: intensity-based image registration
imrotate: rotate image
imsharpen: sharpen image using unsharp masking
imshowpair: compare differences between images
imsubtract: subtract one image from another
imtool: open image viewer app
imtophat: top-hat filtering
imtranslate: translate image
imwarp: apply geometric transformation to image
ind2gray: convert indexed image to grayscale image
lab2rgb: convert CIE 1976 L*a*b* to RGB
labeloverlay: overlay label matrix regions on 2-D image
makecform: create color transformation structure
mat2gray: convert matrix to grayscale image
mean2: average or mean of matrix elements
medfilt2: 2-D median filtering
montage: display multiple image frames as rectangular montage
multithresh: multilevel image thresholds using Otsu's method
normxcorr2: normalized 2-D cross-correlation
ntsc2rgb: convert NTSC values to RGB color space
ordfilt2: 2-D order-statistic filtering
otsuthresh: global histogram threshold using Otsu's method
phantom: create head phantom image
plotChromaticity: plot color reproduction on chromaticity diagram

projective2d: 2-D projective geometric transformation
psnr: peak Signal-to-Noise Ratio (PSNR)
qtdecomp: quad-tree decomposition
qtgetblk: get block values in quad-tree decomposition
qtsetblk: set block values in quad-tree decomposition
regionprops: Measure properties of image regions
rgb2lab: convert RGB to CIE 1976 L*a*b*
rgb2ntsc: convert RGB values to NTSC color space
rgb2xyz: convert RGB to CIE 1931 XYZ
rgb2ycbcr: convert RGB values to YCbCr color space
roifilt2: filter region of interest (ROI) in image
roipoly: specify polygonal region of interest (ROI)
ssim: structural Similarity Index (SSIM) for measuring image quality
std2: standard deviation of matrix elements
strel: morphological structuring element
subimage: display multiple images in single figure window
superpixels: 2-D superpixel oversegmentation of images
viscircles: create circle
warp: display image as texture-mapped surface
wiener2: 2-D adaptive noise-removal filtering
xyz2rgb: convert CIE 1931 XYZ to RGB
ycbcr2rgb: convert YCbCr values to RGB color space

Audio System Toolbox (AST) Functions

audioDeviceReader: record from sound card
audioDeviceWriter: play to sound card
audioOscillator: generate sine, square, and sawtooth waveforms
compressor: dynamic range compressor
crossoverFilter: audio crossover filter
expander: dynamic range expander
getAudioDevices: list available audio devices
integratedLoudness: measure integrated loudness
loudnessMeter: standard-compliant loudness measurements
mfcc: extract MFCC from audio signal
mididevice: send and receive MIDI messages
mididevinfo: MIDI device information
midimsg: create MIDI message
midisend: send MIDI message to MIDI device
noiseGate: dynamic range noise gate
pitch: estimate fundamental frequency of audio signal
reverberator: add reverberation to audio signal
visualize: visualize filter characteristics
voiceActivityDetector: detect presence of speech in audio signal
wavetableSynthesizer: generate a periodic signal with tunable properties

Computer Vision System Toolbox (CVST) Functions

configureKalmanFilter: create Kalman filter for object tracking
detectBRISKFeatures: detect binary robust invariant scalable keypoints (BRISK)
detectFASTFeatures: detect features from accelerated segment test (FAST) algorithm
detectHarrisFeatures: detect corners using Harris–Stephens algorithm
detectKAZEFeatures: detect features using the KAZE algorithm
detectMinEigenFeatures: detect corners using minimum eigenvalue algorithm
detectMSERFeatures: detect maximally stable extremal regions (MSER) features
detectSURFFeatures: detect features using speeded-up robust features (SURF) algorithm
estimateFlow: estimate optical flow
extractFeatures: extract interest point descriptors
extractHOGFeatures: extract histogram of oriented gradients (HOG) features
extractLBPFeatures: extract local binary pattern (LBP) features
insertObjectAnnotation: annotate image or video
insertShape: insert shape in image or video
insertText: insert text in image or video
ocr: recognize text using optical character recognition
opticalFlowFarneback: estimate optical flow using Farneback method
opticalFlowHS: estimate optical flow using Horn–Schunck method
opticalFlowLK: estimate optical flow using Lucas–Kanade method
opticalFlowLKDoG: estimate optical flow using Lucas–Kanade derivative of Gaussian
predict: predict image category from classifier output
step: play one video frame at a time
vision.BlobAnalysis: properties of connected regions
vision.BlockMatcher: estimate motion between images or video frames
vision.CascadeObjectDetector: detect objects using the Viola–Jones algorithm
vision.DeployableVideoPlayer: display video
vision.ForegroundDetector: foreground detection using Gaussian mixture models
vision.HistogramBasedTracker: histogram-based object tracking
vision.KalmanFilter: kalman filter for object tracking
vision.PeopleDetector: detect upright people using HOG features
vision.PointTracker: track points in video using Kanade–Lucas–Tomasi (KLT) algorithm
vision.VideoFileReader: read video frames from video file
vision.VideoFileWriter: write video frames to video file
vision.VideoPlayer: play video or display image

Statistics and Machine Learning Toolbox (SMLT)

cluster: construct agglomerative clusters from linkages
confusionmat: create confusion matrix
cvpartition: partition data for cross-validation
dendrogram: generate dendrogram plot

evalclusters: evaluate clustering solutions
fitcdiscr: fit discriminant analysis classifier
fitcnb: train multiclass naive Bayes model
fitcsvm: train binary support vector machine (SVM) classifier
fitctree: fit binary classification decision tree for multiclass classification
fitgmdist: fit gaussian mixture model to data
gscatter: scatter plot by group
kmeans: apply *k*-means clustering
kmedoids: apply *k*-medoids clustering
knnsearch: find *k*-nearest neighbors
linkage: hierarchical cluster tree
mahal: mahalanobis distance
mvnrnd: multivariate normal random numbers
pca: principal component analysis
pdf: probability density function for Gaussian mixture distribution
pdist: pairwise distance between pairs of observations
predict: predict labels using classification model
silhouette: silhouette plot
statset: create statistics options structure
tabulate: frequency table
test: test indices for cross-validation
training: training indices for cross-validation

Signal Processing Toolbox (SPT)

designfilt: design digital filters
spectrogram: spectrogram using short-time Fourier transform
chirp: generate signals with variable frequencies
dct: discrete cosine transform
idct: inverse discrete cosine transform
window: create a window function of specified type

DSP System Toolbox (DSPST)

dsp.AudioFileReader: stream from audio file
dsp.AudioFileWriter: stream to audio file
dsp.TimeScope: time-domain signal display and measurement
dsp.SineWave: generate discrete sine wave
dsp.SpectrumAnalyzer: display frequency spectrum of time-domain signals
dsp.ArrayPlot: display vectors or arrays
fvtool: visualize frequency response of DSP filters

Neural Network Toolbox (NNT)

feedforwardnet: create a feedforward neural network
patternnet: create a pattern recognition network
perceptron: create a perceptron
train: train neural network
view: view neural network

Wavelet Toolbox (WT)

appcoef2: 2-D approximation coefficients
detcoef2: 2-D detail coefficients
dwt2: single level discrete 2-D wavelet transform
idwt2: single level inverse discrete 2-D wavelet transform
wavedec2: 2-D wavelet decomposition
waverec2: 2-D wavelet reconstruction

Fuzzy Logic Toolbox (FLT)

gaussmf: generate Gaussian function of specified mean and variance

References

Alcantarilla, P.F., A. Bartoli, and A.J. Davison. KAZE features, ECCV 2012, Part VI, LNCS 7577 pp. 214, 2012.

Barron, J.L., D.J. Fleet, S.S. Beauchemin, and T.A. Burkitt, Performance of optical flow techniques, CVPR, 1992.

Bay, H., T. Tuytelaars, L.V. Gool, SURF: Speeded Up Robust Features, *Computer Vision and Image Understanding*, vol. 110, no. 3, pp. 346–359, 2008.

Belongie, S., et al., Color and Texture based image segmentation using EM and its application to content-based image retrieval, *Proceedings of the 6th International Conference on Computer Vision*, (IEEE Cat. No.98CH36271), Bombay, India, 1998, pp. 675–682.

Bishop, C.M., *Neural Networks for Pattern Recognition*. New York: Oxford University Press, 2005.

Bishop, C.M., *Pattern Recognition and Machine Learning*. New York: Springer, 2006.

Bradski, G.R., Computer vision face tracking for use in a perceptual user interface, *IEEE Workshop on Applications of Computer Vision*, Princeton, NJ, 1998, pp. 214–219.

Breiman, L., J.H. Friedman, R.A. Olshen, and C.J. Stone. *Classification and Regression Trees*. Boca Raton, FL: Chapman & Hall, 1984.

Burt, P. and T. Andelson, The laplacian pyramid as a compact image code, *IEEE Transactions on Communications*, vol. 9, no. 4, pp. 532–540, 1983.

Calinski, T. and J. Harabasz. A dendrite method for cluster analysis, *Communications in Statistics*, vol. 3, no. 1, pp. 1–27, 1974.

Christianini, N., and J.C. Shawe-Taylor, *An Introduction to Support Vector Machines and Other Kernel-Based Learning Methods*. Cambridge, UK: Cambridge University Press, 2000

CIE, *Commission Internationale de l'Éclairage Proceedings*. Cambridge: Cambridge University Press, 1931.

CIE, Methods for re-defining CIE D illuminants, Technical Report 204, 2013, ISBN: 978-3-902842-45-9 (http://cie.co.at/publications/methods-re-defining-cie-d-illuminants).

Comaniciu, D. and P. Meer, Mean Shift : A robust approach toward feature space analysis, *IEEE Transactions on Pattern Analysis and Machine Intelligence (PAMI)*, vol.24, no. 5, pp. 603–619, May 2002.

Coppersmith, D., S.J. Hong, and J.R.M. Hosking. Partitioning nominal attributes in decision trees. *Data Mining and Knowledge Discovery*, Vol. 3, pp. 197–217, 1999.

Crow, F., Summed-area tables for texture mapping, *Proceedings of SIGGRAPH*, vol. 18, no. 3, pp. 207–212, 1984.

Dalal, N. and B. Triggs. Histograms of oriented gradients for human detection, *IEEE Computer Society Conference on Computer Vision and Pattern Recognition*, vol. 1, pp. 886–893, 2005.

Dasiopoulou S., et al., Knowledge-assisted semantic video object detection, *IEEE Transactions on Circuits and Systems for Video Technology* vol. 15, no. 10, pp. 1210–1224, 2005.

Davis S.B. and P. Mermelstein, Comparison of parametric representations for monosyllabic word recognition in continously spoken sentences, *IEEE Transactions on Acoustics, Speech and Signal Processing*, vol. 28., no. 4, pp. 357–366, 1980.

Deza, M.M. and E. Deza, *Encyclopedia of Distances*. Berlin: Springer-Verlag, 2009.

European Broadcasting Union, R 128- Loudness Normalisation and Permitted Maximum Level of Audio Signals, EBU R 128. 2014.

Farneback, G., Two-frame motion estimation based on polynomial expansion, *Proceedings of the 13th Scandinavian Conference on Image Analysis*, Gothenburg, Sweden, 2003.

Fisher R.A., The use of multiple measurements in taxonomic problems, *Annals of Eugenics*, vol. 7, no. 2, pp. 179–188, 1936.

Fourier, J.B.J., *Theorie Analytique de la Chaleur*. Paris: Firmin Didot, 1822.

Gabor, D., Theory of communication, *Journal of the Institution of Electrical Engineers Part III, Radio and Communication*, vol. 93, p. 429, 1946.

Gonzalez, R.C., R.E. Woods, S.L. Eddins, *Digital Image Processing Using MATLAB*. Upper Saddle River, New Jersey: Prentice Hall, 2003.

Guild, J., The colorimetric properties of the spectrum, *Pholosophical Transactions*, vol. 230, no. 685, pp. 149–187, 1931.

Haar, A., Zur Theorie der orthogonalen Funktionensysteme, *Mathematische Annalen*, vol. 69, pp. 331–371, 1910.

Hagan, M.T. and M. Menhaj, Training feed-forward networks with the Marquardt algorithm, *IEEE Transactions on Neural Networks*, vol. 5, no. 6, pp. 989–993, 1994.

Haralick, R.M., K. Shanmugan, and I. Dinstein, Textural features for image classification, *IEEE Transactions on Systems, Man, and Cybernetics*, Vol. SMC-3, pp. 610–621, 1973,.

Harris, C., and M. Stephens, A combined corner and edge detector, *Proceedings of the 4th Alvey Vision Conference*, August 1988, pp. 147–151.

Hastie, T., R. Tibshirani, and J. Friedman, *The Elements of Statistical Learning*, Second Edition. New York: Springer, 2008.

Hopfield, J.J., Neural networks and physical systems with emergent collective computational abilities, *Proceedings National Academy of Science, USA*, vol. 79, no. 8, pp. 2554–2558, 1982.

Horn B.K.P. and B.G. Schunck, Determining optical flow. *Artificial Intelligence*, vol. 17, pp. 185–203, 1981.

Hunter, R., Photoelectric color-difference meter, *Journal of Optical Society of America*, vol. 38, no. 7, p. 661, 1948.

International Telecommunication Union - Radiocommunication, Algorithms to Measure Audio Programme Loudness and True-Peak Audio Level, ITU-R BS.1770-4. 2015.

Jain, A.K., *Fundamentals of Digital Image Processing*. Englewood Cliffs, NJ: Prentice Hall, 1989, p. 439.

Kalman, R.E., A new approach to linear filtering and prediction problems, *Journal of Basic Engineering*, vol. 82, pp. 35–45, 1960.

Kaufman, L. and P.J. Rousseeuw, *Finding Groups in Data: An Introduction to Cluster Analysis*. Hoboken, NJ: John Wiley & Sons, Inc., 1990.

Kendall, M.G., *The Advanced Theory of Statistic*, Fourth Edition. New York: Macmillan, 1979.

Lam, E.Y., Iterative statistical approach to blind image deconvolution, *Journal of the Optical Society of America (JOSA)*, vol. 17, no. 7, pp. 1177–1184, 2000.

Leutenegger, S., M. Chli, and R. Siegwart. BRISK: Binary Robust Invariant Scalable Keypoints. *Proceedings of the IEEE International Conference. ICCV*, 2011.

Lewis, J.P., *Fast Normalized Cross-Correlation*. San Francisco, CA: Industrial Light & Magic, 1995.

Lim, J.S., *Two-Dimensional Signal and Image Processing*. Englewood Cliffs, NJ: Prentice Hall, 1990.

Lowe, D.G., Distinctive image features from scale-invariant keypoints, *International Journal of Computer Vision*, vol. 60, no. 2, pp. 91–110, 2004.

Lucas B.D. and T. Kanade, An iterative image registration technique with an application to stereo vision, *Proceedings of Image Understanding Workshop*, 1981, pp. 121–130.

Lucas, B.D. and T. Kanade. An Iterative Image Registration Technique with an Application to Stereo Vision, *Proceedings of the 7th International Joint Conference on Artificial Intelligence*, April 1981, pp. 674–679.

Lucy, L.B., An iterative technique for the rectification of observed distributions, *Astronomical Journal*, vol. 79, no. 1, pp. 745–754, 1974.

Mallat, S.G., A theory for multiresolution signal decomposition: The wavelet representation, *IEEE Transactions on Pattern Analysis and Machine Intelligence*. Vol. 11, no. 7, pp. 674–693, July 1989.

Marquardt, D., An algorithm for least-squares estimation of nonlinear parameters, *SIAM Journal on Applied Mathematics*, vol. 11, no. 2, pp. 431–441, 1963.

Matas, J., O. Chum, M. Urban, and T. Pajdla, Robust wide baseline stereo from maximally stable external regions, *British Machine Vision Conference*, 2002, pp. 384–396.

Nyquist, H., Certain topics in telegraph transmission theory, *Transactions of American Institute of Electrical Engineers (AIEE)*, vol. 47, no. 2, pp. 617–644, Apr. 1928.

Ojala, T., M. Pietikainen, and T. Maenpaa. Multiresolution gray scale and rotation invariant texture classification with local binary patterns. *IEEE Transactions on Pattern Analysis and Machine Intelligence*, Vol. 24, no. 7, pp. 971–987, July 2002.

Otsu, N., A threshold selection method from gray-level histograms. *IEEE Transactions on Systems, Man, and Cybernetics*, vol. 9, no. 1, pp. 62–66, 1979.

Perona, P. and J. Malik, Scale-space and edge detection using annisotropic diffusion. *IEEE Transactions on Pattern Analysis and Machine Intelligence*, vol. 12, pp. 1651–1686, 1990.

Poynton, C., A guided tour of color space, *Proceedings of the SMPTE Advanced Television and Electronic Imaging Conference*, 1995, pp. 167–180.

Poynton, C., *A Technical Introduction to Digital Video*. Hoboken, NJ: John Wiley & Sons Inc., 1996, p. 175.

Richardson, W., Bayesian based iterative method of image restoration, *Journal of the Optical Society of America (JOSA)*, vol. 62, no. 1, pp. 55–59, 1972.

Rosenblatt, F., The perceptron: A probabilistic model for information storage and organization in the brain, *Psychological Review*, vol. 65, no. 6, pp. 386–408, 1958.

Rosten, E. and T. Drummond, Machine learning for high speed corner detection, *9th European Conference on Computer Vision*, vol. 1, 2006, pp. 430–443.

Roussopoulos, N., S. Kelley, F.D.R. Vincent, Nearest neighbor queries, *Proceedings of 1995 ACMSIGMOD International Conference on Management of Data – SIGMOD '95*, p. 71.

Shepp, L., and B. Logan, The fourier reconstruction of a head section, *IEEE Transactions on Nuclear Science*, vol. NS21, no. 3, pp. 21–43, 1974.

Shi, J. and C. Tomasi, Good features to track, *IEEE Conference on Computer Vision and Pattern Recognition (CVPR)*, Seattle, 1994.

Sohn, J., N.-S. Kim, and W. Sung. A statistical model-based voice activity detection. *Signal Processing Letters IEEE*, vol. 6, no. 1, pp. 1–3, 1999.

Stauffer, C. and W.E.L. Grimson, Adaptive background mixture models for real-time tracking, *IEEE Computer Society Conference on Computer Vision and Pattern Recognition*, vol. 2, 06 August 1999, pp. 2246–252.

Stevens, S.S., J. Volkmann, E.B. Newman, A scale for the measurement of the psychological magnitude pitch, *Journal of the Acoustic Society of America (ASA)*, vol. 8, pp. 185–190, 1937.

Taylor, B. Methodus Incrementorum Directa et Inversa (Direct and Reverse Methods of Incrementation) (in Latin), London. pp. 21–23, 1715.

Tomasi, C. and T. Kanade, Detection and tracking of point features, Computer Science Department, Carnegie Mellon University Technical Report CMU-CS-91-132, April 1991.

Viola, P. and M.J. Jones, Rapid object detection using a boosted cascade of simple features, *Proceedings of the 2001 IEEE Computer Society Conference on Computer Vision and Pattern Recognition (CVPR)*, vol. 1, 2001, pp. 511–518.

Wiener, N., *Extrapolation Interpolation and Smoothing of Stationary Time Series*. New York: John Wiley & Sons Inc., 1949.

Witkin, A.P. Scale-space filtering, *Proceedings of the 8th International Joint Conference on Artificial Intelligence*, Karlsruhe, Germany, 1983, pp. 1019–1022.

Wright, W.D., A re-determination of the trichromatic coefficients of the spectral colours, *Transactions of the Optical Society*, vol. 30, no. 4, pp. 141–164, 1929.

Young, T., Bakerian Lecture: On the theory of light and colors, *Philosophical Transactions, Royal Society of London*, vol. 92, pp. 12–48, 1802.

Zhang, S., Single shot refinement neural network for object detection, *Proceedings of the IEEE Conference on Computer Vision and Pattern Recognition (CVPR)*, 2018, pp. 4203–4212.

Zhou, W., A.C. Bovik, H.R. Sheikh, and E.P. Simoncelli. Image qualifty assessment: From error visibility to structural similarity, *IEEE Transactions on Image Processing*, Vol. 13, no. 4, pp. 600–612, April 2004.

Index

AdobeRGB color space 35
Affine transformation 64
Arithmetic operations 99
Artificial neural network (ANN) 339
Audio I/O 210
Audio system toolbox (AST) 195

Band-pass filter 249
Band-stop filter 252
Binarization threshold 17
Blind deconvolution 108
Blob detector 282
Block matcher 296
Block processing 95
BRISK algorithm 321

Calinski-Harabasz (C-H) criterion 366
Checkerboard 45
Chessboard distance 121, 326
Chirp signal 242
Chromaticity values 34
CIE L*a*b* 41
CIE XYZ 34
Classification 337
Classification learner app 356
Clustering 324
CMY color model 32
Color conversions 37
Color look up table (CLUT) 21
Colormap 25
Component video 260
Composite video 260
Computer Vision System Toolbox (CVST) 263, 304
Confusion matrix 359, 367
Contrast adjustment 86
Contrast stretching 50, 87
Convolution 74
Convolution theorem 103
Correlation 74
Covariance 68
Cropping 58
Cross-over filter 224

Data acquisition 306
Decision tree classifier 346
Dendrogram 332
DFT basis functions 203

DFT coefficients 202
Discrete Cosine Transform (DCT) 135
Discrete Fourier Transform (DFT) 132, 202
Discrete Wavelet Transform (DWT) 138
Discriminant analysis classifier 348
Dithering 21
DSP System Toolbox (DSPST) 196
Dynamic range compressor 218
Dynamic range expander 219

Edge detection 80
Eigenvalue and Eigenvector 313
Euclidean distance 326

Face detector 286
FAST algorithm 315
Feature extraction 312
Filter design 244
Finite impulse response (FIR) 237, 238
Fisher-Iris dataset 307
Foreground detector 284
44,100 Hz 44.1 kHz 282
Fourier theorem 200

Gabor filter 78
Gamma curve 86
Gaussian function 48
Gaussian Mixture Model (GMM) 335
Geometric transformations 58
GMM based clustering 335
Gray level co-occurrence matrix (GLCM) 126

Harris corner detector 314
Hierarchical clustering 332
High-pass filter 147
High-pass FIR filter 246
High-pass IIR filter 249
Histogram 57
Histogram based tracker 289
Histogram equalization 91
Histogram of Oriented Gradients (HOG) 324
Hough transform 114
Hue saturation value (HSV) 41

Image arithmetic 99
Image blurring 74
Image filtering 73

Image fusion 52
Image gradient 82, 286
Image import and export 14
Image processing toolbox (IPT) 9
Image quality 129
Image read and write 14
Image registration 67
Image segmentation 109
Image sharpening 80
Image type conversion 16
Image warping 54
Indexed image 21
Infinite impulse response filter (IIR) 237, 238
Interactive exploration 54
Interactive tools 57
Interpolation 62
Inverse filtering 103

Kalman filter 294
KAZE algorithm 320
Kernel 73
k-means clustering 328
k-medoids clustering 329
k-nearest neighbor classifier 338

Laplacian of Gaussian (LoG) 83
Local Binary Pattern (LBP) 322
Logical operations 101
Loudness 226
Low-pass filter 145
Low-pass FIR filter 244
Low-pass IIR filter 246
Lucy-Richardson deconvolution 107

Mahalanobis distance 327
Manhattan distance 326
Maximally Stable Extremal Regions
    (MSER) 316
Median filter 77
Mel frequency cepstral coefficient (MFCC) 229
Minimum eigenvalue method 312
Morphological operations 92
Motion tracking 289
Multi-layer perceptron (MLP) 344
Musical instrument digital interface
    (MIDI) 235, 236

Naïve Bayes classifier 353
National Television Systems Committee
    (NTSC) 279
Neuron 339
Noise 47, 77, 102, 129
Noise filter 76

Noise gate 217
Normalized cross-correlation 71
Nyquist sampling theorem 197

Object analysis 111
Object detection 282
Optical character recognition (OCR) 288
Optical flow 291
Order statistic filter 77
Otsu's method 18
Overfitting 337

People detector 285
Perceptron 339
Phantom head 45
Phase Alternation Lines (PAL) 280
Pitch 225
Pixelation 62
Pixel connectivity 121
Point spread function (PSF) 102
Point tracker 293, 294
Pre-processing 311
Projective transformation 64

Quad-tree decomposition 116
Quantization 20

Reflection 64
Region of interest (ROI) 95
Reverberation 216
RGB color model 31
Rotation 60

Salt and pepper noise 77
Scale Invariant Feature Transform (SIFT) 317
Scale space 318
Scaling 61
Signal Processing Toolbox (SPT) 13, 196
Silhouette value 365
Similarity metrics 324
Simulink 149, 254, 297
Single-layer perceptron (SLP) 343
Sound waves 197
Spectral filters 241
Spectrogram 242
Speeded Up Robust Features (SURF)
    algorithm 317
sRGB color space 35
Statistics and Machine Learning Toolbox
    (SMLT) 305
Supervised grouping 325, 337
Support vector machine (SVM) classifier 354
Synthesizer 214
Synthetic image 44

Temporal filter 237
Texture analysis 125
3-D plotting functions 182
Translation 60
Tristimulus values 34
2-D Gaussian function 49, 77
2-D plotting functions 155

Unsigned integer 8-bits 14
Unsupervised grouping 325
User-defined functions 98

Validation set 338
Video color spaces 278
Video I/O 264
Video processing 259
Voice activity detection (VAD) 225

White point illuminant 34
Wiener deconvolution 106
Window function 239

YCbCr 261

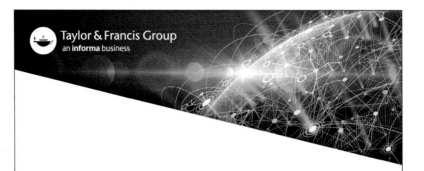